# Lecture Notes in Artificial Intelligence 10259

Subseries of Lecture Notes in Computer Science

T0171820

More information about this series at http://www.springer.com/series/1244

Annette ten Teije · Christian Popow
John H. Holmes · Lucia Sacchi (Eds.)

# Artificial Intelligence in Medicine

16th Conference on Artificial Intelligence
in Medicine, AIME 2017
Vienna, Austria, June 21–24, 2017
Proceedings

 Springer

*Editors*
Annette ten Teije ⓘ
Vrije Universiteit Amsterdam
Amsterdam
The Netherlands

John H. Holmes ⓘ
University of Pennsylvania
Philadelphia, PA
USA

Christian Popow ⓘ
Medical University of Vienna
Vienna
Austria

Lucia Sacchi ⓘ
University of Pavia
Pavia
Italy

ISSN 0302-9743 ISSN 1611-3349 (electronic)
Lecture Notes in Artificial Intelligence
ISBN 978-3-319-59757-7 ISBN 978-3-319-59758-4 (eBook)
DOI 10.1007/978-3-319-59758-4

Library of Congress Control Number: 2017943002

LNCS Sublibrary: SL7 – Artificial Intelligence

This Springer imprint is published by Springer Nature
The registered company is Springer International Publishing AG
The registered company address is: Gewerbestrasse 11, 6330 Cham, Switzerland

# Preface

The European Society for Artificial Intelligence in Medicine (AIME) was established in 1986 following a very successful workshop held in Pavia, Italy, the year before. The principal aims of AIME are to foster fundamental and applied research in the application of artificial intelligence (AI) techniques to medical care and medical research, and to provide a forum at biennial conferences for discussing any progress made. For this reason, the main activity of the society is the organization of a series of biennial conferences, which have been held in Marseilles, France (1987), London, UK (1989), Maastricht, The Netherlands (1991), Munich, Germany (1993), Pavia, Italy (1995), Grenoble, France (1997), Aalborg, Denmark (1999), Cascais, Portugal (2001), Protaras, Cyprus (2003), Aberdeen, UK (2005), Amsterdam, The Netherlands (2007), Verona, Italy (2009), Bled, Slovenia (2011), Murcia, Spain (2013), and Pavia, Italy (2015). This volume contains the proceedings of AIME 2017, the 16th Conference on Artificial Intelligence in Medicine, held in Vienna, Austria, June 21–24, 2017.

The AIME 2017 goals were to present and consolidate the international state of the art of AI in biomedical research from the perspectives of theory, methodology, systems, and applications. The conference included two invited lectures, full and short papers, tutorials, workshops, and a doctoral consortium.

In the conference announcement, authors were invited to submit original contributions regarding the development of theory, methods, systems, and applications for solving problems in the biomedical field, including AI approaches in biomedical informatics, molecular medicine, and health-care organizational aspects. Authors of papers addressing theory were requested to describe the properties of novel AI models potentially useful for solving biomedical problems. Authors of papers addressing theory and methods were asked to describe the development or the extension of AI methods, to address the assumptions and limitations of the proposed techniques, and to discuss their novelty with respect to the state of the art. Authors of papers addressing systems and applications were asked to describe the development, implementation, or evaluation of new AI-inspired tools and systems in the biomedical field. They were asked to link their work to underlying theory, and either analyze the potential benefits to solve biomedical problems or present empirical evidence of benefits in clinical practice.

AIME 2017 received 141 abstract submissions; 113 thereof were eventually submitted as complete papers. Submissions came from 35 countries, including 13 outside Europe. All papers were carefully peer-reviewed by experts from the Program Committee with the support of additional reviewers. Each submission was reviewed in most cases by three reviewers, and at least by two reviewers. The reviewers judged the overall quality of the submitted papers, together with their relevance to the AIME conference, originality, impact, technical correctness, methodology, scholarship, and quality of presentation. In addition, the reviewers provided detailed written comments on each paper, and stated their confidence in the subject area.

A small committee consisting of the AIME 2017 scientific chair, Annette ten Teije, the local organization chair, Christian Popow, John H. Holmes, doctoral consortium chair and AIME 2015 scientific chair, and Lucia Sacchi, AIME 2015 local organization co-chair, made the final decisions regarding the AIME 2017 scientific program. This process began with virtual meetings held monthly starting in March 2016. The process ended with a two-day face-to-face meeting of the committee in Vienna to assemble the final program.

As a result, 21 long papers (an acceptance rate of 22%) and 24 short papers (including demo papers) were accepted; one short paper was withdrawn. Each long paper was presented in a 25-minute oral presentation during the conference. Each regular short paper was presented in a five-minute presentation and by a poster. Each demo short paper was presented in a five-minute presentation and by a demo during the demo session. The papers were organized according to their topics in the following main themes: (1) Ontologies/Knowledge Representation (2) Bayesian Methods; (3) Temporal methods; (4) Nature Language Processing; (5) Health Care Processes; (6) Machine Learning; and (7) Demo's.

AIME 2017 had the privilege of hosting two invited speakers: Stefan Schulz, from the University of Graz, Austria, and Kenneth J. Barker, from T.J. Watson Research Center, IBM Research, New York, USA. In his keynote entitled "SNOMED CT: The Thorny Way Towards Interoperability of Clinical Routine Data" Stefan Schulz discussed the crucial role of the quality of the vocabularies and the annotation process for achieving data interoperability. The quality of terminology-annotated clinical data should be considered with realism, and the automated annotation approaches have to take into account human inter-annotator disagreement.

Ken Barker's keynote focused on intelligent question answer (QA) systems to support professionals in medicine and health care to explore the medical literature. In their approach the three main dimensions are context analysis, content management, and answer management. Furthermore, the collaborative setting plays a role in the learning capabilities of the adaptable QA system.

The doctoral consortium provided an opportunity for six PhD students to present their research goals, proposed methods, and preliminary results. A scientific panel consisting of experienced researchers in the field (Riccardo Bellazzi, Mor Peleg, David Riaño, Lucia Sacchi, Yuval Shahar, and Allan Tucker) provided constructive feedback to the students in an informal atmosphere. The doctoral consortium was chaired by John H. Holmes.

Four workshops were organized after the AIME 2017 main conference. These included the 9th International Workshop on Knowledge Representation for Health Care (KRH4C) and the 10th International Workshop on Process-Oriented Information Systems in Health Care (ProHealth), joined together for the second time at AIME. This workshop was chaired by David Riaño, Richard Lenz, Mor Peleg, and Manfred Reichert. A second full-day workshop was the Second Workshop on Extracting and Processing of Rich Semantics from Medical Texts, chaired by Kerstin Denecke, Yihan Deng, Thierry Declerck, and Frank van Harmelen. The third workshop was the Second Workshop on Artificial Intelligence for Diabetes, chaired by Clare Martin, Beatriz López, and Pau Herrero Vinas. The fourth workshop was the Workshop on Advanced

Predictive Models in Health Care organized by Niels Peek, Gregor Štiglic, Nophar Geifman, Petra Povalej Brzan, and Matthew Sperrin.

In addition to the workshops, five interactive half-day tutorials were presented prior to the AIME 2017 main conference:

(1) Natural Language Processing for Clinical Information Extraction (Stéphane Meystre, Meliha Yetisgen, Scott DuVall, Hua Xu); (2) Latest Speech and Signal Processing for Affective and Behavioral Computing in mHealth, (Bjorn Schuller, Bodgan Vlasenko, Hesam Sagha), (3) Evaluation of Prediction Models in Medicine (Ameen Abu-Hanna); (4) Medical Decision Analysis with Probabilistic Graphical Models (Francisco Javier Diez, Manuel Luque); (5) Clinical Fuzzy Control Systems and Fuzzy Automata with HL7's Clinical Decision Support Standard: The Fuzzy Arden Syntax (Jeroen de Bruin, Klaus-Peter Adlassnig).

We would like to thank everyone who contributed to AIME 2017. First of all, we would like to thank the authors of the papers submitted and the members of the Program Committee together with the additional reviewers. Thanks are also due to the invited speakers as well as to the organizers of the workshops, the tutorials and doctoral consortium. Many thanks go to the local Organizing Committee, who managed all the work making this conference possible. The free EasyChair conference system (http://www.easychair.org/) was an important tool supporting us in the management of submissions, reviews, selection of accepted papers, and preparation of the overall material for the final proceedings. We would like to thank our sponsors, who so generously supported the conference: the American Medical Informatics Association (AMIA/KDDM), the European Association for Artificial Intelligence (EurAI), and Springer. We thank IMIA for the recent endorsement of the AIME conference. Finally, we thank the Springer team for helping us in the final preparation of this LNAI book.

June 2017

Annette Ten Teije
Christian Popow
John H. Holmes
Lucia Sacchi

# Organization

## AIME Organization Team

| | |
|---|---|
| Annette ten Teije | Vrije Universiteit Amsterdam, The Netherlands (Chair) |
| Christian Popow | Medical University of Vienna, Austria (Local chair) |
| John H. Holmes | University of Pennsylvania, USA (Doctoral Consortium Chair) |
| Lucia Sacchi | University of Pavia, Italy (Co-chair) |

## Artificial Intelligence in Medicine Board

| | |
|---|---|
| Amar Das | The Dartmouth Institute, USA |
| Stefan Darmoni | University of Rouen, France |
| Milos Hauskrecht | University of Pittsburgh, USA |
| John Holmes | University of Pennsylvania, USA |
| Jose M. Juarez | University of Murcia, Spain |
| Mar Marcos | Universitat Jaume I, Castellón, Spain |
| Roque Marín Morales | University of Murcia, Spain |
| Stefania Montani | Università del Piemonte Orientale, Italy |
| Barbara Oliboni | University of Verona, Italy |
| Niels Peek | The University of Manchester, UK (Chair) |
| Mor Peleg | University of Haifa, Israel |
| Christian Popow | Medical University of Vienna, Austria |
| David Riaño | Universitat Rovira i Virgili, Spain |
| Lucia Sacchi | University of Pavia, Italy |
| Annette Ten Teije | Vrije Universiteit Amsterdam, The Netherlands |
| Paolo Terenziani | Università del Piemonte Orientale, Italy |
| Samson Tu | Stanford University, USA |
| Allan Tucker | Brunel University London, UK |
| Szymon Wilk | Poznan University of Technology, Poland |
| Blaz Zupan | University of Ljubljana, Slovenia |

## Program Committee

| | |
|---|---|
| Ameen Abu-Hanna | AMC-UvA, The Netherlands |
| Klaus-Peter Adlassnig | Medical University of Vienna, Austria |
| Laura Barnes | University of Virginia, USA |
| Riccardo Bellazzi | University of Pavia, Italy |
| Henrik Boström | Stockholm University, Sweden |
| Carlo Combi | Università degli Studi di Verona, Italy |
| Arianna Dagliati | University of Pavia, Italy |
| Stefan Darmoni | University of Rouen, France |

Niels Peek                The University of Manchester, UK
Mor Peleg                 University of Haifa, Israel
Christian Popow           Medical University of Vienna, Austria (Local Chair)
Cédric Pruski             Luxembourg Institute of Science and Technology,
                             Luxembourg
Silvana Quaglini          University of Pavia, Italy
David Riaño               Universitat Rovira i Virgili, Spain
Pedro Pereira Rodrigues   University of Porto, Portugal
Lucia Sacchi              University of Pavia, Italy (Co-chair)
Aleksander Sadikov        University of Ljubljana, Slovenia
Stefan Schulz             Medical University of Graz, Austria
Brigitte Seroussi         Assistance Publique, Hôpitaux de Paris, France
Yuval Shahar              Ben Gurion University, Israel
Erez Shalom               Ben Gurion University, Israel
Constantine               NCSR Demokritos, Greece
   Spyropoulos
Gregor Štiglic            University of Maribor, Slovenia
Annette Ten Teije         Vrije Universiteit Amsterdam, The Netherlands (Chair)
Paolo Terenziani          Università del Piemonte Orientale, Italy
Allan Third               The Open University, UK
Samson Tu                 Stanford University, USA
Allan Tucker              Brunel University London, UK
Ryan Urbanowicz           University of Pennsylvania, USA
Frank Van Harmelen        Vrije Universiteit Amsterdam, The Netherlands
Alfredo Vellido           Universitat Politècnica de Catalunya, Spain
Szymon Wilk               Poznan University of Technology, Poland
Blaz Zupan                University of Ljubljana, Slovenia
Pierre Zweigenbaum        LIMSI, CNRS, Université Paris-Saclay, France

## Additional Reviewers

Elias Alevizos            Tianyong Hao              Luca Piovesan
Mawulolo Ameko            Xiang Ji                  Vid Podpecan
Luca Anselma              Aida Kamisalic            Vassiliki Rentoumi
Michael Barrowman         Lefteris Koumakis         Fabrizio Riguzzi
Elena Bellodi             Thomas Kupka              Carla Rognoni
Miguel Belmonte           Siqi Liu                  Stelios Sfakianakis
Asim Bhuta                Daniela Loreti            Matthew Sperrin
Alessio Bottrighi         Begoña                    Grigorios Tzortzis
Marcos Luiz                  Martinez-Salvador      Natalia Viani
   de Paula Bueno         Dragana Miljkovic         Ute von Jan
Joana Ferreira            Christopher Ochs          Yonghui Wu
Paolo Fracçaro            Bruno Oliveira            Jinghe Zhang
Vida Groznik              Pierpaolo Palumbo

## Doctoral Consortium Committee

Riccardo Bellazzi        University of Pavia, Italy
John Holmes              University of Pennsylvania, USA (Chair)
Mor Peleg                University of Haifa, Israel
David Riaño              Universitat Rovira i Virgili, Spain
Lucia Sacchi             University of Pavia, Italy
Yuval Shahar             Ben Gurion University, Israel
Allan Tucker             Brunel University London, UK

## Workshops

### 9th International Workshop on Knowledge Representation for Health Care (KRH4C) and the 10th International Workshop on Process-oriented Information Systems in Health Care (ProHealth)

#### Co-chairs

David Riaño              Universitat Rovira i Virgili, Spain
Richard Lenz             University of Erlangen-Nuremberg, Germany
Mor Peleg                University of Haifa, Israel
Manfred Reichert         University of Ulm, Germany

### Second International Workshop on Extraction and Processing of Rich Semantics from Medical Texts

#### Co-chairs

Kerstin Denecke          Bern University of Applied Sciences, Switzerland
Yihan Deng               Bern University of Applied Sciences, Switzerland
Thierry Declerck         Saarland University and German Research Center
                           for Artificial Intelligence, Germany
Frank van Harmelen       Vrije Universiteit Amsterdam, The Netherlands

### Second Workshop on Artificial Intelligence for Diabetes

#### Co-chairs

Clare Martin             Oxford Brookes University, UK
Beatriz López            University of Girona, Spain
Pau Herrero Vinas        Imperial College London, UK

**Workshop on Advanced Predictive Models in Health Care**

**Co-chairs**

| | |
|---|---|
| Niels Peek | The University of Manchester, UK |
| Gregor Štiglic | University of Maribor, Slovenia |
| Nophar Geifman | The University of Manchester, UK |
| Petra Povalej Brzan | University of Maribor, Slovenia |
| Matthew Sperrin | The University of Manchester, UK |

# Tutorials

**Natural Language Processing for Clinical Information Extraction**

| | |
|---|---|
| Stéphane Meystre | Medical University of South Carolina, USA |
| Meliha Yetisgen | University of Washington, USA |
| Scott DuVall | University of Utah and Department of Veterans Affairs Salt Lake City Health Care System, USA |
| Hua Xu | University of Texas, USA |

**Latest Speech and Signal Processing for Affective and Behavioral Computing in mHealth**

| | |
|---|---|
| Björn Schuller | Imperial College London, UK |
| Bodgan Vlasenko | University of Passau, Germany |
| Hesam Sagha | audEERING GmbH, Germany |

**Evaluation of Prediction Models in Medicine**

| | |
|---|---|
| Ameen Abu-Hanna | University of Amsterdam, The Netherlands |

**Medical Decision Analysis with Probabilistic Graphical Models**

| | |
|---|---|
| Francisco Javier Díez | UNED, Spain |
| Manuel Luque | UNED, Spain |

**Clinical Fuzzy Control Systems and Fuzzy Automata with HL7's Clinical Decision Support Standard: The Fuzzy Arden Syntax**

| | |
|---|---|
| Jeroen de Bruin | Medical University of Vienna, Austria |
| Klaus-Peter Adlassnig | Medical University of Vienna, Austria |

## Sponsors

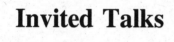

**Invited Talks**

# SNOMED CT: The Thorny Way Towards Interoperability of Clinical Routine Data

Stefan Schulz (ID)

Institute of Medical Informatics, Statistics and Documentation,
Medical University of Graz, Graz, Austria
stefan.schulz@medunigraz.at

**Abstract.** To achieve the goal of data interoperability, the quality of the vocabularies used as well as the coding/annotation processes involved are crucial. Inter-coder agreement is a major quality criterion. In the project ASSESS-CT, the clinical terminology SNOMED CT was assessed for its usefulness regarding manual and machine annotations of clinical texts. Due to the low inter-annotator agreement obtained a manual analysis of disagreements was done.

**Keywords:** Biomedical ontologies · SNOMED CT · Semantic annotation

The EU support action ASSESS CT (Assessing SNOMED CT for Large Scale eHealth Deployments in the EU) [1] aimed at collecting empirical evidence for the fitness of SNOMED CT, compared to other terminology scenarios. A series of manual and machine annotation experiments was performed in order to measure terminology coverage [2] in clinical models and narratives. One striking result was a generally low inter-annotator agreement when clinical text samples from different domains and languages were annotated by domain experts. Although detailed annotation guidelines had been elaborated before, Krippendorf's Alpha was only 37% when annotating English texts with SNOMED CT.

An in-depth analysis of disagreements between annotators yielded several factors that affect consistent. In the following, frequent reasons of disagreement are listed:

- **Human factors**: ignoring the guideline, slips because of haste or carelessness;
- **Tooling**: Retrieval problems due to poor tool support or lacking entry terms;
- **Lack of textual definitions**. This is a typical problem in SNOMED CT, especially when concepts are semantically close (e.g. *Significant* vs. *Severe*, or *Worried* vs. *Anxiety*);
- **Lack of formal definitions**. Fully defined SNOMED CT concepts could support ex-post reconciliation of annotation disagreements. However, definitional axioms were often found missing, e.g. in the primitive concept *Diabetic monitoring (regime/therapy)*, despite the existence of *Diabetes mellitus (disorder)* and *Monitoring - action (qualifier value)*;
- **Logical polysemy** [3], e.g. *Malignant lymphoma (disorder)* vs. *Malignant lymphoma (morphological abnormality)* or *Liver structure (body structure)* vs. *Liver (body structure)*;

- **Navigational concepts**, e.g. *Finding of measures of palpebral fissure (finding)* is a common parent of the pre-coordinated findings, whereas *Measure of palpebral fissure (observable entity)* is to be completed by a qualifier.

These factors can be mitigated by better annotator training, refinement of the guideline, completion of SNOMED CT by textual and formal definitions as well as by flagging of concepts with purely navigational character. Retrieval performance of term browsers can be enhanced, e.g. by fuzzy match and more synonyms. Logical polysemy can be dealt with by postprocessing (addition of inferred concepts).

More difficult are those factors that derive from the characteristics of clinical language, especially as a consequence of its high compactness:

- **Context** (deixis, anaphora) such as in "These ailments have substantially increased since October". Here, "increased" belongs to something in an earlier sentence, which require some reasoning by the reader. The same applies to
- **Co-ordination and negation** such as in "normal factors 5, 9, 10, and 11", or "no tremor, rigidity or bradykinesia";
- **Lexical ambiguity**: One text sample started with "IV:", which could mean the fourth item on a list, the abbreviation for "intravenous" or the 4th cerebral nerve. Here also, the understanding of the context is fundamental;
- **Scope**: in the previous example, even after disambiguation, there was disagreement regarding *Trochlear nerve structure (body structure)* vs. *Exploration of trochlear nerve (IV) (procedure)*.

These latter factors require a real understanding of the text, which constitutes an ongoing challenge for automated (NLP-based) annotation. Regarding the scope of annotation, an enhanced guideline could recommend, e.g., that anatomical entities, devices and organisms are, if possible, always seen as modifiers of a "head" concept, normally a procedure or a finding, which however, requires an advance parsing not only of the whole discourse structure.

One conclusion of this work is that the quality of terminology-annotated clinical data should be considered with realism, especially if they trigger intelligent systems. Another conclusion is that the creation of annotation gold standards is inherently difficult. The assessment of the performance of automated (NLP) annotation approaches must therefore always take human inter-annotator disagreement into account.

# References

1. SNOMED International. What is SNOMED CT? (2017). http://www.snomed.org/snomed-ct/what-is-snomed-ct
2. Kalra, D., Schulz, S., Karlsson, D., et al.: ASSESS CT Recommendations (2016). http://assess-ct.eu/fileadmin/assess_ct/final_brochure/assessct_final_brochure.pdf
3. Arapinis, A., Vieu, L.: A plea for complex categories in ontologies. Appl. Ontology, **10**(3–4), 285–296 (2015)

# Collaborative, Exploratory Question Answering Against Medical Literature

Ken Barker

IBM Research, Thomas J. Watson Research Center, Yorktown Heights,
NY 10598, USA
kjbarker@us.ibm.com

## Background

The amount of text being generated in technical domains such as medicine and healthcare has skyrocketed, making broad, individual, independent mastery of such domains impossible. More than ever, intelligent information gathering tools are essential to support any task that requires consuming and acting on such knowledge. Decision makers require more intelligent Question Answering (QA) systems to support decision making. In particular, professionals in medicine and healthcare could benefit from advanced tools to help explore literature intelligently in the context of specific, complex cases. Existing QA systems suffer from several weaknesses:

- They assume that the decision problem can be directly formulated as a question and that the answer can directly inform the decision to be made. But users may need to supply much context and explore relevant information before even knowing what questions to ask.
- They assume questions are independent. But the decision making process may require a series of questions, influenced by the answers to prior questions. Users' questions and the problem state may change through interacting with the system.
- They assume that questions have definitive answers in a corpus. But the question may be ill-specified, requiring the system to generate passages of text. These are broader than a simple phrase (the focus of factoid-based QA systems) but narrower than an entire document (the focus of web search).
- They do not expose their interpretation of the question or how the QA system arrives at a particular answer. Without this information, the decision maker is left to guess where the QA system may have gone wrong. This guessing at opaque system behavior leads to a painful trial-and-error approach to information gathering.
- There is no way to guide their behavior directly. Even if the interpretation were transparent, there is no way for the user to change or correct the interpretation.
- They do not adapt to the decision maker's needs or learn from interactions. QA systems are generally trained prior to deployment using sets of questions and corresponding answers (ground truth).

To address these weaknesses, we conduct research on technology for QA systems that are *Contextual*, *Collaborative*, *Transparent*, *Guidable*, *Adaptable*, and *Proactive*. More simply, we are investigating "Collaborative, Exploratory Question Answering".

# Research in Collaborative, Exploratory Question Answering

Our approach to Question Answering involves research in several traditional areas of automated language analysis. We group the research into three main dimensions: *Context Analysis*, *Content Management*, and *Answer Management*. Furthermore, in a collaborative setting, adaptable systems should use all user utterances, gestures and feedback as instances for learning.

## Context Analysis

We expect users of our Collaborative, Exploratory QA systems to supply background context and questions in natural language. Context analysis includes applying Natural Language Processing and Learning to understand the user's utterances, determine appropriate actions to take, and formulate the queries needed by the agents who will carry out those actions. Specific research projects include utterance classification, semantic interpretation, concept clustering, dialog, and hypothesis generation.

## Content Management

Content management involves finding appropriate unstructured text and structured knowledge resources for the information goals that users have. It includes defining models of the content, crafting appropriate ingestion and indexing strategies, and search. Our research projects in Content Management include combining structured and unstructured sources, term and concept weighting, conceptual query expansion, and lexical, syntactic and semantic indexing and search,

## Answer Management

Non-factoid Question Answering poses unique challenges for Answer Management: creating relevant, coherent units from search results, aggregating and visualizing them, and providing answer justification. We conduct specific research in passage segmentation, passage similarity, results clustering, emergent concept induction, and learning from user manipulation of result sets.

# An Interactive Platform for Exploratory QA

We have built an interactive Exploratory Question Answering system to serve as a platform for investigating the research challenges described above. The chat-based system allows users to specify background context, to examine and edit the results of interpretation, to state hypotheses and ask questions, and to explore and interact with the results of question answering.

# Contents

## Temporal Methods

## Natural Language Processing

## Health Care Processes

## Machine Learning

## Demo's

# Ontologies and Knowledge Representation

# Studying the Reuse of Content in Biomedical Ontologies: An Axiom-Based Approach

Manuel Quesada-Martínez and Jesualdo Tomás Fernández-Breis<sup>(✉)</sup>

Facultad de Informática, Universidad de Murcia, IMIB-Arrixaca,
30100 Murcia, Spain
{manuel.quesada,jfernand}@um.es

**Abstract.** The biomedical community has developed many ontologies in the last years, which may follow a set of community accepted principles for ontology development such as the ones proposed by the OBO Foundry. One of such principles is the orthogonality of biomedical ontologies, which should be based on the reuse of existing content. Previous works have studied how ontology matching techniques help to increase the number of terms reused. In this paper we investigate to what extent the reuse of terms also mean reuse of logical axioms. For this purpose, our method identifies two different ways of reusing terms, reuse of URIs (implicit reuse) and reuse of concepts (explicit reuse). The method is also able of detecting hidden axioms, that is, axioms associated with a reused term but that are not actually reused. We have developed and applied our method to a corpus of 144 OBO Foundry ontologies. The results show that 75 ontologies implicitly reuse terms, 50% of which also explicitly does it. The characterisation based on reuse enables the visualisation of the corpus as a dependency graph that can be clustered for grouping ontologies by their reuse profile. Finally, the application of a locality-based module extractor reveals that roughly 2 000 terms and 20 000 hidden axioms, on average, could be automatically reused.

**Keywords:** Biomedical ontologies · Ontology axiomatisation · Reuse

## 1 Introduction

The biomedical community has now developed a significant number of ontologies. The curation of biomedical ontologies is a complex task and they evolve rapidly, so new versions are regularly and frequently published in ontology repositories. Ontologies should play a critical role in the achievement of semantic interoperability in healthcare, as it was stated by the Semantic Health Net[1]. Therefore, the *quality assurance* of the content of biomedical ontologies is important, but it is becoming harder and harder due to the increasing number and size of biomedical ontologies. Briefly speaking, ontologies describe a domain using terms/classes, properties and instances that are implemented using a formal language.

---

[1] http://www.semantichealthnet.eu/.

© Springer International Publishing AG 2017
A. ten Teije et al. (Eds.): AIME 2017, LNAI 10259, pp. 3–13, 2017.
DOI: 10.1007/978-3-319-59758-4_1

Ontology entities have natural language annotations that make them understandable by humans, but such meaning is provided to the machines in the form of logical axioms.

The OBO Foundry [10] promotes as a set of principles for building ontologies. One of these principles promotes the reuse of terms for building an orthogonal set of ontologies[2]. Orthogonality could be used when terms can be jointly applied to describe complementary but distinguishable perspectives on the same biological or medical entity. The reuse in biomedical ontologies has been studied in works like [3,7–9]. In [9] the analysis of prominent case studies on ontology reuse was performed, discussing the need for methodologies that optimally exploit human and computational content when terms are reused. Later, in [3] the level of explicit term reuse among the OBO foundry ontologies was studied. Recently, a systematic analysis of term reuse and overlap has been performed in (1) Gene Ontology and (2) between other biomedical ontologies [7,8]. However, those works mainly focused on analysing and promoting term reuse but did not analyse the reuse of axioms. In general, the more axioms the ontology has, the more inferencing capability it has. Hence, the goal of this work is to provide insights in how the reuse of logical axioms can be improved.

## 2 Methods

### 2.1 Types of Term Reuse in Biomedical Ontologies

The reuse of content is a best practice included in methodologies for building ontologies [9] and it is one of the principles proposed by the OBO Foundry. As mentioned, orthogonality permits ontology developers to focus on the creation of the content specific of a given subdomain, and to include content from other subdomains by reusing properties or axioms. According to the OBO Foundry principle, ontology terms can be reused in different ways:

- **Explicit reuse of full ontologies:** options for importing ontologies of languages such as OWL permits to have access to their entities and axioms[3]. The `owl:imports` operation is transitive, which means that if an ontology $\theta_1$ imports the ontology $\theta_2$, and $\theta_2$ imports $\theta_3$, then $\theta_1$ imports the content of $\theta_2$ and $\theta_3$. The *import closure* of an ontology $\theta$ is the smallest set containing the axioms of $\theta$ and all the axioms of the ontologies imported by $\theta$ [2]. For an ontology $\theta$ we define two sets of classes $\theta C$ and $\theta C_{IC}$ where $\theta C$ contains all the classes directly defined by $\theta$ and $\theta C_{IC}$ the classes imported from external ontologies. We consider that a term is explicitly reused when it comes from an imported ontology.
- **Implicit reuse of individual terms:** this can be done by reusing the term URI (Uniform Resource Identifier) without importing the ontology.

The reuse of ontology content requires a *source ontology* and an *external* one. Figure 1 shows the axiomatic definition of the term *Cleavage: 16-cell*[4]. This

---

[2] http://www.obofoundry.org/principles/fp-001-open.html.
[3] https://www.w3.org/TR/owl2-syntax/#Imports.
[4] http://purl.obolibrary.org/obo/ZFS_0000005.

term is originally defined in the Zebrafish Developmental Stages Ontology (ZFS) (Fig. 1 right) and it is implicitly reused in the Zebrafish Anatomy and Development Ontology (ZFA) (Fig. 1 left). In this example, ZFA plays the role of *source ontology* and ZFS is the *external* one. In this example only the URI is reused, since the axioms defined in ZFS are not available in ZFA. Thus, the implicit reuse of ZFS_0000005 does not imply reusing the axioms: `part of some cleavage` or `immediately_preceded_by some Cleavage:8-cell`. This means that a tool using ZFA could not use these two axioms to make inferences. In this work, we will refer to these axioms as *hidden axioms*.

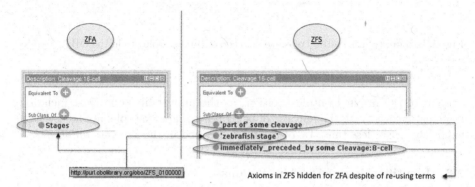

**Fig. 1.** Axiomatic definition of the term *Cleavage: 16-cell* (ZFS_0000005). (Right) Axioms associated with the term in the original ontology (ZFS). (Left) Axioms associated with the term in the ZFA ontology, which implicitly imports the term through its URI.

## 2.2   Characterisation of Ontologies Based on Reuse

Ontologies can be characterised according to the type of reuse they exhibit. The relation between a *source ontology* and *external ontologies* is usually 1:m. Figure 2 shows three examples of the behaviour followed by three ontologies extracted from the OBO Foundry repository: ZFA, the Comparative Data Analysis Ontology (CDAO) and the Cephalopod Ontology (CEPH). Dark circles represent the *source ontologies*, and the number of terms with local URI are shown in brackets. White circles represent the *external ontologies*, dotted circle lines mean implicit reuse and solid circle lines mean explicit reuse. For example, CEPH defines 325 terms, and it reuses terms from the Uberon Multi-Species Anatomy Ontology (UBERON): 72 implicitly and 408 explicitly reused.

Therefore, an ontology can be classified in one of the following groups: (1) no reuse, (2) implicit reuse, (3) explicit reuse, and (4) implicit and explicit reuse. In the running example, ZFA, CDAO and CEPH belong to groups 2, 3 and 4 respectively. The explicit importation of one ontology does not necessarily imply that the content of one ontology is reused. This does not mean to reuse the whole

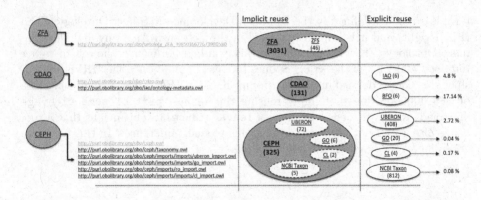

**Fig. 2.** Example of the method of reuse between of three ontologies in the OBO Foundry repository.

content of the original ontology either, as the import file could just include a fragment created with the purpose of being reused. For example, in Fig. 2, CEPH explicitly reuses less than 3% of the terms defined in the *external ontologies*.

## 2.3 Identification of Hidden Axioms

We propose a method to measure how much of the potentially reusable logical knowledge is actually reused. For this, we follow the next steps:

1. **Analysis of content driven by URIs:** Analysis of the identifiers of the *source ontology* entities, assuming that the OBO Foundry principle of URI / Identifier Space[5] is followed. This principle defines that the URI of each term is the concatenation of the ontology base URI (prefix) and an identifier. For example, *Cleveage: 16-cell* in Fig. 1 has the prefix ZFS and the identifier 0000005. We process term URIs by applying a regular expression[6]. The analysis groups terms in *reused sets*, which are groups of terms defined in a *source ontology* or in its *import closure* and that share the prefix (white circles in Fig. 2).
2. **Retrieval of the *external ontologies*:** The method needs the complete ontologies that are reused in order to calculate how much content is actually reused. For example, if ZFA implicitly reuses ZFS, then the method needs to process the complete ZFS ontology.
3. **Creating axioms sets:** For each *reused set* we create two sets of axioms, one for the axioms included in the *source ontology*, and another one for the axioms included in the *complete external ontology*.
4. **Finding hidden axioms:** For each *reused set*, the axioms of the *complete external ontology* that are not included in the *source ontology* are considered *hidden axioms*.

---

[5] http://www.obofoundry.org/principles/fp-003-uris.html.
[6] ^http://purl.obolibrary.org/obo/([A-Za-z]+)_(\\d+).

## 2.4   A Modular Strategy for Increasing the Amount of Knowledge that is Already Being Reused

Finally, we want to propose an automatic mechanism that exploits the information provided by our method to increase the amount of knowledge that is already being reused. We propose the use of mechanisms for the automatic extraction of ontology modules [4,6]. In particular, we propose to use locality-based modules[7]. A locality-based module $M$ is a subset of the axioms in an ontology $\theta$, and is extracted from $\theta$ for a set $S$ of terms (class or property names). The set $S$ is called a *seed signature* of $M$. Informally, everything the ontology $\theta$ knows about the topic consisting of the terms in $S$ and $M$ is already known by its module $M$. The remainder of $O$ knows nothing non-trivial about this topic.

We propose to extract modules of the *complete external ontologies* using as *seed signature* the classes reused by the *source ontology*. This will axiomatically enrich the *source ontology* using the minimum amount of logical content linked to the reused terms. The module could include new axioms but also new terms. For example, if the axiom `part of` of Fig. 1 right is reused, then the term *cleavage* from ZFS would be included too.

# 3   Results

## 3.1   Experimental Setup

We analysed the OBO Foundry ontologies publicly available at[8]. The corpus was formed by 144 ontologies. For each ontology, we processed the latest version available in BioPortal [11]. In case such ontology was not available in BioPortal we tried to download the file through the PURL[9] address. The ontologies were downloaded in January 2017. Our automatic process was not able to obtain 3 out of the 144 ontologies in OWL format. We used the OWL API [5] for the manipulation of the ontologies. The method was implemented in Java by using a shared memory algorithm. The method was executed using 64 processors and 300 GB RAM. The processing time was 2.5 h (download time not included). 18 out of 141 ontologies could not been loaded by the OWL API due to inaccessible import references or unparseable content. As a result, we analysed 123 ontologies.

Next, the major results are described. The complete description of the corpus and further results can be found at our website[10].

## 3.2   Analysis of the Reused Terms URIs

63 ontologies correctly applied the OBO principle explained in Sect. 2.3 to define their URIs. 60 ontologies contained terms that do not follow the principle: 55

---

[7] http://owl.cs.manchester.ac.uk/research/modularity/.

[8] http://www.obofoundry.org/.

[9] https://github.com/OBOFoundry/purl.obolibrary.org/.

[10] http://sele.inf.um.es/ontoenrich/projects/reuse/aime2017/.

ontologies had such cases only for implicitly-reused terms, 5 for only explicitly-reused ones, and 5 ontologies had cases for both types of reuse. Table 1 shows 5 examples of such situations.

**Table 1.** Example of URIs that do not follow the format proposed by the URIs/Identifiers principle of the OBO Foundry.

|         | Source ontology     | Incorrect URI                                                              |
|---------|---------------------|----------------------------------------------------------------------------|
| E.g. 1  | CDAO                | http://www.geneontology.org/formats/oboInOwl#DbXref                        |
| E.g. 2  | Chemical Inf. Ont.  | http://semanticscience.org/resource/CHEMINF_000318                         |
| E.g. 3  | Chemical Inf. Ont.  | http://www.ifomis.org/bfo/1.1/snap#GenericallyDependentContinuant          |
| E.g. 4  | Cell Line Ont.      | http://www.ebi.ac.uk/cellline#cervical_carcinoma_cell_line                 |
| E.g. 5  | Cell Line Ont.      | http://ncicb.nci.nih.gov/xml/owl/EVS/Thesaurus.owl#Iliac_Vein              |

### 3.3   Analysis by the Type of Reuse

Figure 3 shows the distribution of ontologies by the type of reuse. 49 ontologies did not reuse terms. The remaining 75 ontologies imported at least one term, with implicit reuse being the prominent strategy. The explicit reuse of terms is commonly combined with the implicit one, so ontology developers integrate the reused terms in the *source ontology*, and perform some enrichment with external content.

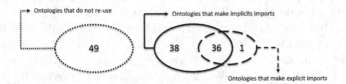

**Fig. 3.** Distribution of the ontologies according to the type of reuse that they perform.

We used the data obtained by our method as input of the The Open Graph Viz Platform[11]. We built a graph, which can be visualised and explored from different focuses by using filters, for example: (1) Fig. 4 left highlights those ontologies that reuse OBI and other ontologies and relations can be observed in the background; (2) Fig. 4 right shows the filtered graph based on a clustering algorithm that is explained next.

It should be pointed out that the nodes represent ontologies. The directed edges between two nodes means that the node from which the edge departs is the *source ontology* and the other is the *external* one. The weight of each edge represents the number of terms reused, which is represented by the thickness of the edges. The size of each node represents the number of times that the ontology

---

[11] https://gephi.org/.

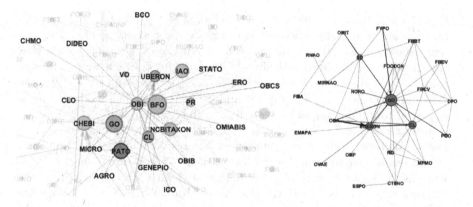

**Fig. 4.** Graphs that represent the reuse between the ontologies in our corpus. Generated with Gephi using Fruchterman Reingold as layout algorithm to minimise overlap.

is reused. Using Fig. 4 right as example, the Gene Ontology (GO) is reused by the Ontology of Biological Attributes (OBA), GO is reused more times than OBA, and OBA reused more terms from UBERON than from GO. Finally, we performed a cluster analysis of the ontologies using as parameter the weight of the edges (see report at[12]). Clusters are represented by colours in the graph.

The cluster analysis returned 51 clusters. More than 60% of the ontologies were classified in 9 clusters; the reminder clusters had just one member, what means that they do not reuse content. Conceptually, the clusters can be used, for example, to visualise: (1) groups of ontologies that reuse a similar number of terms between them (Fig. 4 right), (2) groups of ontologies that are frequently reused by others, or (3) a small set of ontologies with a high reuse between them in comparison with the members of other clusters (see more figures with the clusters in our webpage). Visualisations like these might contribute to the understanding of the reuse among a large set of ontologies, and they offer different perspectives of analysis to ontology developers.

## 3.4 Analysis of Hidden Axioms and Terms Already Reused

Finally, we analysed the existence of *hidden axioms* associated with relations already reused in our corpus. Figure 5 summarizes to what extent the reuse of axioms is performed and how the application of the modularity algorithm could be used to increase the reuse of terms and axioms. This result comes from analysing both the implicit and explicit reuse.

– *Terms reuse:* Fig. 5 left compares the mean number of terms that are reused and the potentially reusable ones from the *external ontologies*. The mean number of terms implicitly reused by the analysed ontologies is 855 and the number of explicitly reused ones is 1 210 terms. This difference makes us think

---

[12] http://sele.inf.um.es/ontoenrich/projects/reuse/aime2017/cluster.

that the `owl:import` operation is not including all the content from the original ontology but a simplified version (e.g. see the percentage of the explicit reuse shown in Fig. 2). The application of our modular strategy finds that the signature of the automatically obtained modules, which were extracted using as *seed signature* already reused terms, contains a mean of 2016 and 2376 terms respectively for implicit and explicit reuse. The modules can be imported containing terms logically link to the one reused.

– *Axioms reuse:* Fig. 5 right performs a similar analysis, but focused on axioms instead of terms. The mean number of axioms associated with implicitly reused terms in *source ontologies* (also existing in the external ones) is 2390, whereas the mean number of axioms associated with such terms only in the *external ontologies* is 22680; this means that, on average, each ontology has 20290 *hidden axioms*. The results for explicitly reused terms is, respectively, 2690 reused axioms and 27710 hidden ones.

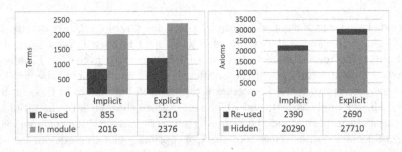

**Fig. 5.** (Left) Comparison between the number of reused terms and those included in the locality-module extracted. (Right) Comparison between those axioms reused and *hidden axioms* in the complete *external ontology*.

Finally, Fig. 6 shows the most frequent axioms linked with terms reused in the *source ontologies* (left), and that are hidden in the *external* ones (right).

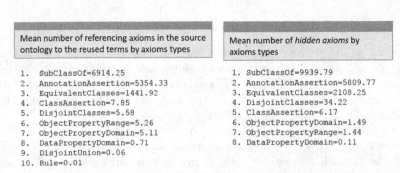

**Fig. 6.** Ranking of different types of axioms related with the reused terms.

# 4   Discussion and Conclusions

How much and which content is necessary to reuse is an open discussion in the ontology community. One option is to import (using `owl:import`) the complete *external ontology* when at least one term is reused. This option may require high computational resources when reasoning is required, since even the content that is not reused should be processed by a reasoner. This motivated the development of the MIREOT principle [1] promoting the URIs reuse. MIREOT is likely to be the main reason of the implicit reuse to avoid working with too large ontologies and to not worry about the potential unintended inferences if the complete ontology is imported, what can be criticized from a formal point of view.

The goal of our method is to study the amount of logical knowledge in the *external ontologies* that could be used to axiomatically enrich the *source ontologies*. We have designed a strategy that complements both implicit and explicit reuse. For this reason, we decided to start by analysing the already reused terms. It is worth pointing out that the number of terms shown in Fig. 5 represents less than 2% of all the terms implicitly and explicitly defined in the *external ontologies* (see the graphical representation in our webpage). Increasing the number of terms reused, which could be in line with works such as [7,8], is out of the scope of this work, except for those linked to *hidden axioms*.

Our method requires us to find the ontology to which each reused term belongs. This is currently performed through URI analysis, but this would exclude all the terms that do not follow the URI principle. For example, the URI in the row 3 of Table 1 is quite close to the OBO proposed format; row 4 uses an old reference to the updated term http://purl.obolibrary.org/obo/BFO_0000031. This is a limitation of our current implementation as the method could be improved to use heuristics to overcome such issues or to handle XREF references. Moreover, for all the ontologies associated with terms, the method needs to process their complete implementation. Otherwise, the method could not compute the module or have the information about the potential amount of knowledge that could be reused. Therefore, the results presented here must be contextualised to the set of 123 OBO Foundry ontologies that were successfully processed.

Concerning the impact of our method in current ontologies, the extracted modules could be reused through `owl:import` operations, which would include all the mentioned *hidden axioms*/related terms. This would contribute to the *quality assurance* of *source ontologies* from a logical point of view, and reasoners could use this new content to make inferences. Despite those terms in the modules are selected because they are linked through logical relations with terms already reused in the *source ontology* (used as a *seed signature*), it should be measured if they are conceptually of interest for the *source ontology*. Moreover, once the modules are explicitly imported a reasoner should be used to check the consistency of the enriched ontology. Therefore, our method can be used as a complementary and automatic approach to the application of the MIREOT principle with the Ontofox [12] tool, where ontology developers manually configure what to import using, e.g., a SPARQL-based ontology term retrieval algorithm.

In conclusion, we believe that our method contributes to the *quality assurance* of biomedical ontologies. The paper describes the application of the method to characterise the reuse within the OBO Foundry ontologies. This corpus has been selected because this community builds ontologies by applying a set of shared principles. The findings are that 49 ontologies do not make any type of reuse and 75 do reuse terms. Implicit reuse is the predominant action, that being complemented in a 50% of the cases with explicit reuse. The study of the reused terms has permitted us to visualise the dependencies between ontologies and to cluster them according to the number of ontologies and terms reused. Finally, the exploration of axioms that reference to already reused terms, has revealed that the combination of the content currently being reused with our modular extraction strategy might contribute to increase the axiomatic content of current ontologies, with both new terms and axioms. As future work, we propose the analysis of a larger set of ontologies, improving the mechanism for linking terms with the *source ontology*, and studying the impact of axiomatic richer ontologies in tools that exploit the semantics of biomedical ontologies like [13].

**Acknowledgements.** This work has been partially funded by to the Spanish Ministry of Economy, Industry and Competitiveness, the FEDER Programme and by the Fundación Séneca through grants TIN2014-53749-C2-2-R and 19371/PI/14.

# References

1. Courtot, M., Gibson, F., Lister, A.L., Malone, J., Schober, D., Brinkman, R.R., Ruttenberg, A.: MIREOT: the minimum information to reference an external ontology term. Appl. Ontol. **6**(1), 23–33 (2011)
2. Delbru, R., Tummarello, G., Polleres, A.: Context-dependent OWL reasoning in sindice - experiences and lessons learnt. In: Rudolph, S., Gutierrez, C. (eds.) RR 2011. LNCS, vol. 6902, pp. 46–60. Springer, Heidelberg (2011). doi:10.1007/978-3-642-23580-1_5
3. Ghazvinian, A., Noy, N.F., Musen, M.A.: How orthogonal are the OBO foundry ontologies? J. Biomed. Semant. **2**(2), S2 (2011)
4. Grau, B.C., Horrocks, I., Kazakov, Y., Sattler, U.: Modular reuse of ontologies: theory and practice. J. Artif. Int. Res. **31**(1), 273–318 (2008)
5. Horridge, M., Bechhofer, S.: The OWL API: a JAVA API for OWL ontologies. Semant. Web **2**(1), 11–21 (2011)
6. Jiménez-Ruiz, E., Grau, B.C., Sattler, U., Schneider, T., Berlanga, R.: Safe and economic re-use of ontologies: a logic-based methodology and tool support. In: Bechhofer, S., Hauswirth, M., Hoffmann, J., Koubarakis, M. (eds.) ESWC 2008. LNCS, vol. 5021, pp. 185–199. Springer, Heidelberg (2008). doi:10.1007/978-3-540-68234-9_16
7. Kamdar, M., Tudorache, T., Musen, M.A.: A systematic analysis of term reuse and term overlap across biomedical ontologies. Semant. Web, 1–19 (2016). http://content.iospress.com/articles/semantic-web/sw238
8. Quesada-Martínez, M., Mikroyannidi, E., Fernández-Breis, J.T., Stevens, R.: Approaching the axiomatic enrichment of the gene ontology from a lexical perspective. Artif. Intell. Med. **65**(1), 35–48 (2015)

9. Simperl, E.: Reusing ontologies on the semantic web: a feasibility study. Data Knowl. Eng. **68**(10), 905–925 (2009)

10. Smith, B., Ashburner, M., Rosse, C., Bard, J., Bug, W., Ceusters, W., Goldberg, L.J., Eilbeck, K., Ireland, A., Mungall, C.J., et al.: The OBO foundry: coordinated evolution of ontologies to support biomedical data integration. Nat. Biotechnol. **25**(11), 1251–1255 (2007)

11. Whetzel, P.L., Noy, N.F., Shah, N.H., Alexander, P.R., Nyulas, C., Tudorache, T., Musen, M.A.: BioPortal: enhanced functionality via new web services from the national center for biomedical ontology to access and use ontologies in software applications. Nucleic Acids Res. **39**(Suppl 2), W541–W545 (2011)

12. Xiang, Z., Courtot, M., Brinkman, R.R., Ruttenberg, A., He, Y.: OntoFox: web-based support for ontology reuse. BMC Res. Notes **3**(1), 175 (2010)

13. Znaidi, E., Tamine, L., Latiri, C.: Answering PICO clinical questions: a semantic graph-based approach. In: Holmes, J.H., Bellazzi, R., Sacchi, L., Peek, N. (eds.) AIME 2015. LNCS, vol. 9105, pp. 232–237. Springer, Cham (2015). doi:10.1007/978-3-319-19551-3_30

# Ontological Representation of Laboratory Test Observables: Challenges and Perspectives in the SNOMED CT Observable Entity Model Adoption

Mélissa Mary[1,2(✉)], Lina F. Soualmia[2,3], Xavier Gansel[1],
Stéfan Darmoni[2,3], Daniel Karlsson[4], and Stefan Schulz[5] (iD)

[1] Department of System and Development, bioMérieux,
La-Balme-Les-Grottes, France
{melissa.mary,xavier.gansel}@biomerieux.com

[2] LITIS EA 4108, Normandy University, University of Rouen, Rouen, France
{lina.soualmia,stefan.darmoni}@chu-rouen.fr

[3] French National Institutes for Health (INSERM), LIMICS UMR_1142,
Paris, France

[4] Department of Biomedical Engineering/Health Informatics,
Linköping University, Linköping, Sweden
daniel.karlsson@liu.se

[5] Institute of Medical Informatics, Statistics and Documentation,
Medical University of Graz, Graz, Austria
stefan.schulz@medunigraz.at

**Abstract.** The emergence of electronic health records has highlighted the need for semantic standards for representation of observations in laboratory medicine. Two such standards are LOINC, with a focus on detailed encoding of lab tests, and SNOMED CT, which is more general, including the representation of qualitative and ordinal test results. In this paper we will discuss how lab observation entries can be represented using SNOMED CT. We use resources provided by the Regenstrief Institute and SNOMED International collaboration, which formalize LOINC terms as SNOMED CT post-coordinated expressions. We demonstrate the benefits brought by SNOMED CT to classify lab tests. We then propose a SNOMED CT based model for lab observation entries aligned with the BioTopLite2 (BTL2) upper level ontology. We provide examples showing how a model designed with no ontological foundation can produce misleading interpretations of inferred observation results. Our solution based on a BTL2 conformant formal interpretation of SNOMED CT concepts allows representing lab test without creating unintended models. We argue in favour of an ontologically explicit bridge between compositional clinical terminologies, in order to safely use their formal representations in intelligent systems.

**Keywords:** Biomedical ontologies and terminologies · LOINC · SNOMED CT · BioTopLite2

© Springer International Publishing AG 2017
A. ten Teije et al. (Eds.): AIME 2017, LNAI 10259, pp. 14–23, 2017.
DOI: 10.1007/978-3-319-59758-4_2

# 1 Introduction

The emergence of Electronic Health Records has raised interoperability challenges in (i) the establishment of common data structures and (ii) the definition of semantic standards to represent clinical information. The representation of *in vitro* diagnostic observation (Table 1) in laboratory reports follows a global tendency by health care providers and public health institutions [1] towards two semantic standards, *viz.* the Logical Observation Identifiers Names and Codes (LOINC) terminology [2, 3] and the ontology-based clinical terminology SNOMED CT [4–6]. Whereas LOINC provides precise, compositional encodings of lab tests and other clinical observables, SNOMED CT provides codes for nominal and ordinal scale result values. For three years, the respective maintenance organisations, Regenstrief Institute and SNOMED International have worked together in order to elaborate a first representation of 13,756 LOINC tests (over a total of 79,000) as post-coordinated SNOMED CT expressions [7, 8]. The main advantage of this LOINC–SNOMED CT interoperation is to enable the representation of observation results (pairs of lab test observables with result values) within SNOMED CT, using its post-coordination mechanism.

**Table 1.** Definitions of main notions and examples of a naïve observation model.

| Term | Definition | Example |
|------|-----------|---------|
| Observable | a plan for an observation procedure to observe a feature (quality, disposition, or process quality) of an entity | *Non-invasive systolic blood pressure, measured on upper left arm.* |
| Lab test observable | *in vitro* diagnostic tests represented by LOINC or SNOMED CT concepts in the Observable Entity subhierarchy | *17279-1 Bacteria identified: Prid:Pt:Plr fld:Nom:Aerobic Culture* |
| Result value | immaterial, information-like outcomes of an action | *Present, Absent* |
| Observation | realization of an Observable yielding a Result value, typically described as a Observable – Result vale pair | |

This article focuses on the use of SNOMED CT to describe lab test observables and observations. We present a hierarchical reorganisation of lab test observables computed by the ELK inference engine [9] and discuss their usability into observation context. We demonstrate that hierarchical lab test structures combined into naïve ad-hoc observation models raise misleading interpretations. Motivated by unintended results, we propose a new model to represent observation entries, compatible with the SNOMED CT lab test observables formalization under the biomedical upper level ontology BioTopLite2 [10].[1]

---

[1] In the following, we will abbreviate BioTopLite by BTL2 and SNOMED CT by SCT. In lower case, these acronyms will also be used as namespace identifiers.

This article is organized as follows. Section 2 presents current resources and the classification method used. Section 3 presents our results. In the last section, we discuss the results obtained and give an outlook to further work.

## 2 Materials and Methods

### 2.1 Terminologies and Ontologies

**LOINC** is a terminology created in 1994 by the Regenstrief Institute to represent clinical and lab tests [2, 3]. Its version 2.54, used in this study, describes 79,000 tests, of which approximately 50,000 are lab tests. Test models in LOINC are defined by six main dimensions that represent information about protocols (Analyte, Method, Time and System), together with the type of result values expected (Scale and Property). Test descriptions can be refined by the addition of three optional items of information (Challenge, Divisor and Adjustment).

**SNOMED CT** (SCT) is an increasingly ontology-based clinical terminology created in 2002 to formally represent the wide range of terms used of the clinical domain [5]. It is maintained by SNOMED International (formerly IHTSDO), and distributed in a relational file format (RF2) [6], from which an OWL EL ontology can be programmatically derived [4]. SNOMED CT's January 2016 international release is constituted by around 300,000 concepts, thematically arranged by 18 mostly disjoint subhierarchies. A SNOMED International working group has, since then, invested considerable effort in improving SCT's *Observable Entity* subhierarchy by ontology patterns [11], which sets clinical and lab tests description on formal-ontological grounds and allows enough flexibility to mimic the granularity of LOINC. The new Observable entity model was introduced in the international SCT release in January 2017.

**The LOINC – SCT resource** described in this paper is the third technical preview release provided by Regenstrief and SNOMED International [7, 8] in April 2016, using the SCT January 2016 release, as an outcome of an interinstitutional cooperation agreement signed in July 2013. It represents the representation of 13,786 LOINC (28% of lab) codes into SCT post-coordinated expressions using Observable Entity ontology patterns. The LOINC – SCT alignment release is distributed in three formats: RF2, OWL and Excel ("Human Readable"). In this study we used the OWL format.

**BioTopLite2** (BTL2) is a top-domain ontology [10] that intends to address the need for clear-cut upper-level classes (55) and relations (37), thus improving and facilitating ontology engineering and inter-operation. A preliminary bridge between SCT and BTL2 has been proposed in 2015 [12], addressing the problem that SCT's basic ontological assumptions are partly hidden in the documentation and partly underpinned by OWL examples scattered across publications.

### 2.2 Lab Test Observable Classification

The reason for the automatic classification of LOINC concepts based on SCT expressions was to use description logics inference in order to add new subclass links to the hitherto flat LOINC structure. It was performed on the OWL file of the third release of the LOINC – SCT resource as described in the previous paragraph. The classification process is composed of two steps:

- Firstly, the OWL version of the SCT January 2016 international release is imported.
- Secondly, the ELK reasoner [9], which computes description logics inferences on OWL EL with a good scalability behaviour, is used to reclassify the merged ontologies.

## 3 Results

### 3.1 ELK Classification Metrics

We first observed that 45 lab test observables were inferred as pairwise equivalent. The ELK inference engine also infers subsumption relationships for half (6,789) of the SCT post-coordinated expressions that represent LOINC codes. Among them, we observed that 16.6% were classified into poly-hierarchies. The resulting taxonomy of lab test observables had an average depth of 1.5, with a maximum of five levels below the top concept *363787002 | Observable entity*. We distinguished two reasons of the obtained subsumption: definition increment (31% of inference) and definition refinement (78%).

▼    **'31718-0 Astrovirus Ag:ACnc:Pt:Stool:Ord'**        *(A)*
      **'7810-5 Astrovirus Ag:ACnc:Pt:Stool:Ord:EIA'**

*'7810-5 Astrovirus Ag:ACnc:Pt:Stool:Ord:EIA'* equivalentTo       *(B)*
    sct:*Observable entity* and
          **sct:Component** some sct:*AstrovirusAntigen* and
          **sct:Scale type** some sct:*OrdinalValue* and
          **sct:Time aspect** some sct:*SinglePointInTtime* and
          **sct:Property type** some sct:*ArbitraryConcentration* and
          **sct:Inheres in** some sct:*GastrointestinalTractMaterial* and
          **sct:Direct site** some sct:*StoolSpecimen* and
          <u>**sct:Technique** some sct:*EnzymeImmunoassayTechnique*</u>

**Fig. 1.** ELK reclassification of a LOINC post-coordinated concept: definition increment. (A) Hierarchical relation inferred by ELK; (B) *'7810-5 Astrovirus Ag: ACnc:Pt:Stool:Ord:EIA* definition (the restriction added to the '31718-0 *Astrovirus Ag:ACnc:Pt:Stool:Ord*' definition is <u>underlined</u>).

We understand by "definition increment" computed subclass inferences entailed by additional restrictions in the formal definition of more specific lab test observable concepts (Fig. 1). For instance definition of the post-coordinated concept *'7810-5 Astrovirus Ag:ACnc:Pt:Stool:Ord:EIA'* specifies the technique used (underlined in Fig. 1B, *viz.* lab test using enzyme immunoassay technique), whereas the definition of *'31718-0 Astrovirus Ag:ACnc:Pt:Stool:Ord* ' does not (lab test using any technique). We observed that 91.3% of the definition increment subsumptions were caused by the addition of the technique information to the subtest definition, in this particular case, the technique enzyme immunoassay.

By "definition refinement" we mean classification inferences entailed by hierarchical relationships between existing SCT concepts, used in the same part of lab test observables definitions (Fig. 2). For instance the LOINC post-coordinated concept *'17279-1 Bacteria identified:Prid:Pt:Plr fld:Nom:Aerobic Culture'* is computed as a subclass of *'618-9 Bacteria identified:Prid:Pt:Plr fld:Nom:Culture'*, because the concept *Aerobic culture* is a child of *Culture* (i.e. any culture) in SCT. We observed that such definition refinement assertions were mainly (4,878 - 90.1%) due to concepts representing the *Component* dimension in LOINC code, i.e. the specific component of the material analyzed like a bacteria of pleural fluid in this case.

```
'618-9 Bacteria identified:Prid:Pt:Plr fld:Nom:Culture'
  '17279-1 Bacteria identified:Prid:Pt:Plr fld:Nom:Aerobic culture'
  '38393-5 Legionella sp identified:Prid:Pt:Plr fld:Nom:Organism specific culture'
  '53909-8 Mycobacterium sp identified:Prid:Pt:Plr fld:Nom:Organism specific culture'
```

**Fig. 2.** ELK classification of organism culture in pleural fluids sample lab test observables, example of definition refinement.

## 3.2   Lab Test Classification Issue

In ontologies, the subsumption relationship (**rdf:subClassOf** [13]) is transitive (1) and expresses that every individual member of subsumed class is also member of the corresponding superclass(es) (2).

$$c_1 \text{ \textbf{rdfs:subClassOf} } c_2 \ \wedge \ c_2 \text{ \textbf{rdfs:subClassOf} } c_3 \ \Rightarrow \ c_1 \text{ \textbf{rdfs:subClassOf} } c_3 \qquad (1)$$

$$c_1 \text{ \textbf{rdfs:subClassOf} } c_2 \ \Rightarrow \ \text{Instance}(c_1) \subset \text{Instance}(c_2) \qquad (2)$$

Figure 2 illustrates the classification of observables on organism cultures from pleural fluids. From a logical perspective, the classification is consistent. Indeed, *Aerobic culture* is a kind of *Culture*, as well as the assertion that a lab test for an anaerobic bacteria culture (*'17279-1 Bacteria identified:Prid:Pt:Plr fld:Nom:Aerobic Culture'*) is more specific than a lab test applying any bacteria culture test technique (*'618-9 Bacteria identified:Prid:Pt:Plr fld:Nom:Culture'*). In a laboratory report context, lab test observables described by LOINC are used to represent observation entries (see Table 1). In the next section we will present how the lab tests classification can mislead users in the interpretation of inferred observation statements.

**Problem Statement.** In this section we consider a naïve interpretation of *in vitro* diagnostic observations (Table 1 and Fig. 3A), in which an observation is expressed by a direct relationship between a lab test observable and test result value, linked by the relation **hasResultValue**. In the following example, $i_1$ and $i_2$ are two instances of the observation result concept (Fig. 3B, C) representing the lab test observable *'17279-1 Bacteria identified:Prid:Pt:Plr fld:Nom:Aerobic Culture'* on pleural fluid, together with the result values *Present* and *Absent*.

First, we consider the $i_1$ observation result pattern representing the presence of bacteria in aerobic culture condition from pleural fluid. Under the interpretation of $i_1$ being an instance of *'17279-1 Bacteria identified:Prid:Pt:Plr fld:Nom:Aerobic Culture'* and **hasResultValue** some *Present* (literally the **Presence** *of bacteria in **aerobic** culture in pleural fluid).* $i_1$ is inferred as being also an instance of the expression *618-9 Bacteria identified:Prid:Pt:Plr fld:Nom:Culture'* and **hasResultValue** some *Present* (***Presence** of bacteria in **some** culture of pleural fluid)* because $i_1$ is also an instance of the *'618-9 Bacteria identified:Prid:Pt:Plr fld:Nom:Culture'* test (Eq. 2, Fig. 2). In this assertion the *some* constructor is inherited from the formal definition and thus is not explicitly stated in the observation. Nevertheless, it plays a key role in the interpretation of the inferred observation. This explains why a naïve interpretation would be misleading, because the existentiality notion (*i.e.* ∃) would be intuitively ignored, as we will see later. As long as the $i_1$ observation is positive ("presence"), the entailment of *"presence of bacteria in pleural fluid culture"* seems plausible and straightforward.

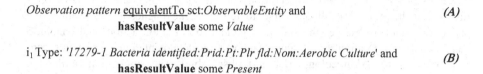

*Observation pattern* <u>equivalentTo</u> sct:*ObservableEntity* and    *(A)*
            **hasResultValue** some *Value*

$i_1$ Type: *'17279-1 Bacteria identified:Prid:Pt:Plr fld:Nom:Aerobic Culture'* and    *(B)*
            **hasResultValue** some *Present*

$i_2$ Type: *'17279-1 Bacteria identified:Prid:Pt:Plr fld:Nom:Aerobic Culture'* and    *(C)*
            **hasResultValue** some *Absent*

**Fig. 3.** Simple model of observation entry pattern (A) and two examples of instances.

Contrary to $i_1$, the instance $i_2$ of the observation class represents a negative assertion (***Absence** of bacteria in pleural fluid **aerobic** culture*) and can produce a misleading interpretation due to the subsumption of the observation result. The individual $i_2$ is also an instance of *'618-9 Bacteria identified:Prid:Pt:Plr fld:Nom:Culture'*, equivalent to *'618-9 Bacteria identified:Prid:Pt:Plr fld:Nom:Culture'* and '**hasResultValue**' some *Absence* (literally "***Absence** of bacteria in **some** culture of pleural fluid sample"*). In our example, this means that the statement "***Absence** of bacteria in some culture of pleural fluid sample"* is not sufficient to infer the general statement "***Absence** of bacteria in pleural fluid culture"*. In other words, this would not contradict that in another pleural fluid culture, e.g. for anaerobic bacteria, the finding is positive.

This example shows that the 'naïve' representation of the observation model (Fig. 3A), as it might be interpreted by non-ontologists, misleads the interpretation of inferred lab test observation results. The phenomenon, exemplified regarding *in vitro* diagnostic observations with binary result values (*Present, Absent*), can easily be generalized to ordinal and quantitative result values. In the next section we propose a solution to this issue by formalizing observation entry and lab tests with general categories proposed by the BTL2 upper-level ontology [10].

**Lab Test Redefinition with BioTopLite2.** Lab tests as described in LOINC, as well as in SCT under *Observable Entity* are meant to be representations, i.e. information about processes but not the processes themselves. The Regenstrief Institute's characterisation of LOINC codes as "universal names and ID codes for identifying laboratory and clinical test results" [14] suggests exactly this. This distinction is also justified by the fact that a given LOINC code can be assigned to different *in vitro* diagnosis products from different manufacturers, each with a different laboratory process.

The distinction between information and process is materialized in BTL2 by the disjointness between the categories *btl2:InformationObject* and *btl2:Process*. The characterization of lab test observables as being information objects (*btl2: InformationObject)* and not processes sheds light on misleading interpretations of the observation pattern as exposed in the previous section (Fig. 3A). That lab test observables are information objects in the sense of BTL2 is fully coherent with the fact that they are complemented by observation result values. Result values are (immaterial, information-like) outcomes of an action (in our example represented by the lab test processes proper). The word "observation" (Table 1), in this context, is rather confusing than helpful, as it alludes, first, to the classical diagnostic observation of a patient: a clinician observes the skin of a patient (observation action) and concludes that it is pale (observation result value). This value is documented in the patient record, next to the entry "skin colour" (lab test observable). The summation of the two information entities "skin colour" and "pale", makes the information complete.

In parallel, a machine "observes" (actually measures) a blood sample for haemoglobin. The outcome "9 mg/ml" (observation result value), is, in this case, more precisely a piece of information produced by a machine, which completes the lab observable "Haemoglobin concentration in blood" (again, an information object). Both composite information objects, *viz.* "Skin colour: pale" and "Haemoglobin concentration in blood: 9 mg/dl" then represent some medical condition like *Anaemia*. Note that this does not mean that there is always an instance of anaemia, because the results of observations and measurements, as such, bear the possibility of being non-referring, e.g. due to the clinicians' diagnostic error, due to inappropriate light conditions, or due to a technical error in the machine.

The discussed implausible inference also highlights difficulties in interpreting ontologies for practical applications. A clarification of the intended meaning, and, in consequence the prevention of implausible interpretations, can be achieved by reference to an upper-level ontology like BTL2, as we have demonstrated. We will therefore propose a consistent modelling pattern for representing observables extracted from laboratory reports, placing SCT and LOINC under BTL2. Model requirements and the definition of the main concepts are presented in the following section.

## 3.3    Representation of Observation Using BTL2

The above issue is addressed by proposing a new SCT approach to represent laboratory observables called "Observation entry". This model intends to complete SNOMED International's work on the formalization of lab test observables [11] rather than competing with it. Indeed, SNOMED International started to address this issue

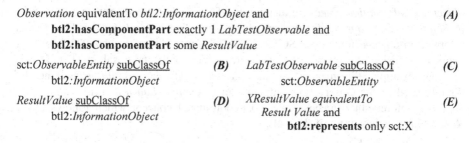

*Observation* equivalentTo *btl2:InformationObject* and                    *(A)*
    **btl2:hasComponentPart** exactly 1 *LabTestObservable* and
    **btl2:hasComponentPart** some *ResultValue*

sct:*ObservableEntity* <u>subClassOf</u>   *(B)*   *LabTestObservable* <u>subClassOf</u>   *(C)*
   btl2:*InformationObject*               sct:*ObservableEntity*

*ResultValue* <u>subClassOf</u>   *(D)*   *XResultValue* equivalentTo   *(E)*
   btl2:*InformationObject*         *Result Value* and
                      **btl2:represents** only sct:X

**Fig. 4.** Observation model main class definitions. *XResult* according to [15].

especially with the Observables working group, which formalized an ontology pattern for lab test observables within the *Observable Entity* hierarchy and proposes a formalization of the *result value* (named *Observation Result* in the SCT document) according to existing concepts and relations in SCT [11]. Our contribution to the conceptualization of observation is to offer a different point of view using BTL2 as ordering principle.

We define *Observation* (Fig. 4A, Table 1) as an *Information object* composed by a lab test observable and associated result value information: it is formalized by the mereological sum (**btl2:hasComponentPart**) of the *Lab Test Observable* (a specific type of information object) and the test result value (*Result value*). These definitions, in addition of being consistent with BTL2, solve misleading interpretations. Indeed, as observation result concepts (Table 1, Fig. 4A) are no longer subsumed by lab test observable concepts, the entailment of instances due to the **rdfs:subClassOf** definition will no longer occur. The *Lab test observable* concept (Fig. 4C) is an *Observable Entity* post-coordinated expression (Fig. 4B) as defined in the LOINC–SNOMED CT harmonization resource. *Lab test observable* is modelled as indirectly subsumed by *btl2:InformationObject*.

The *Result Value* concept (Fig. 4D) is instantiated by information produced by observation or analysis processes. In laboratory reports, we distinguish between two kinds of results and each has its own formalization: Literal result values are alphanumeric symbols or concatenations thereof, which form numbers or strings of characters like DNA sequences. *Literal Result Value* is a new concept linked to *rdfs: Literal* with the **hasValue** datatype property. It is disjoint from *Result Value*. We refrain from any further ontological account of literal results, especially numbers, due to the inherent intricacy of the ontology of mathematical objects, and the lack of relevance for most use cases. Opposed to *Literal result values*, *X result values* (Fig. 4E) are "conceptual" outcomes, which correspond to concepts in ontologies like SCT. Interestingly, *X result values* cannot be directly represented by the terminology codes. Considering *3092008 | Staphylococcus aureus,* this SCT concept can be used in SCT to define a disease (*i.e. 441658007 | Pneumonia caused by Staphylococcus aureus*) as well as in lab reports to point to the result of a bacteria identification test. Whereas in the first case, the definitions implies that *Staphylococcus aureus* instances, i.e. real bacterial organisms (under *btl2:Material object*), in the case of lab reports the target concept must be *Information object*, according to our stipulations. Because *material* *object* and *information object* are disjoint in BTL2, we here need another way to refer to *Staphylococcus aureus*. So we propose to formalize *X result values* (Fig. 4E) as information objects linked to a SCT concept (not an instance) by the BTL2 relation **represents**, using the quantifier "only", according to the proposal in [16]. In our previous example, this means that the concept *Staphylococcus aureus* will not be directly used in the Observation Entry model to express *Lab test result values*. We would therefore rather create a new post-coordinated concept which follows the *X Result Value* pattern [15], cf. Fig. 4E.

## 4   Conclusion

This paper elaborates on the representation of laboratory observables with SNOMED CT. We first studied the LOINC – SNOMED CT harmonization resource, which proposes a representation of LOINC lab tests by post-coordinated SNOMED CT expressions and observed that the classification of lab test observables was enhanced due to a formal representation in OWL-EL and the SNOMED CT concept hierarchy. A previous study [16] on the LOINC – SNOMED CT resource had also demonstrated benefits of SCT to enhance lab test queries.

We then focused on the formalization of lab test observables. We analysed a typical pattern representing information stored in laboratory reports and demonstrated how it might be interpreted by lab staff. We showed how naïve interpretations of lab test observation results is misleading because they blend the meaning of represented and representing entities. An implementation of this naïve model in clinical decision support system could, in the worst case, infer wrong observation results and affect patient safety if included in a medical decision support pipeline.

We finally formalized a new observation model constrained by the BioTopLite (BTL2) upper level ontology. Bridging the observation model to BTL2 clarifies the intended meaning of lab tests and observations. This shows the normative value of a strict upper level ontology, which would also be helpful for guiding the development of other ontologies like SNOMED CT. By that means, the ontology could evolve in a more principled way, avoiding the risk of competing ontological commitments [17]. SNOMED CT would thus gain more reliability in coding clinical information like lab results, which impacts on decision support and data analytics use cases. Standardization of patient data, especially with SNOMED CT, opens up new opportunities for implementing new clinical decision support tools putting *in vitro* diagnostic observation into a global patient context. In clinical microbiology [18], experts systems (as Vitek2 AES [19]) or ontologies [20] propose therapeutic corrections and antibiotic stewardship implementing rules extracted from guidelines. A principled observation model addressing the representation of *in vitro* diagnosis, compatible with a worldwide clinical terminology like SNOMED CT would therefore be a cornerstone for reliable decision support. Further steps will be to enhance and evaluate its capacity to perform decision support based on *in vitro* diagnostic data.

# References

1. Blumenthal, D.: Launching HITECH. N. Engl. J. Med. **362**, 382–385 (2010)
2. Logical Observation Identifiers Names and Codes (LOINC®) — LOINC. https://loinc.org/
3. McDonald, C.J., Huff, S.M., Suico, J.G., Hill, G., Leavelle, D., Aller, R., Forrey, A., Mercer, K., DeMoor, G., Hook, J., Williams, W., Case, J., Maloney, P.: LOINC, a universal standard for identifying laboratory observations: a 5-year update. Clin. Chem. **49**, 624–633 (2003)
4. Schulz, S., Suntisrivaraporn, B., Baader, F., Boeker, M.: SNOMED reaching its adolescence: ontologists' and logicians' health check. Int. J. Med. Inf. **78**(Suppl. 1), S86–S94 (2009)
5. Cornet, R., de Keizer, N.: Forty years of SNOMED: a literature review. BMC Med. Inform. Decis. Mak. **8**, S2 (2008)
6. SNOMED CT Document Library - SNOMED CT Document Library - IHTSDO Confluence. https://confluence.ihtsdotools.org/display/DOC/SNOMED+CT+Document+Library
7. Santamaria, S.L., Ashrafi, F., Spackman, K.A.: Linking LOINC and SNOMED CT: a cooperative approach to enhance each terminology and facilitate co-usage. In: ICBO 2014, pp. 99–101 (2014)
8. Regenstrief: Alpha (phase 3) Edition of Draft LOINC-SNOMED CT Mappings and Expression Associations. http://loinc.org/news/alpha-phase-3-edition-of-draft-loinc-snomed-ct-mappings-and-expression-associations-now-available.html/
9. Kazakov, Y., Krötzsch, M., Simančík, F.: ELK: a reasoner for OWL EL ontologies. Technical report, University of Oxford (2012)
10. Beisswanger, E., Schulz, S., Stenzhorn, H., Hahn, U.: BioTop: an upper domain ontology for the life sciences. Appl. Ontol. **3**, 205–212 (2008)
11. Spackman, K., Karlsson, D.: Observables and investigation procedures redesign. SNOMED International (2015)
12. Schulz, S., Martínez-Costa, C.: Harmonizing SNOMED CT with BioTopLite: an exercise in principled ontology alignment. In: MedInfo, pp. 832–836 (2015)
13. Smith, B., Ceusters, W., Klagges, B., Köhler, J., Kumar, A., Lomax, J., Mungall, C., Neuhaus, F., Rector, A.L., Rosse, C.: Relations in biomedical ontologies. Genome Biol. **6**, R46–R61 (2005)
14. LOINC Committee: LOINC User's Guide. Regenstrief Institute, Indianapolis (2016)
15. Schulz, S., Martínez-Costa, C., Karlsson, D., Cornet, R., Brochhausen, M., Rector, A.L.: An ontological analysis of reference in health record statements. In: FOIS, pp. 289–302 (2014)
16. Mary, M., Soualmia, L.F., Gansel, X.: Projection des propriétés d'une ontologie pour la classification d'une ressource terminologique. Journée Francophones sur les Ontologies, Bordeaux, 1–12 (2016)
17. Schulz, S., Cornet, R., Spackman, K.: Consolidating SNOMED CT's ontological commitment. Appl. Ontol. **6**, 1–11 (2011)
18. Rhoads, D.D., Sintchenko, V., Rauch, C.A., Pantanowitz, L.: Clinical microbiology informatics. Clin. Microbiol. Rev. **27**, 1025–1047 (2014)
19. Barry, J., Brown, A., Ensor, V., Lakhani, U., Petts, D., Warren, C., Winstanley, T.: Comparative evaluation of the VITEK 2 Advanced Expert System (AES) in five UK hospitals. J. Antimicrob. Chemother. **51**, 1191–1202 (2003)
20. Bright, T.J., Furuya, E.Y., Kuperman, G.J., Cimino, J.J., Bakken, S.: Development and evaluation of an ontology for guiding appropriate antibiotic prescribing. J. Biomed. Inform. **45**, 120–128 (2012)

# CAREDAS: Context and Activity Recognition Enabling Detection of Anomalous Situation

Hela Sfar[1]([⊠]), Nathan Ramoly[1], Amel Bouzeghoub[1], and Beatrice Finance[2]

[1] CNRS Paris Saclay, Telecom SudParis, SAMOVAR, Paris, France
{hela.sfar,nathan.ramoly,amel.bouzeghoub}@telecom-sudparis.eu
[2] DAVID, University of Versailles Saint-Quentin-en-Yvelines, Versailles, France
beatrice.finance@uvsq.fr

**Abstract.** As the world population is growing older, more and more peoples are facing health issues. For elderly, leaving alone can be tough and risky, typically, a fall can have serious consequences for them. Consequently, smart homes are becoming more and more popular. Such sensors enriched environment can be exploited for health-care applications, in particular Anomaly Detection (AD). Currently, most AD solutions only focus on detecting anomalies in the user daily activities while omitting the ones from the environment itself. For instance the user may have forgotten the pan on the stove while he/she is phoning. In this paper, we present a novel approach for detecting anomaly occurring in the home environment during user activities: CAREDAS, We propose a combination between ontologies and Markov Logic Network to classify the situations to anomaly classes. Our system is implemented, tested and evaluated using real data obtained from the Hadaptic platform. Experimental results prove our approach to be efficient in terms of recognition rate.

**Keywords:** Smart home · Anomaly Detection · Ontology · Markov Logic Network

## 1 Introduction

According to a survey from the British Office for National Statistics[1], in 2011–2030, around 53% of elderly persons in nine european countries will be living independently. Accordingly, less carefulness and heed are provided to the elderly from their family members. This missing care may lead to several problems, since senior population is minded to have a more sensitive health and physical conditions. Over the past few years, technological progress in pervasive computing has enabled the concept of smart homes. Smart homes are aiming to provide an environment for assisted living, which enables monitoring of the home contextual information and the resident's in-home activities.

---

[1] http://www.cairn-int.info/focus-E_POPU_704_0789–who-will-be-caring-for-europe-s-dependen.htm.

© Springer International Publishing AG 2017
A. ten Teije et al. (Eds.): AIME 2017, LNAI 10259, pp. 24–36, 2017.
DOI: 10.1007/978-3-319-59758-4_3

Reliable Anomaly Detection (AD) in daily in-home situations is the most important component of many home health care applications. In literature, an anomaly is defined as a deviation from the normal behavior [7–9]. Research has emphasized the daily user activities for the normal behavior learning. The normal behavior is, indeed, a model of the usual user activities classified under features using different machine learning techniques [10]. The anomaly is then detected as a deviation from this model either by the application of logic rules defined by experts, or through the exploitation of methods [11]. As a result, the detected anomaly can be an unusual activity or group of activities that can be analyzed considering some context. However, the anomalies produced by the environment context during an activity occurrence are not tackled in the state of the art. The following scenario illustrates this problematic:

**Scenario 1:** *Patrick is an old poker player living alone in his smart home. It is Monday, 10am, after watching TV and before lunch, Patrick is cooking in the kitchen as usual. The vent is shut down and all the kitchen windows are closed.*

In scenario 1, Patrick risks a 'suffocation' as the smoke induced by cooking can't be evacuated: this is an anomalous situation. State of the art approaches for AD are not able to detect such an anomaly since the anomaly is not in the activity (Patrick has done a usual activity) but in the environment. As a matter of fact, classical AD solutions mostly analyze one or a sequences of activities and detect deviation from expected activities, not on the context itself. Hence, the activity is considered as the main cause of anomaly.

In order to tackle this problematic, we propose a new method for AD named CAREDAS. CAREDAS is aimed to detect anomalous situation. A situation is a combination of the environment contextual data with the user activity. The situation is considered anomalous when the contextual data can identify anomalies while an activity is occurring, such as the suffocation anomaly in Scenario 1. CAREDAS reaches this objective through two contributions: firstly, it combines the ontological modeling and reasoning with a machine learning method: Markov Logic Network (MLN) [5]. Secondly, CAREDAS proposes an improvement of MLN to enable more flexibility.

The rest of the paper is organized as follows: Sect. 2 discusses related work. Section 3 details the contributions of the paper. Then, Sect. 4 introduces the knowledge base in CAREDAS while Sect. 5 presents its reasoning engine. Section 6 reports experimental results. Finally, Sect. 7 concludes the paper.

## 2   Related Work

As our contribution is two-fold, the literature was reviewed according to two axis: the first subsection reviews the existing AD techniques while the second deals with the MLN method.

### 2.1   Anomaly Detection

Previous proposed approaches can be classified into three classes of anomalies: Point, collective, and context anomalies [7].

**Point anomalies** consider each activity independently and decide whether it is anomalous or not regarding the normal behavior. For detecting point anomalies, Han et al. [13] use the mean of different features for different activities and apply a classification method to define regular behavior. Then, it looks for anomalous activities based on predefined thresholds of deviation. In [14–16] authors learn which rooms the resident is in during different times of day. Using circadian rhythms and room location, they monitor anomalies regarding room occupancy. In [8,17], the authors propose new systems of anomaly detection concerning mild cognitive impairment. In order to automatically reason with anomalies, they represent them in propositional logic. Then, according to expert defined rules, the anomaly is detected as an activity containing a deviation from the normal behavior. [9] is an extension of [17] in which the rules describing anomalies are generated automatically through a new classification method.

**Collective anomalies** consider groups of activity instances together to determine whether the group is normal or no. To do so, Anderson et al. [18] use an automata based approach to define sequence of activities as normal behaviors and learn those behaviors. They also support combining multiple days of activities to detect anomalies that occur over the time. In [12] Authors use unsupervised pattern clustering techniques to identify behavior model of the resident. Later, they apply a supervised machine learning method to detect anomalous sequences of activities.

**Contextual anomalies** consider activities under some context. Holmes [7] is a typical approach combining point, context, and collective anomalies detection. The considered context in this paper is the day of week. Holmes starts by constructing a hierarchical normal behavior. At its bottom level are the several regular behaviors classified per day. Then, these latter are gathered to model the temporal correlations between activities. After training, the anomalous activities are detected by computing their distance from the normal behavior.

Based on the papers we reviewed, all works focus on detecting anomalous user activities. They do not consider more global anomalies (e.g. anomalous situations) that involve the environment itself.

## 2.2    Markov Logic Network (MLN)

Unlike other machine learning methods that are purely probabilistic, a Markov Logic Network provides many benefits and allows to deal with first order logic and handle uncertainty. These characteristics make it an appealing tool to reason with anomalous situation classification. A Markov Logic Network (MLN) [5] is a machine learning method that allows to handle uncertainty, imperfection, and contradictory knowledge. Technically, a MLN is a finite set of pairs $(F_i; w_i)$; $1 \leq i \leq n$, where each $F_i$ is function-free first-order logic and $w_i \in R$ is its weight. Together with a finite set of constants $C = c_1 \ldots c_n$ it defines the ground MLN, i.e., the MLN in which logic rules do not contain any free variable.

Hence, a MLN defines a log-linear probability distribution over Herbrand interpretations (possible worlds):

$$P(x) = \frac{1}{Z} exp \left( \sum_{i}^{F} w_i n_i(x) \right) \tag{1}$$

where F is the set of rules in the MLN, $n_i(x)$ is the number of true groundings of $F_i$ in the possible world x, $w_i$ is the weight of $F_i$, and Z is the normalizing constant.

The MLN suffers from three main shortcomings: (1) scalability, (2) dealing with numerical constraints, and (3) staticness of ground MLN. Regarding the first issue, the work in [19] proposes a scalable MLN that can be applied in a big social graph. In order to overcome the second issue, the $MLN_{NC}$ [4] extends MLN with numerical constraints in the logic rules. The third problem of MLN regards the staticness of ground MLN. The rules, from which the ground MLN is built, should be given as input instantiated with all possible constants. Moreover, the rules weights are given as evidence. However, this is unsuitable in our case of smart homes: the rules weights depend on the given weights of the contextual data. Hence, these weights are changing depending on several conditions. Furthermore, the rules can be instantiated only from the given contextual data. This means that the system should be able to detect anomalies despite missing constants.

## 3 Contributions

In this paper we propose CAREDAS, a novel approach that features: (1) a new method for anomalous situation detection, and (2) a Dynamic Ground MLN.

**Anomalous Situations Detection:** As pointed out in the related work section, most works only focus on abnormality of activities. In this paper, we propose a new method to *identify anomalous situation on the whole context* and not only on activities. A situation has a set of context data, including activities. CAREDAS takes as input a situation and infer if it is anomalous or not. As the analysis should be continuous, the method works according to *time windows*. Furthermore, as sensor are not always reliable, *CAREDAS tackles and takes into account uncertainty*. All these features are enabled by using a particular knowledge base and a MLN. However, as it, MLN is not suitable for our needs: this leads to our second contribution.

**The Dynamic Ground MLN:** As mentioned in the related work section, ground MLN staticness is a major flaw for our approach. Classically, a ground MLN, in other word an instantiated MLN, is generated from an abstract MLN and a set of provided rules. But in our case, the sensors do not provide rules but context data. In this work, we propose a enhancement of MLN to *dynamically generate rules instances from context data*. Moreover, *rules' weights are also dynamically computed and take into account the uncertainty of data*.

The next sections will present in detail our contributions.

## 4    CAREDAS Knowledge Base

CAREDAS relies on a knowledge base with particular data structure in order to infer anomalous situations. We will now discuss this knowledge base.

### 4.1    Data Structure Definition

CAREDAS relies on various concepts defined as follows.

**Definition 1.** FUZZY CONTEXT DATA (FCD) *is a data obtained from a sensor observation of the environment. It contains multiple possible values each associated with a trust weight. Formally, it is a 4-tuple (subj, type, wVal, time) where subj is an entity of the environment (i.e. the Freezer), type is the data type (i.e. BinaryProperties), wVal is a set of pair representing all possible values with their trust weight (i.e. [Open (0,5);Close (0,5)]) and time is date of observation of this data.*

FCD can be provided by adapted sensors or by existing solution for context acquisition, such as FSCEP [1,2].

**Definition 2.** STATIC CONTEXT DATA (SCD) *represents invariable, non observable and timeless data about the environment. Unlike FCD, it carries one value that is considered true. Formally, it is a 3-tuple (subj, type, val) where subj is an entity of the environment (i.e. stove), type is the data type (i.e. location) and val its value.*

SCD represent invariant data about the environment that are supposed provided and known. For instance the position of the stove in the house will hardly ever change while the position of the user continuously varies.

**Definition 3.** ACTIVITY (ACT) *is a high level context data representing the action of the user. Formally we define the activity is a 4-tuple: act($l$, $t_s$, $u$, $c$) where: $l$ is its label (e.g. cooking), $t_s$ is its start time, $u$ is its uncertainty value, and $c$ is the set of context data that characterize the activity (e.g. its location).*

**Definition 4.** ANOMALOUS SITUATION CLASS (ASC) *is a class that represents a set of anomalous situations sharing same anomaly (e.g. Suffocation, FireElectricity).*

**Definition 5.** ANOMALOUS SITUATION (AS) *is a situation holding a set of FCD that may produce anomaly considering the user activity. Formally we define an anomalous situation as a 5-tuples sit(act, fcd, asc, unc, win): **act** is the activity of the user during the time window win, **fcd** are the observed fuzzy contextual data during the time window win, **asc** is the anomaly situation class that this situation belongs to, it is the result of the Inference engine (Sect. 5), **win** is the time window for this situation, and **unc** is the uncertainty value of the classification of this situation to the class asc. Actually, a situation is considered anomalous if its uncertainty value, that corresponds to a computed weigh of a predicate in the MLN'rules, is upper than a threshold that is defined by expert knowledge. More explanations are in Sect. 5.*

From a more technical point of view, various technologies can be used to represent these concepts. In our work, we chose to model this knowledge using ontologies: they allows reasoning and interoperability with other systems (including FSCEP [1,2]).

## 4.2   The MLN Model for Anomalous Situation Detection

The MLN model is a non-oriented graph where nodes are predicates and an edge between two nodes means that the two predicates are found in at least one logic rule in the set $F$. For example, a rule about a possible electricity Fire anomaly class must states that: if the user is doing an activity and he is far away from an electric device that has a very high temperature level, then it risks a Electricity fire. $F_1$ in the following example formally expresses this rule. This example shows an MLN that models three rules.

**Example 1.** *Let $F = \{(F_1, w_1), (F_2, w_2), (F_3, w_3)\}$ the set of rules in the MLN model where:*

- *$F_1$: $\forall$ act $\in$ Activity $\exists$ x $\in$ ElectricDevice $\exists$ l1, l2 $\in$ Location*
  *State(x, "ObjectIn", $w_{state}$) $\wedge$ State(x, "Hot+", $w_{state}$) $\wedge$ LocatedIn(x, l1, $w_{loc}$) $\wedge$ OccuresIn(act, l2, $w_{occures}$) $\wedge$ Different(l1, L2, $w_{diff}$) $\rightarrow$ Prediction( act, FireElectricity, $p_{red}$)*
- *$F_2$: $\forall$ act $\in$ Activity $\exists$ x $\in$ Entree $\exists$ l1 $\in$ Location*
  *Similar(act, "Sleeping", $w_{sim}$) $\wedge$ OccuresIn(act, l1, $w_{occures}$) $\wedge$ State ("Weather", "Cold", $w_{state}$) $\wedge$ Access (x, l1, $w_{access}$) $\wedge$ State(x, "Open", $w_{state}$) $\rightarrow$ Prediction( $\langle x; hasBinaryProperties; [\langle Open; hasValue; w_{Open}\rangle]\rangle$, Discomfort,$p_{pred}$)*
- *$F_3$: $\forall$ act $\in$ Activity $\exists$ x $\in$ Entree $\exists$ l1 $\in$ Location*
  *Requirement (act, "Vent", $w_{req}$) $\wedge$ OccuresIn(act, l1, $w_{occures}$) $\wedge$ Access(x, l1, $w_{access}$) $\wedge$ State(x, "Close", $w_{state}$) $\wedge$ State("Vent", "Shutdown", $w_{state}$) $\rightarrow$ Prediction(act, $\langle x; hasBinaryProperties; [\langle Close; hasValue; w_{close}\rangle]\rangle$, Suffocation, $p_{pred}$)*

*In the MLN model the values of rules weights ($w_i$) are initially unknown, they are computed through Inference Engine (Sect. 5). The three rules allow the prediction of three anomaly situation classes: FireElectricity, Discomfort, and Suffocation as depicted in Fig. 1.*

These three rules define constraint about three anomalies classes: FireElectricity, Discomfort, and Suffocation. As we can see, the rules rely on particular predicates. However, these predicates do not match at first the format of FCD and SCD. Thus, we define three types of predicates:

**Definition 6.** Probabilistic Evidence Predicate (PEP) *in our MLN is a predicate corresponding to FCD. The distribution truth values (uncertainty value= $w_{predicate}$ ) of the PEP may be given as evidence. It is worth mentioning that when multiple uncertainty values are provided in the predicate definition,*

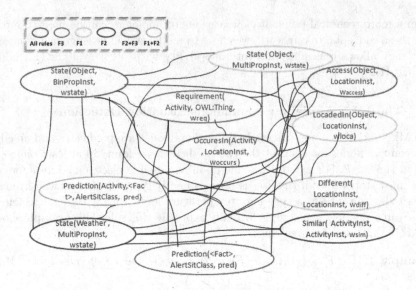

**Fig. 1.** The abstract MLN model of rules $F_1$, $F_2$, and $F_3$

the possibility logic [6] is applied to find the final uncertainty value of the predicate. For instance, this rules State(Object, BinPropInst, $w_{state}$) ← hasBinaryProperties(Object, BinPropInst) ∧ hasValue(BinPropInst, $w_{state}$) (e.g. State (StovePlate, ObjectIn, 0,8)) in its left part creates a PEP, in Example 1, that represents the FCD in the right part of the rule.

**Definition 7.** DETERMINISTIC EVIDENCE PREDICATES (DEP) in our MLN is a predicate corresponding to SCD. A distribution truth value of a DEP is either 0 or 1.

PEP and DEP are generated from the FCD and SCD by applying simple expert provided rules. Let us have an example of PEP considering an ontology: State(Object, BinPropInst, $w_{state}$) ← hasBinaryProperties(Object, Bin-PropInst) ∧ hasValue(BinPropInst, $w_{state}$). This rule generates the predicate 'State' (e.g. State (StovePlate, ObjectIn, 0,8)) from a FCD that have 'Stove-Plate' as a subject which is stored in the ontology as an Object type.

PEP and DEP are predicates representing the context, as such, they are in most cases present in the left part of the rules in the MLN. The right part of the rules relies on a different type of predicate:

**Definition 8.** Probabilistic Hidden Predicate (PHP) in the MLN can be referred as the query node: the right side of the rule. Its distribution truth ($p_{pred}$) is unknown. Prediction (Activity, AnomalyClass, $p_{pred}$) is one of the considered PHP in CAREDAS.

# 5    CAREDAS Inference Engine

The main role of this step is to find the most probable anomaly situation class of the current situation. To do so, four main steps are required: (1) Situation construction, (2) Dynamic Ground MLN creation (3) Rules weights calculus, and (4) Hidden Predicates (PHP) weights calculus.

## 5.1    Situation Construction

CAREDAS is conceived to recognize anomaly situations from real time data. It receives as input activities and FCD from state-of-the-art solutions such as [3] and [1,2]. From these inputs, CAREDAS builds **situations**. This construction process is applied periodically on time window (see Definition 4). This time window contains the activity and the FCD that occur during its duration. Once the situation is built, the system analyses it to find the most probable anomaly class. In the case the current situation doesn't match any anomaly class, CAREDAS repeats the inference but take into consideration further data from the next time window. We suppose the activity recognition process to be very efficient, meaning it detects all the user activities exhaustively. Therefore, the inference keeps going until one or two conditions are verified: 1- anomaly situation class is detected or 2- new activity is received. In the condition 1: the situation is considered as anomalous (Definition 4) with $asc$ as the most probable anomaly class, $unc$ as its computed probability, and $win$ as the sum of duration of time window until the detection of anomaly. As for the condition 2: the reception of a new activity means this situation is not anomalous; A a new situation is constructed and the inference process described above restarts. Figure 2 illustrates these steps through an example.

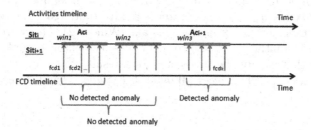

**Fig. 2.** Regarding the situation $Sit_1$ about the activity $ac_{i+1}$, during the time window $wind_1$ the system does not detect anomaly, so it continues inference by adding the fcd in $wind_2$. Similarly, no anomaly detected with $wind_1 + wind_2$. The system stop the inference and constructs a new situation $Sit_2$ since it receives a new activity $ac_{i+1}$

---

**Algorithm 1.** Dynamic Ground MLN

---

**Input:** Situation sit, Ontologies ont; Set ⟨*Predicates*⟩ P, Graph mln
**Output :** Graph groundMLN

1: fcd ← GetFCD(sit);act ← GetAct(sit); actCarac ← GetActCarac(sit);
2: ⟨*InstP*⟩ ← InstantiateP(P, fcd, act, actCarac);{    Instantiation of the predicates in P according to the perceived situation
3: *MiInstP* ← GetPartPred(InstP) { Extraction of partially valued predicates } }
4: **FOR** each miInstp ∈ MiINSTp **DO**
    miInstP ← ReplaceVar(miInstP, onto);{    The ontology onto is used to instantiate the variables in the miInstansiated predicates}
    **IF** Instansiated(miInstP) **THEN**
    InstP ← InstP ⋃ miInstP;
    **ENDIF**
5: **ENDFOR**
6: groundMLN ← CreateGroundMLN(InstP, mln)

---

## 5.2  Dynamic Ground MLN Creation

As we mentioned in Sect. 2, in this work we propose the Algorithm 1 called Dynamic Ground MLN for the creation of the ground MLN based on the situation and the ontology in the knowledge base. The algorithm proceeds as a first step to instantiate the predicates defined in the knowledge base (see Sect. 4) from the FCD, the activity ACT, and the activity characteristics ACTCARAC in the situation SIT (Line 1 and 2). The result of the instantiation function INSTANTI-ATEP is three kinds of predicates: INSTP, MIINSTP, and UNINSTP as described in the Algorithm 1. Afterwards, the algorithm focuses on the MIINSTP predicates by trying, for each predicate ∈ MIINSTP to instantiate its variables from the ontology (Line 3...4) to enrich the ground MLN with more instantiated predicates. It is worth mentioning that only the truth value of the deterministic evidence predicates (DEP) can be deduced from the ontology. For example, after execution of the Dynamic Ground MLN algorithm we obtain the predicate $Access(Window2, x, w_{access})$. The latter is a miInstantiated DEP (∈ MIINSTP) since it contains a variable (x) and a constant (Window2). Afterwards, the algorithm tries to instantiate x. To do so, it browses the ontology and tries to find the instances that satisfy the predicate. As a result, it obtain theses instantiated predicates (∈ INSTP): *Access(Window2, Kitchen, 1)* and *Access(Window2, Garden, 0)*.

## 5.3  Rules Weights Calculus

After the creation of the Ground MLN, we obtain the set **IPEP**$_w$ (weighted **I**nstanciated **P**robabilistic **E**vidence **P**redicate) where: ∀ instP$_w$ ∈ IPEP$_w$ ↔ instP$_w$ ∈ InstP & the truth value of the predicate is known

**Example 2.** Let *State(StovePlate, Hot, 0.9), (Stove, Hot, –)* ∈ *InstP. These two instantiated predicates are obtained from the Algorithm 1. Obviously, State(StovePlate, Hot, 0.9)* ∈ *IPEP$_w$ since its truth value is known and equal to 0.9. However, (Stove, Hot, –)* ∉ *IPEP$_w$ since its truth value is unknown.*

Then, to compute the weight ($w_i$) of logic rule $F_i \in F$ of the MLN model (see Example 1) we apply this proposed formula:

$$w_i = ln\frac{(\sum_j w_{satP_{ij}})/n_i}{1 - \sum_j w_{satP_{ij}}/n_i} \qquad (2)$$

where $satP_{ij} \in IPEP_w$ & satisfies $F_i$ and $w_{satP_{ij}}$ is its weight. $n_i$ is the number of predicates $\in PEP \cap F_i$. For instance, from a given situation, after the execution of Algorithm 1, we can obtain State(StovePlate, hasObject, 0.95), State(StovePlate, hasNoObject, 0.05) $\in IPEP_w$. $F_1$ in Example 1 is satisfied only with State(StovePlate, hasObject, 0.95). This is because in $F_1$ the predicate State(Object, BinPropInst, $w_{state}$) is satisfied only with $BinPropInst = hasObject$. Through the application of the Eq. 2, the rule's weight increases with the increased number of the predicates that satisfy the corresponding rule. As a matter of fact, a high number of predicates satisfy a rule means that the contextual data and the activity are about the anomaly class in this rule. Therefore, the rule gets a high weight.

### 5.4 Computation of the Weight of Probabilistic Hidden Predicates

CAREDAS can now compute the weights of the hidden predicates. Let $k \in PHP$ in the ground MLN, $P_{pred_k}$ is its weight to be computed as follows:

$$P_{pred_k} = Av(P(x = const))\forall const \in Constant_k \qquad (3)$$

where $Constant_k$ is the set of constants in the Ground MLN that are in the predicates which have edge with the PHP $k$. $P(x = const)$ is computed by the application of the formula in Eq. 1. The final classification of a given situation is the anomaly situation class that has the maximum weight value in the corresponding PHD. If this weight is upper than a threshold then the situation is considered anomalous. This weight will be its uncertainty value $unc$ (Definition 4), $asc$ is the selected anomaly class.

## 6 Experimental Evaluation

In order to evaluate CAREDAS, we developed a prototype of it. We have extensively evaluated the proposed method with a dataset acquired out of more of 2 h of elderly-like routine in the Hadaptic platform[2]. The dataset contains approximatively 600 events and 10 different activities that occurred 75 times for 15 different scenarios. The smart lab is equipped with motion sensors, beacons, switches, thermometers and more. CAREDAS was integrated with a FSCEP [1,2] implementation, for data acquisition, and a state-of-the-art activity recognition system [3]. For this evaluation, a set of logic rules has been defined for three anomalies classes: Intrusion, Suffocation, and ElectricityFire. Before execution,

---

[2] http://hadaptic.telecom-sudparis.eu/.

CAREDAS requires the preliminary step in which the value of parameter *win*, that corresponds to the time window duration of situation, is experimentally chosen. Therefore, we have tested the method with different values of *win* ∈ [60 s. . . 300 s]. The constant Z, in Eq. 1 was set to 10. CAREDAS was evaluated by comparing its output against expected results. For each time window, the precision and recall of the system was computed. Furthermore, we computed the correctness, which is simply the rate of correct (expected) answers of the system. Figure 3 shows the obtained result. In this experiments the focus is in the evaluation of CAREDAS's efficiency. However the study of its computational performance and the complexity of its process will be addressed in a future work. As we can see in Fig. 3, CAREDAS has a high precision for all time windows. This means our system rarely detects untimely events, this is an important feature for the comfort of the user. However, as depicted by the recall, it sometimes misses anomalies, but relatively rarely for time window shorter than 4 min. On the overall, we can see that correctness decreases with the size of the time window. This can be explained by the increase of data volume and the time window overlapping multiple activities. This activities overlapping sometimes leads the system to make wrong decisions. As a conclusion, with a suited time window (2 min), CAREDAS is highly reliable with the proposed improvement of MLN, e.g. Dynamic Ground MLN and rules weights, to recognize anomalous situations.

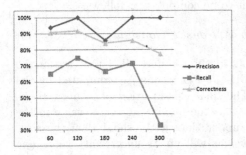

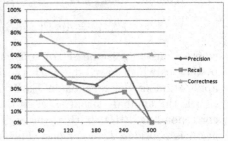

**Fig. 3.** Precision, recall, and correctness of CAREDAS for different time window *win*

**Fig. 4.** Precision, recall, and correctness of CAREDAS with static ground MLN for different time window *win*

In order to prove the efficiency of the Dynamic Ground MLN regarding the Static one, we have also implemented the MLN with a static ground MLN [5] with the same dataset and rules. However, as the static ground MLN [5] requires, all the rules are instantiated with all possible constants from the beginning, and rules weights values are fixed to 1. This version of CAREDAS was tested and evaluated on the same dataset and metrics. The evaluation results are presented in Fig. 4. As we can see in Fig. 4, the static ground MLN [5] does not work well for anomalous situation detection as it shows poor result compared to a dynamic one. It particularly encounters difficulty for 5 min time windows, where no anomalies were detected at all. In fact, it is not able to handle a large variety of context data.

# 7  Conclusion and Future Work

In this paper, we proposed CAREDAS a new method for AD. Unlike previous AD methods, CAREDAS is able to detect anomalies through contextual data during the user activity occurrence. To do so, it features combination of ontology and MLN as a machine learning method. Moreover, as a second contribution we proposed a Dynamic Ground MLN in the MLN model allowing it to deal with dynamic and online data. Our experiments underlines the viability of CAREDAS and its high precision level. Studying the computational performance of CAREDAS and the complexity of its process is one of the future directions.

**Acknowledgements.** This work has been partially supported by the project COCAPS (https://agora.bourges.univ-orleans.fr/COCAPS/) funded by Single Interministrial Fund N20 (FUI N20).

# References

1. Jarraya, A., Ramoly, N., Bouzeghoub, A., Arour, K., Borgi, A., Finance, B.: FSCEP: a new model for context perception in smart homes. In: Debruyne, C., et al. (eds.) OTM 2016. LNCS, vol. 10033, pp. 465–484. Springer, Cham (2016). doi:10.1007/978-3-319-48472-3_28
2. Jarraya, A., Ramoly, N., Bouzeghoub, A., Arour, K., Borgi, A., Finance, B.: A fuzzy semantic CEP model for situation identification in smart homes. In: ECAI (2016)
3. Sfar, H., Bouzeghoub, A., Ramoly, N., Boudy, J.: AGACY monitoring: a hybrid model for activity recognition and uncertainty handling. In: ESWC (2017)
4. Melisachew, C., Jakob, H., Christian, M., Heiner, S.: Markov logic networks with numerical constraints. In: ECAI (2016)
5. Matthew, R., Pedros, D.: Markov logic networks. Mach. Learn. **62**, 107–136 (2006)
6. Dubois, D., Lang, J., Prade, H.: Automated reasoning using possibilistic logic: semantics, belief revision, and variable certainty weights. In: TKDE, vol. 6 (1994)
7. Hoque, E., Dickerson, F.R., Preum, S.M.: Holmes: a comprehensive anomaly detection system for daily in-home activities. In: DCOSS (2016)
8. Riboni, D., Bettini, C., Civitares, G., Janjua, Z.H.: SmartFABER: recognizing fine-grained abnormal behaviors for early detection of mild cognitive impairment. Artif. Intell. Med. **67**, 57–74 (2016)
9. Janjua, Z.H., Riboni, D., Bettini, C.: Towards automatic induction of abnormal behavioral patterns for recognizing mild cognitive impairment. In: SAC (2016)
10. Ye, J., Dobson, S., McKeever, M.: Situation identification techniques in pervasive computing: a review. Pervasive Mob. Comput. **9**, 36–66 (2012)
11. Huang, J., Zhu, Q., Feng, L.Y.J.: A non-parameter outlier detection algorithm based on Natural Neighbor. Knowl.-Based Syst. **92**, 71–77 (2016)
12. Jakkula, V., Cook, D.J.: Detecting anomalous sensor events in smart home data for enhancing the living experience. In: AIII (2011)
13. Han, Y., Han, M., Lee, S., Sarkar, A.M.J., Lee, Y.K.: A framework for supervising lifestyle diseases using long-term activity monitoring. Sensors **12**, 5363–5379 (2012)
14. Lot, A., Langensiepen, C., Mahmoud, S.M., Akhlaghinia, M.J.: Smart homes for the elderly dementia suerers: identication and prediction of abnormal behavior. J. Ambient Intell. Humaniz Comput. **3**, 205–218 (2012)

15. Novak, M., Binas, M., Jakab, F.: Unobtrusive anomaly detection in presence of elderly in a smart-home environment. In: ELEKTRO (2012)
16. Novak, M., Jakab, F., Lain, L.: Anomaly detection in user daily patterns in smart-home environment. In: JSHI, vol. 3 (2013)
17. Riboni, D., Bettini, C., Civitarese, G., Janjua, Z.H., Helaoui, R.: Fine-grained recognition of abnormal behaviors for early detection of mild cognitive impairment. In: PerCom (2015)
18. Anderson, D.T., Ros, M., Keller, J.M., Cuellar, M.P., Popescu, M., Delgado, M., Vila, A.: Similarity measure for anomaly detection and comparing human behaviors. Int. J. Intell. Syst. **27**, 733–756 (2012)
19. Chen, H., Ku, W.S., Wang, H., Tang, L., Sun, M.T.: Scaling up Markov logic probabilistic inference for social graphs. In: TKDE, vol. 29 (2016)

# Using Constraint Logic Programming for the Verification of Customized Decision Models for Clinical Guidelines

Szymon Wilk[1]($\boxtimes$), Adi Fux[2], Martin Michalowski[3], Mor Peleg[2], and Pnina Soffer[2]

[1] Poznan University of Technology, Poznan, Poland
szymon.wilk@cs.put.poznan.pl
[2] University of Haifa, Haifa, Israel
[3] MET Research Group, Ottawa, Canada

**Abstract.** Computer-interpretable implementations of clinical guidelines (CIGs) add knowledge that is outside the scope of the original guideline. This knowledge can customize CIGs to patients' psycho-social context or address comorbidities that are common in the local population, potentially increasing standardization of care and patient compliance. We developed a two-layered contextual decision-model based on the PROforma CIG formalism that separates the primary knowledge of the original guideline from secondary arguments for or against specific recommendations. In this paper we show how constraint logic programming can be used to verify the layered model for two essential properties: (1) secondary arguments do not rule in recommendations that are ruled out in the original guideline, and (2) the CIG is complete in providing recommendation(s) for any combination of patient data items considered. We demonstrate our approach when applied to the asthma domain.

## 1 Introduction

Clinical practice guidelines (CPGs) include evidence-based recommendations intended to optimize patient care [1]. When physicians use CPGs in practice to make clinical decisions, they often take into consideration additional aspects related to the patient's personal context (e.g., the patient's level of family support, the degree to which his/her daily schedule is routine) [2]. Additional clinical aspects not contained in the CPG, such as comorbidities that the patient may have, are also considered during decision making. Typically there is no evidence-based recommendation for weighing in personal considerations into clinical decision-making. Moreover, CPGs cannot address all possible comorbidities that patients may have and such considerations are usually left to the discretion of the physician.

It follows that when a decision support system (DSS) is developed based on computer-interpretable clinical guidelines (CIGs) [9], it may be desirable to customize the CIG by also including arguments (conditions that provide support for and against specific recommendations) that are based on comorbidities and personal considerations which are common in the local settings. Customization aims to achieve more standard

© Springer International Publishing AG 2017
A. ten Teije et al. (Eds.): AIME 2017, LNAI 10259, pp. 37–47, 2017.
DOI: 10.1007/978-3-319-59758-4_4

management by physicians, given that they better address relevant secondary considerations not mentioned in the CPG. Moreover, patients may be more compliant to recommendations that address their personal context [3].

Nevertheless, it is important to acknowledge that arguments associated with the considerations not contained in the original CPGs should be secondary to the recommendations found in the CPGs, as they are generally not evidence-based. Hence any customized decision-making model should obey the *secondarity* property: *Secondary arguments should only modulate existing primary recommendations, while not suggesting recommendations that are not clinically indicated.* Modulation includes re-ranking of decision options, changes in dose or frequency of treatment or monitoring, changes in treatment or monitoring schedule, etc. Moreover, we would like the decision-model to obey the *completeness* property: *for any valid combination of primary and secondary decision parameters (data items and results of previous decisions), at least one recommendation is indicated.* This property guarantees that the customized CIG will not encounter a situation where no valid candidate exists.

In this paper we introduce a two-layered contextual decision-model based on the PROforma CIG formalism [4] and operationalized within the Tallis enactment engine (http://www.cossac.org/tallis). We then present logic-based methods for verifying the secondarity and completeness properties of the two-layered model. We use a case-study from the asthma domain to demonstrate our approach.

## 2   Related Work

### 2.1   Customization of CPGs

Other researchers have developed methodologies that allow customizing CIG models so that they can be personalized at run time. Riaño et al.'s methodology [5] uses algorithms that manipulate domain ontologies to yield a personalized view of the healthcare knowledge to support clinical decisions for chronically ill comorbid patients. The methodology uses domain ontologies to provide decision support for adjusting a patient's condition based on disease profiles; these profiles are consulted to suggest additional signs and symptoms that the patient is likely to exhibit and which could be used to generate a more complete record. Grandi and coauthors [6, 7] suggest efficient management of multi-version CIGs collections by representing, in a knowledge base, multi-version clinical guidelines and domain ontologies in XML or in relational schemas. Personalized CIGs can be created by building from the knowledge base an on demand version that is tailored to the patient's current time (or desired temporal perspectives) and to the patient's disease profile (i.e., set of comorbidities). Finally, Michalowski et al. [8] expanded their mitigation framework based on first-order logic to account for patient preferences related to treatment. These preferences are represented in the form of preference-related revision operators that describe undesired circumstances (e.g., a sequence of treatment actions) that the patient would like to avoid, and specify changes that should be introduced to CPGs in order to make them consistent with patient preferences.

On one hand our work shares some similarities with the above approaches – the secondary layer can be seen as a very complex revision operator that expands the primary layer and brings additional data items into consideration. On the other hand, unlike other approaches, it explicitly verifies the validity of the obtained model to ensure it maintains the required properties.

## 2.2 Automatic Verification and Evaluation of CIGs

CIG verification techniques fall into three categories [9]: (1) proving that the CIG specification is internally consistent and free of anomalies, (2) proving that the CIG specification satisfies a set of desired formally defined properties, using model checking or theorem proving, and (3) checking inconsistencies between CIGs that are concurrently applied to a patient with comorbidities [10]. The approach presented in [11] uses two techniques: model checking to verify guidelines against semantic errors and inconsistencies in CIG definition, and model-driven development to automatically process manually-created CIGs against temporal logic statements that should hold for these CIG specifications. Another technique is described in [12] where theorem proving explores logical derivations of a theory representing a CIG to confirm whether a formal CIG protocol complies with certain protocol properties.

Our approach relies on model checking – specifically we use constraint logic programming (described in Sect. 3.2) to ensure that the required properties hold for a given two-layered decision model. Generally, model checking is easier and more efficient than theorem proving [13]. Moreover, applied techniques (e.g., constraint propagation) further facilitate representation and processing of CIGs.

# 3   Methods

## 3.1   Two-Layered Contextual Decision Model

Following the definitions used in PROforma, a *plan* (*task network*) is a network composed of *tasks* and *scheduling constraints*. Tasks are specialized into a *plan, enquiry, action*, and *decision*. A decision has at least two *candidates* (or *recommendations*). Each *candidate* has at least one *argument* – a condition that refers to patient data items and an associated numerical weight (support) for or against the candidate. Our two-layered contextual decision-model extends PROforma's CIG model by distinguishing between *primary* and *secondary* arguments. Primary arguments are formalizations of evidence-based recommendations found in a CPG and refer to *clinical (primary) data items*. Secondary arguments extend the CPG by constructing arguments that relate to additional *secondary data items* that are not part of the CPG. In our two-layered decision model, the *primary layer* is a plan where all arguments associated with decision candidates are primary arguments; the *secondary layer* includes a set of secondary arguments and their secondary data items that are associated with the decision candidates of a given primary layer. Weights for primary arguments correspond to the grades of evidence used by the CPG, while weights for secondary arguments are established by clinical experts based on their knowledge and experience.

Completeness is satisfied if for each decision there is at least one candidate with total support (i.e., sum of argument weights) in both layers that is equal to or greater than a threshold defined in the CIG; secondarity is satisfied if there is no decision candidate for which total support in the primary layer is lower than 0 and support is greater or equal to the threshold in both layers.

We created software that combines PROforma models representing primary and secondary layers into a single two-layered model. The software integrates the primary and secondary arguments into their respective decisions. Upon enactment (using Tallis), all arguments are evaluated and decision candidates are ranked accordingly.

### 3.2    Constraint Logic Programming and MiniZinc

Constraint logic programming (CLP) unifies logic programming (LP) and a constraint satisfaction problem (CSP) by using LP as a constraint programming language to solve a CSP [14]. A CLP model is made up of a set of variables with finite domains, a set of clauses with constraints, and a goal to be satisfied. The clauses in the model capture the relationships between variables and they restrict the possible combinations of values assigned to variables. Solving a CLP model entails satisfying the goal given the set of constraints, where a value is assigned to each variable such that no constraints are violated (i.e., bodies of all clauses are satisfied). It is also possible to expand the goal with a goal function and look for solutions that optimize it (maximize or minimize this function) while preserving all constraints. This is usually referred to as a constraint optimization problem (however, in this work we are not considering this variant).

There are specialized solvers for CSPs that employ various finite domain and linear programming techniques and use different and often incompatible modeling languages. MiniZinc is a medium-level constraint modeling language that has been widely accepted as a standard for CLP models [15] and we use it as our modeling language.

### 3.3    Using CLP to Check Properties of Two-Layered Decision Models

In this study we use CLP to verify the completeness and secondarity of two-layered decision models and to control the process of introducing revisions necessary to ensure these properties. More specifically, decision models given in PROforma are translated into MiniZinc models, which in turn are verified and revised. Finally, the resulting MiniZinc models are translated back to PROforma.

Our overall goal is to ensure that a given PROforma model satisfies the properties of completeness and secondarity for all possible patient cases (i.e., all clinically valid combinations of primary and secondary data items). We achieve this goal indirectly by creating and solving a corresponding MiniZinc model with constraints to identify *problematic* cases that violate at least one of these properties. Thus, if there are no solutions to the MiniZinc model, this indicates that there are no such cases and the validity of the underlying PROforma model has been positively verified.

Our approach is outlined in more details below:

1. Create a MiniZinc model from the initial PROforma model. The MiniZinc model contains three groups of variables:

   - Variables corresponding to primary and secondary data items defined in the enquiry steps (for each data item there is a unique variable),
   - Variables corresponding to *intermediate* decision candidates, i.e., these decision candidates that affect other decision candidates (for each intermediate candidate there is a unique variable),
   - Variables corresponding to support for specific decision candidates (for each candidate there are two unique variables corresponding to support in each layer).

   Moreover, it contains the following groups of constraints:

   - Constraints enforcing (or computing) support for individual candidates (for each variable corresponding to the support there is a unique constraint),
   - A single constraint that is a disjunction of two "sub-constraints" – one that enforces the violation of the completeness property, and the other that enforces the violation of the secondarity property,
   - Optional constraints corresponding to domain knowledge that exclude combinations of variable values representing clinically invalid solutions. Unlike the earlier constraints, the optional ones need to be specified manually by a clinical domain expert.

2. Solve the MiniZinc model. If there are no solutions, then go to step 4.
3. Revise the MiniZinc model to avoid problematic cases by (1) modifying conditions and weights in existing arguments, (2) removing existing arguments and/or (3) adding new arguments, and then go to step 2. Revisions may be applied to both layers, thus following the principles of evidence-based medicine, we allow experts to adjust CPGs recommendations according to their experience.
4. Translate the MiniZinc model to the final PROforma model focusing on constraints corresponding to arguments, as they are the ones that have been revised.

Currently revisions in step 3 are introduced manually using problematic patient cases from step 2 to direct the search for appropriate modifications, however, we plan to automate the revision step and thus to minimize the need for manual intervention.

### 3.4    Analysis of Property Violations and Revisions of the PROforma Model

We carried out a theoretical analysis of the reasons for violations of both properties and proposed corresponding revisions. The prevalent reason was incorrect argument weights. This calls for rescaling weights either in the primary or secondary layer to increase the difference between their orders of magnitude. Another reason was certain clinically infeasible combinations of data items that were initially missed by CIG modelers. These need to be explicitly excluded. Introducing these revisions requires changing the clinical flowchart and updating of CIG both layers.

## 4    Case Study Example

We use an asthma guideline adapted from [16, 17] to which additional decision arguments based on personal context variables were added as a secondary layer. The additional arguments were based on interviews with 15 clinical experts from Israel. The asthma guideline starts with a decision regarding the clinical goal: an aggressive goal which tries to improve clinical indicators, or a basic goal, which tries to maintain their levels. Once a treatment goal is selected, three decisions are made regarding medication type (steroid or not), dose (high, medium, low), and treatment intervals (frequency: daily or weekly). Decision arguments in the clinical guideline refer to four clinical (primary) data items: number of monthly attacks ($\leq 4$, 4–8, >8), severity of attacks (low, moderate, severe), forced expiratory volume (FEV1, which can be <60%, 60–80%, >80%) and daily limitation level (minor, medium, severe).

The secondary layer provides additional arguments for the existing decisions that are based on personal considerations. These include the routineness of the patient's daily life (routine, semi-routine, no-routine), his/her communication level (low, medium, high) and his/her level of family support (frequent, medium, low).

## 5    Results

### 5.1    A Decision Model for Asthma with the Secondary Personal Domains

The two-layered model developed for this case study is given in Fig. 1a. The first decision in the model ("treatment_goal") is an intermediate decision that does not lead directly to any action, but influences other decisions. Moreover, the model invokes an external plan for patients with a large number of monthly attacks (exceeding 20) as required by the asthma guideline. Figure 1b displays arguments from the primary and secondary layers associated with the "aggressive_goal" decision candidate of the "treatment_goal" decision. For example, the aggressive goal is preferred for patients with high communications skills (who understand directions) and frequent family support (who may commit effort).

### 5.2    Verification and Revision of the Decision Model for Asthma

The PROforma model from Fig. 1 was the initial model for our verification and revision procedure described in the *Methods* section. We started by constructing a corresponding MiniZinc model – its representative parts are given in Figs. 2 and 3.

Figure 2 shows how we represent arguments (for brevity we focus on primary arguments, secondary ones are defined similarly) and how we compute support associated with specific candidates in the primary and secondary layers. Specifically, each argument is a conditional *if...then...else...endif* expression associating conditions on data items with a weight. For brevity in the MiniZinc model we encoded symbolic values of specific data items with numbers (e.g., low, moderate, and severe attack levels are represented as 1, 2, and 3 respectively). A quick comparison of Figs. 1b and 2 highlights the close correspondence between MiniZinc and PROforma.

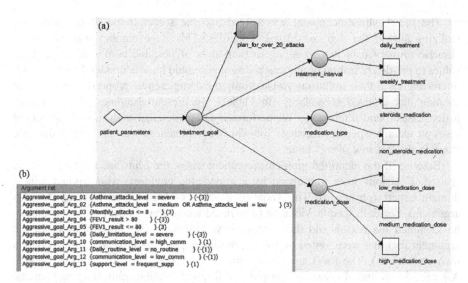

**Fig. 1.** (a) PROforma asthma CIG, where diamonds denote enquiries, circles – decisions, squares – actions and ovals – invocation of an external plan. (b) Primary (01..06) and secondary arguments (10..13) for the "aggressive_goal" candidate of the "treatment_goal" decision.

```
var int: pl_aggressive_goal_support;
constraint
  pl_aggressive_goal_support =
    (if asthma_attacks_level == 3 then -3 else 0 endif) +
    (if asthma_attacks_level == 2 \/ asthma_attacks_level == 1 then 3 else 0 endif) +
    (if monthly_attacks <= 8 then 3 else 0 endif) +
    (if FEV1_result > 80 then -3 else 0 endif) +
    (if FEV1_result <= 80 then 3 else 0 endif) +
    (if daily_limitation_level == 3 then -3 else 0 endif);
```

**Fig. 2.** Constraints that introduce arguments and calculate support for the "aggressive_goal" candidate in the primary layer.

For brevity, Fig. 3 presents selected parts of the constraint that enforces the violation of the completeness and secondarity properties. It is formulated over variables representing support for specific decision candidates. All support thresholds in the initial PROforma model were equal to 1.0, and the same value was used in the MiniZinc model. The constraint in Fig. 3 enables the solver to identify problematic patient cases for whom these properties are violated. Moreover, the MiniZinc model contains constraints to eliminate solutions that are clinically invalid (e.g., that combine more than 8 monthly attacks with minor daily limitation level). The structure of these latter constraints is relatively simple and as such we do not present them here.

```
constraint
  ((pl_plan_for_over_20_attacks_support + sl_over_20_attacks_support_plan_support < 1 /\
    pl_aggressive_goal_support + sl_aggressive_goal_support < 1 /\
    pl_basic_goal_support + sl_basic_goal_support < 1) \/ [...] /\          (a)
\/
  ((pl_plan_for_over_20_attacks_support < 0 /\
    pl_plan_for_over_20_attacks_support + sl_plan_for_over_20_attacks_support >= 1) \/ [...]    (b)
```

**Fig. 3.** Disjunctive constraint that identifies patient cases that violate the completeness property (a) or the secondarity property (b) in the two-layer model. Prefixes "pl_" and "sl_" identify variables associated with primary and secondary layers, respectively

The initial MiniZinc model was solved and the solver found 28,656 solutions violating any of the two considered properties. This large number was caused by numeric primary data items leading to thousands of possible combinations of their values (a pre-discretization of numerical data items could have addressed this problem). Interestingly, all these solutions violated only the completeness property; there was no solution that violated secondarity. In Table 1 we present examples of problematic patient cases found at this stage. We performed a more detailed analysis of these patient cases to identify specific decisions for which completeness was violated – they are listed in the last row of this table.

Because of the identified problematic patient cases, the MiniZinc model had to be revised. Revisions were introduced by a knowledge engineer, who worked with a domain expert (physician). The knowledge engineer focused on primary decision arguments. For each specific violation (associated with a specific decision or candidate) he identified the reason and then proposed several fixes (i.e., modifications of the argument list) that were vetted by the expert (the expert was also able to provide his own corrections). The most appropriate fix was introduced to the MiniZinc model. After the first round of revisions, the solver still found solutions indicating problematic patients, thus the revision and verification steps needed to be repeated. Overall, it took 9 iterations to arrive at the final MiniZinc model where no patient cases violating any of the properties were found. The knowledge engineer and the domain expert accepted this model and it was translated to the final PROforma model.

The differences between the initial and final PROforma models are summarized in Table 2. The changes were focused on the three decisions in the primary layer – "treatment_goal", "treatment_interval" and "medication type." The most extensive revisions were associated with candidates of the first decision, where more than 20 new arguments were added to the model. Significant changes were also introduced for the "steroids" candidate of "medication_type". On the contrary, no revisions were made to the secondary layer and to the "medication_dose" decision in the primary layer.

**Table 1.** Sample problematic patient cases violating the completeness property of the two-layered PROforma model.

| Data item | Case 1 | Case 2 | Case 3 |
|---|---|---|---|
| asthma_attack_level | low | moderate | low |
| daily_limitation_level | severe | minor | minor |
| FEV1_result | 81 | 81 | 81 |
| monthly_attacks | 5 | 1 | 5 |
| daily_routine_level | no_routine | semi_routine | routine |
| communication_level | medium_comm | medium_comm | medium_comm |
| support_level | low_supp | medium_supp | medium_supp |
| Decision with violated completeness | treatment_goal, treatment_interval, medication_type | treatment_interval | medication_type |

In Fig. 4 we present arguments associated with the "non_steroids" candidate of the "medication_type" decision in the initial and final PROforma models. The introduced

**Table 2.** Differences between the initial and final PROforma models in terms of the number of arguments associated with specific decision candidates.

| Decision | Candidate | # Arguments | | | | | |
|---|---|---|---|---|---|---|---|
| | | Primary layer | | | Secondary layer | | |
| | | Initial model | Final model | Change | Initial model | Final model | Change |
| treatment_goal | basic_goal | 4 | 27 | 23 | 2 | 2 | 0 |
| | aggressive_goal | 6 | 37 | 31 | 4 | 4 | 0 |
| | plan_for_over_20_attacks | 1 | 1 | 0 | 0 | 0 | 0 |
| treatment_interval | weekly_interval | 7 | 18 | 11 | 3 | 3 | 0 |
| | daily_interval | 4 | 20 | 16 | 1 | 1 | 0 |
| medication_type | non_steroids | 4 | 14 | 10 | 1 | 1 | 0 |
| | steroids | 6 | 25 | 19 | 2 | 2 | 0 |
| medication_dose | low_dose | 6 | 6 | 0 | 3 | 3 | 0 |
| | medium_dose | 5 | 5 | 0 | 2 | 2 | 0 |
| | high_dose | 8 | 8 | 0 | 2 | 2 | 0 |

revisions not only modified weights or expanded conditions in existing arguments (compare argument 03 in Fig. 4a and 10 in Fig. 4b, or argument 04 in Fig. 4a and arguments 02...04 and 11..12 in Fig. 4b), but also removed existing arguments (01 in Fig. 4a) and added multiple new ones. We note that the list of arguments in the final PROforma model could be shortened by combining some arguments (e.g. 08 and 09 in Fig. 4b). However the extent of changes would still be significant.

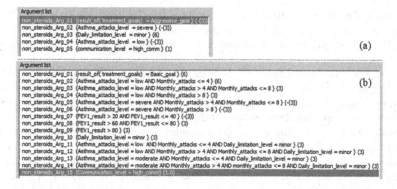

**Fig. 4.** Arguments associated with the "non_steroids" candidate of the "medication_type" decision in the initial PROforma model (a) and in the final PROforma model (b).

## 6  Discussion and Conclusions

In this paper we demonstrated that the application of CLP to two-layered decision models allows for their automatic verification and suggests areas of focus for revision, saving a knowledge engineer significant manual work. Layered models can become quite complex and although the knowledge engineer who created the models is experienced, he could not manually find all errors in the model within a reasonable amount of time. All errors were exposed using the CLP verification approach by iteratively checking the MiniZinc models for satisfiability allowing for easier construction of more complex two-layered models. We plan to further expand this approach to automatically find the appropriate weights for primary and secondary arguments.

By encoding the guideline as a set of constraints in the model, the argument weights as variables, and negating the current version of the secondarity and completeness verification constraint, our approach should solve this extended model for feasibility. A feasible solution will represent an assignment of weights for all arguments such that secondarity and completeness are guaranteed in the decision model.

In this work we focused on one way of modulating the recommendations provided by the primary layer, specifically modulating the support for different candidates by changing their ranking. Our secondary layer also only considered the psycho-social context of patients.

Future research will examine the use of CLP for verifying PROforma models where modulation also involves changes in dose or frequency of treatment or monitoring, changes in treatment or monitoring schedule, etc., and where the secondary context includes other factors such as additional comorbidities and local setting (e.g., organizational resources, local regulations).

# References

1. Institute of Medicine: Clinical Practice Guidelines We Can Trust (2013)
2. Fux, A., Peleg, M., Soffer, P.: How does personal information affect clinical decision making? Eliciting categories of personal context and effects. In: AMIA Annual Symposium, p. 1741 (2012)
3. Quinn, C.C., Gruber-Baldini, A.L., Shardell, M., Weed, K., Clough, S.S., Peeples, M., Terrin, M., Bronich-Hall, L., Barr, E., Lender, D.: Mobile diabetes intervention study: testing a personalized treatment/behavioral communication intervention for blood glucose control. Contemp. Clin. Trials 30(4), 334–346 (2009)
4. Sutton, D.R., Fox, J.: The syntax and semantics of the PROforma guideline modeling language. J. Am. Med. Inform. Assoc. 10(5), 433–443 (2003)
5. Riano, D., Real, F., Lopez-Vallverdu, J.A., Campana, F., Ercolani, S., Mecocci, P., Annicchiarico, R., Caltagirone, C.: An ontology-based personalization of health-care knowledge to support clinical decisions for chronically ill patients. J. Biomed. Inform. 45(3), 429–446 (2012)
6. Grandi, F.: Dynamic class hierarchy management for multi-version ontology-based personalization. J. Comput. Syst. Sci. 82(1 Part A), 69–90 (2016)
7. Grandi, F., Mandreoli, F., Martoglia, R.: Efficient management of multi-version clinical guidelines. J. Biomed. Inform. 45(6), 1120–1136 (2012)
8. Michalowski, M., Wilk, S., Rosu, D., Kezadri, M., Michalowski, W., Carrier, M.: Expanding a first-order logic mitigation framework to handle multimorbid patient preferences. In: AMIA Annual Symposium Proceedings 2015, pp. 895–904 (2015)
9. Peleg, M.: Computer-interpretable clinical guidelines: a methodological review. J. Biomed. Inform. 46(4), 744–763 (2013)

10. Wilk, S., Michalowski, W., Michalowski, M., Farion, K., Hing, M.M., Mohapatra, S.: Mitigation of adverse interactions in pairs of clinical practice guidelines using constraint logic programming. J. Biomed. Inform. **46**(2), 341–353 (2013)
11. Perez, B., Porres, I.: Authoring and verification of clinical guidelines: a model driven approach. J. Biomed. Inform. **43**(4), 520–536 (2010)
12. ten Teije, A., Marcos, M., Balser, M., van Croonenborg, J., Duelli, C., van Harmelen, F., Lucas, P., Miksch, S., Reif, W., Rosenbrand, K., Seyfang, A.: Improving medical protocols by formal methods. Artif. Intell. Med. **36**(3), 193–209 (2006)
13. Halpern, J.Y., Vardi, M.Y.: Model checking vs. theorem proving: a manifesto. In: Vladimir, L. (ed.) Artificial Intelligence and Mathematical Theory of Computation, pp. 151–176. Academic Press Professional, Inc., Cambridge (1991)
14. Dechter, R.: Constraint Processing. MIT Press, Cambridge (1989)
15. Nethercote, N., Stuckey, P.J., Becket, R., Brand, S., Duck, G.J., Tack, G.: MiniZinc: towards a standard CP modelling language. In: Bessière, C. (ed.) CP 2007. LNCS, vol. 4741, pp. 529–543. Springer, Heidelberg (2007). doi:10.1007/978-3-540-74970-7_38
16. British Thoracic Society and Scottish Intercollegiate Guidelines Network, QRG 141 - British Guideline on the Management of Asthma (2014)
17. Israel Medical Association: Clinical Practice Guidelines for Asthma Management. Harefuah (2000). (in Hebrew)

# Constructing Disease-Centric Knowledge Graphs: A Case Study for Depression (short Version)

Zhisheng Huang[1(✉)], Jie Yang[2], Frank van Harmelen[1], and Qing Hu[1,3]

[1] VU University Amsterdam, Amsterdam, The Netherlands
{huang,Frank.van.Harmelen,qhu400}@cs.vu.nl
[2] Beijing Anding Hospital, Beijing, China
jieyangadyy@ccmu.edu.cn
[3] College of Computer Science and Technology,
Wuhan Univesity of Science and Technology, Wuhan, China

**Abstract.** In this paper we show how we used multiple large knowledge sources to construct a much smaller knowledge graph that is focussed on single disease (in our case major depression disorder). Such a disease-centric knowledge-graph makes it more convenient for doctors (in our case psychiatric doctors) to explore the relationship among various knowledge resources and to answer realistic clinical queries.

## 1 Introduction

Major depressive disorder (MDD) has become a serious problem in modern society. Using antidepressants has been considered the dominant treatment for MDD. However, 30% to 50% of the individuals treated with antidepressants do not show a response. Hence, psychiatric doctors confront the challenge to make clinical decision efficiently by gaining a comprehensive analysis over various knowledge resources about depression. In this paper we propose an approach to constructing a knowledge graph of depression using semantic web technology to integrate those knowledge resources, achieving a high degree of inter-operability. With a single semantic query over integrated knowledge resources, psychiatric doctors can be much more efficient in finding answers to queries which currently require them to explore multiple databases and to make a time-consuming analysis on the results of those searches.

The term "Knowledge Graph" is widely used to refer to a large scale semantic network of entities and concepts plus the semantic relationships among them. Medical knowledge graphs typically cover very wide areas of medical knowledge: all proteines (UniProt), as many drugs as possible (Drugbank), as many drug-drug interactions as are known (Sider), and massively integrated knowledge graphs such as Bio2RDF and LinkedLifeData. Such knowledge graphs are very *a-specific* in terms of the diseases that they cover, and are often prohibitively large, hampering both efficiency for machines and usability for people. In this

A. ten Teije et al. (Eds.): AIME 2017, LNAI 10259, pp. 48–52, 2017.
DOI: 10.1007/978-3-319-59758-4_5

abridged paper we propose an approach to the construction of *disease-centric knowledge graphs*. Our claims are (i) that it is indeed possible to make disease-centric subgraphs and (ii) that realistic clinical queries can still be answered over such disease-specific knowledge graphs without substantial loss of recall.

We illustrate our general idea by integrating various knowledge resources about depression (e.g., clinical trials, antidepressants, medical publications, clinical guidelines, etc.). We call the generated knowledge graph *DepressionKG* for short. DepressionKG is represented in RDF/NTriple format [2]. DepressionKG provides a data infrastructure to explore the relationship among various knowledge and data-sources about depression. We show how it provides support for clinical question answering and knowledge browsing.

## 2    Challenges

In order to integrate various knowledge resources of depression, we had to confront the following challenges: (i) *Heterogeneity*: Different knowledge resources are generated by multiple creators. We have to achieve semantic inter-operability among those knowledge resources; (ii) *Free text processing*. Some of our knowledge resources contained a lot of free text. We have to use a natural language processing tool with a medical terminology to extract semantic relations from free text; (iii) *Partiality, inconsistency, and incorrectness*. Knowledge resources often contain some partial, inconsistent, noisy or even erroneous data. We have to develop efficient methods to deal with this; (iv) *Expressive Representation* of Medical Knowledge. The formal languages of Knowledge Graphs (RDF, RFD Schema, OWL) are a decidable fragment of first order logic. Such languages are usually not expressive enough for medical knowledge representation.

In this paper we present some methods for dealing with the first two challenges when constructing a knowledge graph of depression, while leaving the third and fourth challenges for future work. We will show how DepressionKG can be used in some realistic scenarios for clinical decision support. We have implemented a system of DepressionKG aimed at psychiatric doctors with no background knowledge in knowledge graphs.

## 3    Knowledge Resources and Integration

Following commonly used technology, we will construct our knowledge graph as an RDF graph. A summary of DepressionKG is shown in the table below. This shows that the resulting knowledge graph is only of moderate size (8M triples), whereas many of the original knowledge graphs are many times larger than this (10M–100M triples). In Sect. 4 we will illustrate that we can still answer a diversity of clinically relevant questions with such a small disease-centric knowledge graph. Our 8M triples are dominated by SNOMED. We decided to include all of SNOMED (instead of only those parts of the hierarchies relevant to depression) because some of the clinical use-cases presented to us by our psychiatric experts concern comorbidities (e.g., Alzheimer disease is a frequent comorbidity with MDD), and restricting SNOMED would hamper such comorbidity queries.

| Knowledge resource | Number of data item | Number of triple |
|---|---|---|
| ClinicalTrial | 10,190 trials | 1,606,446 |
| PubMed on depression | 46,060 papers | 1,059,398 |
| Medical guidelines | 1 guideline | 1,830 |
| DrugBank | 4,770 drugs | 766,920 |
| DrugBook | 264 antidepressants | 13,046 |
| Wikipedia side effects | 17 antidepressants | 6,608 |
| SNOMED CT | | 5,045,225 |
| Patient data | 1,000 patients | 200,000 |
| Total | | 8,699,473 |

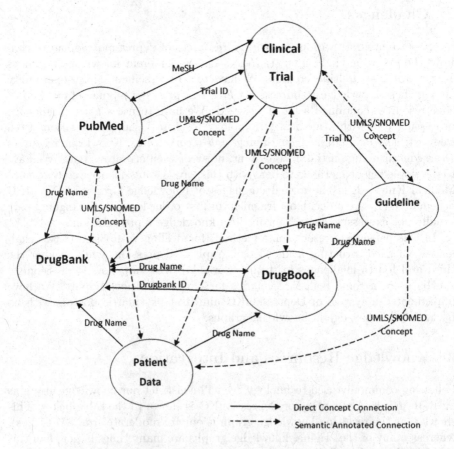

We use the following four methods to integrate the various knowledge resources. (i) *Direct Entity identification*. Some knowledge resources refer to the same entity with identical names, e.g. the PubMed IDs used in both PubMed and the clinical trials. Such entities are obvious links between these knowledge sources; (ii) *Direct Concept identification*. Numerous knowledge resources can be integrated by using direct *concept* identification. For example, both a publication in PubMed and a clinical trial are annotated with MeSH terms. This provides us with a way to detect a relationship between a clinical trial and a publication directly; (iii) *Semantic Annotation with an NLP tool.* We used Xerox's NLP tool XMedlan [1] for semantically annotating medical text (both concept identification and relation extraction) with medical terminologies such as SNOMED CT; (iv) *Semantic Queries with regular expressions.* The previous three approaches are offline approaches to integrate knowledge sources. Semantic Queries with regular expressions are an online approach, because such queries find relationships among knowledge resources at query time. Although online methods lead to more latency, they do provide a method to detect a connection among different knowledge resources based on free text.

The figure shows the connectivity of DepressionKG. An arrow denotes a direct concept connection via a property, and a dashed arrow denotes a concept identification in a medical terminology by using an NLP tool. The figure shows that our set of knowledge resources is well integrated.

# 4    Use Cases

In this section, we will discuss several use cases how the knowledge graph on depression can be used by psychiatric doctors for clinical decision support through SPARQL queries over the knowledge graphs. Because of space constraints, we only give the code of the SPARQL query for one example.

Case 1. Patient A, female, aged 20. She has suffered from MDD for three years. In the past, she took the SSRI antidepressant Paroxetine, however, gained a lot of weight. She wants an antidepressant which has the effect of weight loss. This can be answered with a SPARQL query over a single knowledge source, combining semantic properties and a regular expression on the textual description of symptoms From the result of this query, we learn that taking Bupropion may lead to weight loss, and taking Fluoxetine may lead to a modest weight loss.

Case 2. Patient B, a female adolescent with MDD. She failed to respond to first-line treatment with Fluoxetine. The doctor wants to know the details of any clinical trial which investigates the effect of Fluoxetine and the publications of those trials. The following query searches over two knowledge resources ClincalTrial and PubMed in the knowledge graph:

```
PREFIX ...
select distinct  ?trial ?title ?description ?pmid ?articletitle ?abstract
where {?t sct:BriefTitle ?title.
       FILTER regex(?title,"Fluoxetine")
       ?t sct:NCTID ?trial.
       ?t sct:DetailedDescription ?description.
       ?pmid pubmed:hasAbstractText ?abstract.
       FILTER regex(?abstract,?trial)
       ?pmid pubmed:hasArticleTitle ?articletitle.}
```

This query finds three relevant trials, one with two publications, the others with one publication.

Case 3. Patient C, an adult male, suffers from mood disorder and hopes to try a clinical trial on depression. His clinical doctor wants to find an on-going trial which uses a drug intervention with target "neurotransmitter transporter activity". This requires a search that covers both DrugBank and ClinicalTrial. From DrugBank we find the drug target, and from ClinicalTrial we find which trial has an intervention with the required drugs. This query returns 25 clinical trials whose starting date is in 2016, and which meet the specified condition.

Case 4. Patient D, male, aged 45, has complained a lot that the antidepressant Clomipramine has lead to fatigue. Indeed fatigue is a very common side effect of Clomipramine. The psychiatric doctor wants to know if there exists any other antidepression drug of the same class where fatigue is a rare or uncommon side effect. In this example, we use the predicate `skos:narrower` and `skos:broader` to search over the SNOMED concept hierarchy to find two sibling concepts (i.e. two antidepressants of the same class). The answer for this search is "Dosulepin".

## 5   Implementation, Discussion and Conclusion

We have implemented the DepressionKG system with a graphical user interface, so that psychiatric doctors can use the system to search for the knowledge they need and to explore the relationships among various knowledge resources for clinical decision support. The DepressionKG system supports knowledge browsing and querying, and will be evaluated in Beijing Anding Hospital, one of the biggest psychiatric hospitals in China, for experiments in the Smart Ward project. The objective of the Smart Ward project is to develop a knowledge-based platform for monitoring and analyzing the status of patients and for supporting clinical decision making in a psychiatric ward.

In this paper, we have proposed an approach to making a knowledge graph of depression, and we have shown how various knowledge resources concerning depression can be integrated for semantic inter-operability. We have provided several use cases for such a knowledge graph of depression. From those use cases, we can see that by using a knowledge graph with its semantic search, it is rather convenient for us to detect relationship which cover multiple knowledge resources.

**Acknowledgments.** This work is partially supported by the Dutch national project COMMIT, the international cooperation project No. 61420106005 funded by National Natural Science Foundation of China, and the NWO-funded Project Re-Search. The fourth author is funded by the China Scholarship Council.

## References

1. Ait-Mokhtar, S., Bruijn, B.D., Hagege, C., Rupi, P.: Intermediary-stage ie components, D3.5, Technical report, EURECA Project (2014)
2. Cyganiak, R., Wood, D., Lanthaler, M.: RDF 1.1 concepts and abstract syntax (2014)

# Bayesian Methods

# Implementing Guidelines for Causality Assessment of Adverse Drug Reaction Reports: A Bayesian Network Approach

Pedro Pereira Rodrigues[1,2](✉), Daniela Ferreira-Santos[1], Ana Silva[3],
Jorge Polónia[1,3], and Inês Ribeiro-Vaz[1,3]

[1] CINTESIS - Centre for Health Technology and Services Research,
Rua Dr. Plácido Costa, s/n, 4200-450 Porto, Portugal
[2] MEDCIDS-FMUP, Faculty of Medicine of the University of Porto,
Alameda Prof. Hernâni Monteiro, 4200-319 Porto, Portugal
pprodrigues@med.up.pt
[3] UFN - Northern Pharmacovigilance Centre, FMUP,
Rua Dr. Plácido Costa, s/n, 4200-450 Porto, Portugal

**Abstract.** In pharmacovigilance, reported cases are considered suspected adverse drug reactions (ADR). Health authorities have thus adopted structured causality assessment methods, allowing the evaluation of the likelihood that a medicine was the causal agent of an adverse reaction. The aim of this work was to develop and validate a new causality assessment support system used in a regional pharmacovigilance centre. A Bayesian network was developed, for which the structure was defined by an expert, aiming at implementing the current guidelines for causality assessment, while the parameters were learnt from 593 completely-filled ADR reports evaluated by the Portuguese Northern Pharmacovigilance Centre expert between 2000 and 2012. Precision, recall and time to causality assessment (TTA) was evaluated, according to the WHO causality assessment guidelines, in a retrospective cohort of 466 reports (April to September 2014) and a prospective cohort of 1041 reports (January to December 2015). Results show that the network was able to easily identify the higher levels of causality (recall above 80%), although strugling to assess reports with a lower level of causality. Nonetheless, the median (Q1:Q3) TTA was 4 (2:8) days using the network and 8 (5:14) days using global introspection, meaning the network allowed a faster time to assessment, which has a procedural deadline of 30 days, improving daily activities in the centre.

**Keywords:** Adverse drug events · Causality assessment · Bayesian nets

## 1 Introduction

In pharmacovigilance, most of the reported cases are considered as suspected adverse drug reactions (ADR). Health professionals and consumers are asked

© Springer International Publishing AG 2017
A. ten Teije et al. (Eds.): AIME 2017, LNAI 10259, pp. 55–64, 2017.
DOI: 10.1007/978-3-319-59758-4_6

to report episodes they believe are related with drug intake, but in most of the cases ADR are not particular for each drug and a drug rechallenge rarely occurs. To solve this difficulty, health authorities have adopted structured and harmonized causality assessment methods, in order to classify the ADR reports with one of the causality degrees proposed by the WHO-UMC causality assessment system [20]. Apart from ADR identification, where innovative methods have been proposed [12], causality assessment is an essential tool in the pharmacovigilance system, as it helps the risk-benefit evaluation of commercialized medicines, and is part of the signal detection (being a signal a "reported information on a possible causal relationship between an adverse event and a drug, the relationship being unknown or incompletely documented previously" [20]) performed by health authorities. The Portuguese Pharmacovigilance System has adopted the method of Global Introspection [7], since its creation. During this process, an expert (or a group of experts) expresses judgement about possible drug causation, considering all available data in the ADR report. The decision is based on the expert knowledge and experience, and uses no standardized tools. Although this is the method most widely used [1], it has some limitations related to its reproducibility and validity [2,3,15]. Besides, this method is closely linked with the expert availability which not always allows meeting legal deadlines. Causality assessment can also be done through validated algorithms such as the Naranjo [16], Jones [8] or Karch-Lasagna [9] algorithms. Although these algorithms have better agreement rates than Global Introspection, they also have the disadvantage of not being flexible and, consequently, it is not possible to include more causal factors to be evaluated at the same time [1]. Besides, in our experience, some real cases evaluated by more than one algorithm may give rise to different degrees of causality. Guidelines such as the ones used for causality assessment are several times hard to interpret and to apply, even by experienced practicioners. Moreover, they often result in simple rules or association measures, making their application in decision support somewhat limited, especially in the context of guidelines that are to be computer-interpreted for decision support systems [10]. Bayesian approaches have an extreme importance in these problems as they provide a quantitative perspective and have been successfully applied in health care domains [14]. One of their strengths is that Bayesian statistical methods allow taking into account prior knowledge when analysing data, turning the data analysis into a process of updating that prior knowledge with biomedical and health-care evidence [13]. This way, a possible path for implementing such practice guidelines are Bayesian networks, which are probabilistic graphical models that expose the interdependencies among variables, defining associations based on conditional probabilities (i.e. risks) and enabling a friendly inference interface to users [5]. Bayesian networks can be seen as an alternative to logistic regression, where statistical dependence and independence are not hidden in approximating weights, but rather explicitly represented by links in a network of variables [14]. Generally, a Bayesian network represents a joint distribution of one set of variables, specifying the assumption of independence between them, with the interdependence between variables being represented by a directed acyclic graph. Each variable is represented by a node in the graph, and is dependent on the set of variables represented by its ascendant nodes. This dependence is represented by a conditional probability table that describes the probability distribution of each variable, given their ascendant variables [5]. Given their successful applications in previous healthcare applications, we decided to build (and validate) a Bayesian network model to help in the process of causality assessment carried out in pharmacovigilance centres.

## 2  Materials and Methods

The study is framed as the development of a diagnostic test, where the comparison (gold standard) is the method of expert's global introspection. Three cohorts of suspected ADR were used: a derivation cohort, consisting of the registries of suspected ADR evaluated by a specialist in a regional pharmacovigilance centre between 2000 and 2012; a retrospective validation cohort consisting of all reports of suspected ADR received in the same centre within the initial 6 months of implantation of the system in the centre; and a prospective cohort, consisting of all reports of suspected ADR received in the same centre during the year of 2015. Each suspected ADR was evaluated by the expert of the Northern Pharmacovigilance Centre using global introspection, for causality categories of *Definite*, *Probable*, *Possible* or *Conditional*, according to the WHO causality assessment guidelines. The variables used to develop the network were the usual data needed for common causality assessment algorithms [8,9,16], as explained below:

**Described:** If the ADR was previously reported in other patients so that this event is described in the summary of product characteristics (SPC), it enhances the likelihood that a drug is the cause of the observed event; this variable was slightly enhanced for the prospective cohort, including descriptions also on other sources of published literature.

**Reintroduced/Reappeared:** Data on drug rechallenge is mostly absent, because it is not likely that a patient who has suffered ADR receive the suspected drug again. This data is available when the patient uses the drug for the second time by mistake, or when the first ADR episode has not been interpreted as such.

**Suspended/Improved:** A favorable evolution of the ADR after drug withdrawal increases the likelihood that the suspected drug was the cause of the ADR.

**Concomitant:** The presence of other drugs can represent alternative causes (other than the suspected drug) that could on their own cause the ADR.

**Suspected interaction:** If there is suspicion that an interaction with other drug existed, then the cause of the ADR is less clear.

**Route of Administration:** Some ADR are more likely to occur when the drug is administered intravenously, and others when it is administered orally or topically, for example.

**Notifier:** This variable represents a proxy for quality of information. In most cases, physicians report their ADR suspicions more completely and precisely than other health professionals.

**Pharmacotherapeutic Group:** Although ADR are not specific to drug classes, some events are more related to certain pharmacotherapeutic group(s). The nomenclature used for drug classification was the one adopted by Portuguese Authority of Medicines and Health Products (INFARMED, IP) according to national legislation *(Despacho nº 21844/2004, de 12 de outubro)* which includes a correspondence with the international ATC.

**Ineffectiveness:** Until February 2015, the expert interpretation of this kind of ADR (lack of effect) assumed that they should all receive the causality degree of *Conditional,* as experts needed further information to a complete the assessment. After February 2015, this interpretation has changed into assuming lack of effect is described for all drugs.

Data was collected from INFARMED and completely-filled ADR reports, evaluated by the Portuguese Northern Pharmacovigilance Centre. In case of duplicate reports, an evaluation were performed and the duplicate dismissed. All the reports inserted in the database were cross-checked with the original paper reports. In the cases were we had two or more drugs that had different properties and characteristics in the same report, the classification on the database was unknown for variables *Suspended, Administration* and *PharmaGroup.* All reports where there was no indication of the reaction being described in the literature were classified as not being described. Also, if in the same report described two levels of causality, we considered it having the lowest one, following the recommendations of the experts. Variables *Suspended* and *ImprovedAfterSuspension,* and likewise *Reintroduced* and *Reappeared* have an intimate connection, with the former imposing a "NotApplicable" status to the latter in cases where the former were negative.

## 2.1 Bayesian Network Model Definition

A Bayesian network was developed where the structure of causal dependence was defined towards implementing the current guidelines for causality assessment, in cooperation with the medical expert, whereas the conditional probabilities were induced from the derivation cohort. Figure 1 presents a graphical representation of the development process. ADR reports are sent to pharmacovigilance centres using any available channel (e.g. web service, online form, phone, email, etc.) where a medical expert assesses the causality by the process of global introspection, assigning a causality degree (definite, probable, possible or conditional) to the report, which is then stored in a relational database. Using published guidelines for the causality assessment, and the experienced opinion of the expert, a Bayesian network was built trying to capture the causal interdependences of the ADR. Then, using the historical assessment done by the expert, the network parameters were learnt from data, defining its quantitative model.

### Bayesian Network Model Structure

Figure 2 presents the final structure of the model. Following the guidelines and expert opinion, we separated the relevant variables into four groups of nodes for

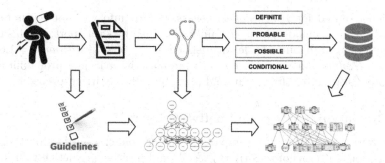

**Fig. 1.** Definition of the expert-informed Bayesian network for causality assessment of adverse drug reaction reports. Both the guidelines and the medical expert opinion were used to define the structures, while parameters were learnt from historical assessment.

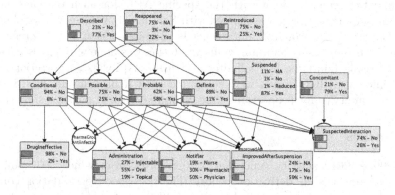

**Fig. 2.** Expert-informed Bayesian network for causality assessment of ADR reports. Monitors show the marginal probabilities for all nodes, except for drug pharmacotherapeutical group, which was hidden for presentation space reasons.

the network: (a) factors that generally influence the occurrence of ADR for the drug in question (i.e. *Described*, *Reintroduced* and *Reappeared*); (b) factors which are related with the particular report in question (i.e. *Suspended*, *ImprovedAfterSupension*, *Concomitant*, *SuspectedInteraction*, *Notifier* and *Administration*); (c) special cases, i.e. *PharmaGroup* which, given the number of possible states, would make the network too complex if modeled in causal ways, and also, *Ineffectiveness* was considered to only influence the *Conditional* degree; and (d) given the peculiarities of some factors, and the limited number of reports for the modeling, we decided to model each causality degree in a separate node.

### Bayesian Network Model Parameters

Since we are trying to model the assessment done by the expert, the model's parameters were learnt from the actual reports and assessment in the derivation cohort. Given the limited quality of the electronic reporting of suspected ADR in years prior to 2012, we only considered complete reports for this step. An exception is noted for node *Ineffectiveness*, as this info was not registered up until 2014; thus, to model the uncertainty described by the expert, for this node, the conditional probability table was defined as $P(Yes|Conditional) = 1.0$ and $P(Yes| \sim Conditional) = .375$. Figure 2 presents the marginal probabilities for each node's state after the conditional probability table fitting procedure.

## 2.2 Evaluation Strategy and Software Used

The network assessement was compared with the gold standard (expert's global introspection) in terms of sensitivity (recall) and positive predictive values (precision). Also evaluated (using the retrospective cohort) was the time to causality assessment (TTA), compared to the manual assessment times recorded in the centre's quality management system. Final validation was done in the prospective validation cohort, along with the specific AUC for each outcome node. Bayesian network structure was defined with *SamIam* [4], while conditional probability tables were learnt from data using R package *bnlearn* [19]. Inference for daily use was done using SamIam, while validation was done with the R package *gRain* [6] using Lauritzen-Spiegelhalter algorithm [11] for exact posterior probability inference. ROC curves were computed with R package *pROC* [18], and confidence intervals for proportions were computed with R package *stats* [17].

## 3 Results

The 2000 to 2012 activity generated 3220 records, from which 593 complete instances were used as derivation cohort. The retrospective validation cohort, collected during 6 months in 2014, included 466 reports. The final prospective validation cohort, collected for the whole year of 2015, included 1041 reports. Over all 2100 ADR, 85% were described, 29% did not include a drug rechallenge, but 77% were suspended leading to patient status improvement in 72% of the cases, 88% considered concomitant medication, although only 4% actually raised suspicion of interaction. The majority of ADR were reported by physicians (61%) and pharmacists (27%), being mainly related to oral (66%) or injectable (31%) drugs. Table 1 presents the descriptive analysis of the three cohorts. The initial analysis of the derivation and validation cohorts gave indications that the network seems better for higher degrees of causality (precision and recall for Probable above 87%). However, in the validation cohort, the network actually tends to overrate causality (96.9% of errors on Possible cases classified as Probable) or give the immediately below level (90.8% of errors on Definite cases classified as Probable; 69.7% of errors on Probable cases classified as Possible). The median (Q1:Q3) time to causality assessment was 4 (2:8) days using the network and 8 (5:14) days using global introspection, meaning the network allowed a faster time to assessment. The prospective validation of the network reinforced the ability to identify higher levels of causality (recall for *Definite* and *Probable* above 80%) while exposing even stronger problems dealing with the lower levels of causality. Table 2 presents the results for that cohort, where the network clearly failed to address *Possible* and *Conditional* levels, although each node, alone, had a specific AUC above 65%.

**Table 1.** Descriptive analysis of the three cohorts used in the study.

| | Derivation n (%) | Validation n (%) | Prospective n (%) | Total n (% [95%CI]) |
|---|---|---|---|---|
| Period | 2000-2012 | Apr-Sep 2014 | Jan-Dec 2015 | |
| **Described** | **593 (28.3)** | **464 (22.1)** | **1041 (49.6)** | **2098 (100)** |
| Yes | 459 (77.4) | 383 (82.5) | 932 (89.5) | 1774 (84.6 [82.9,86.1]) |
| No | 134 (22.6) | 81 (17.5) | 109 (10.5) | 324 (15.4 [13.9,17.1]) |
| **Reintroduced** | **593 (55.6)** | **189 (17.7)** | **284 (26.6)** | **1066 (100)** |
| Yes | 148 (25) | 59 (31.2) | 101 (35.6) | 308 (28.9 [26.2,31.7]) |
| No | 445 (75) | 48 (25.4) | 183 (64.4) | 676 (63.4 [60.4,66.3]) |
| NotApplicable | 0 (0) | 82 (43.4) | 0 (0) | 82 (7.7 [6.2,9.5]) |
| **Reappeared after reintroduction** | **593 (56.6)** | **183 (17.5)** | **271 (25.9)** | **1047 (100)** |
| Yes | 129 (21.8) | 44 (24) | 56 (20.7) | 229 (21.9 [19.4,24.5]) |
| No | 18 (3) | 9 (4.9) | 32 (11.8) | 59 (5.6 [4.4,7.3]) |
| NotApplicable | 446 (75.2) | 130 (71) | 183 (67.5) | 759 (72.5 [69.7,75.2]) |
| **Suspended** | **593 (30.4)** | **418 (21.5)** | **937 (48.1)** | **1948 (100)** |
| Yes | 518 (87.4) | 319 (76.3) | 628 (67) | 1465 (75.2 [73.2,77.1]) |
| Reduced | 3 (0.5) | 5 (1.2) | 24 (2.6) | 32 (1.6 [1.1,2.3]) |
| No | 8 (1.3) | 28 (6.7) | 152 (16.2) | 188 (9.7 [8.4,11.1]) |
| NotApplicable | 64 (10.8) | 66 (15.8) | 133 (14.2) | 263 (13.5 [12,15.1]) |
| **Improved after suspension** | **593 (31.7)** | **399 (21.3)** | **879 (47)** | **1871 (100)** |
| Yes | 486 (82) | 293 (73.4) | 567 (64.5) | 1346 (71.9 [69.8,74]) |
| No | 29 (4.9) | 7 (1.8) | 27 (3.1) | 63 (3.4 [2.6,4.3]) |
| NotApplicable | 78 (13.2) | 99 (24.8) | 285 (32.4) | 462 (24.7 [22.8,26.7]) |
| **Concomitant medication** | **593 (47.7)** | **189 (15.2)** | **460 (37)** | **1242 (100)** |
| Yes | 466 (78.6) | 180 (95.2) | 445 (96.7) | 1091 (87.8 [85.9,89.6]) |
| No | 127 (21.4) | 9 (4.8) | 15 (3.3) | 151 (12.2 [10.4,14.1]) |
| **Suspected interaction** | **593 (34.7)** | **75 (4.4)** | **1041 (60.9)** | **1709 (100)** |
| Yes | 37 (6.2) | 10 (13.3) | 21 (2) | 68 (4 [3.1,5]) |
| No | 556 (93.8) | 65 (86.7) | 1020 (98) | 1641 (96 [95,96.9]) |
| **Route of administration** | **593 (29)** | **443 (21.7)** | **1006 (49.3)** | **2042 (100)** |
| Oral | 429 (72.3) | 267 (60.3) | 654 (65) | 1350 (66.1 [64,68.2]) |
| Injectable | 123 (20.7) | 167 (37.7) | 346 (34.4) | 636 (31.1 [29.2,33.2]) |
| Topical | 41 (6.9) | 9 (2) | 6 (0.6) | 56 (2.7 [2.1,3.6]) |
| **Notifier** | **593 (29.3)** | **466 (23)** | **966 (47.7)** | **2025 (100)** |
| Physician | 372 (62.7) | 295 (63.3) | 565 (58.5) | 1232 (60.8 [58.7,63]) |
| Pharmacist | 175 (29.5) | 91 (19.5) | 283 (29.3) | 549 (27.1 [25.2,29.1]) |
| Nurse | 46 (7.8) | 57 (12.2) | 118 (12.2) | 221 (10.9 [9.6,12.4]) |
| Other | 0 (0) | 23 (4.9) | 0 (0) | 23 (1.1 [0.7,1.7]) |
| **Pharmacotherapeutical group** | **593 (28.8)** | **466 (22.6)** | **1003 (48.6)** | **2062 (100)** |
| AntiallergicMedication | 11 (1.9) | 4 (0.9) | 8 (0.8) | 23 (1.1 [0.7,1.7]) |
| Antiinfectious | 136 (22.9) | 103 (22.1) | 264 (26.3) | 503 (24.4 [22.6,26.3]) |
| AntineoplasticDrugsImmunemodulators | 35 (5.9) | 82 (17.6) | 212 (21.1) | 329 (16 [14.4,17.6]) |
| Blood | 7 (1.2) | 9 (1.9) | 25 (2.5) | 41 (2 [1.4,2.7]) |
| CardiovascularSystem | 72 (12.1) | 29 (6.2) | 42 (4.2) | 143 (6.9 [5.9,8.1]) |
| CentralNervousSystem | 91 (15.3) | 51 (10.9) | 162 (16.2) | 304 (14.7 [13.3,16.4]) |
| DiagnosisMedia | 1 (0.2) | 9 (1.9) | 20 (2) | 30 (1.5 [1,2.1]) |
| DrugsForEyeDisorders | 4 (0.7) | 1 (0.2) | 8 (0.8) | 13 (0.6 [0.4,1.1]) |
| DrugsForSkinDisorders | 23 (3.9) | 3 (0.6) | 3 (0.3) | 29 (1.4 [1,2]) |
| DrugsToTreatPoisoning | 1 (0.2) | 0 (0) | 0 (0) | 1 (0 [0,0.3]) |
| GastrointestinalSystem | 28 (4.7) | 11 (2.4) | 16 (1.6) | 55 (2.7 [2,3.5]) |
| GenitourinarySystem | 13 (2.2) | 1 (0.2) | 10 (1) | 24 (1.2 [0.8,1.8]) |
| Hormones | 17 (2.9) | 14 (3) | 33 (3.3) | 64 (3.1 [2.4,4]) |
| LocomotorSystem | 101 (17) | 77 (16.5) | 84 (8.4) | 262 (12.7 [11.3,14.2]) |
| Nutrition | 3 (0.5) | 2 (0.4) | 2 (0.2) | 7 (0.3 [0.1,0.7]) |
| Otorhinolaryngology | 0 (0) | 2 (0.4) | 0 (0) | 2 (0.1 [0,0.4]) |
| RespiratorySystem | 10 (1.7) | 8 (1.7) | 12 (1.2) | 30 (1.5 [1,2.1]) |
| VaccinesImmunoglobulins | 40 (6.7) | 58 (12.4) | 102 (10.2) | 200 (9.7 [8.5,11.1]) |
| Volaemia | 0 (0) | 2 (0.4) | 0 (0) | 2 (0.1 [0,0.4]) |
| **Expert assessment** | **593 (28.2)** | **466 (22.2)** | **1041 (49.6)** | **2100 (100)** |
| Definite | 60 (10.1) | 37 (7.9) | 36 (3.5) | 133 (6.3 [5.3,7.5]) |
| Probable | 346 (58.3) | 372 (79.8) | 833 (80) | 1551 (73.9 [71.9,75.7]) |
| Possible | 152 (25.6) | 44 (9.4) | 131 (12.6) | 327 (15.6 [14.1,17.2]) |
| Conditional | 35 (5.9) | 13 (2.8) | 41 (3.9) | 89 (4.2 [3.4,5.2]) |
| **Bayesian net assessment** | **593 (28.2)** | **466 (22.2)** | **1041 (49.6)** | **2100 (100)** |
| Definite | 77 (13) | 36 (7.7) | 47 (4.5) | 160 (7.6 [6.5,8.9]) |
| Probable | 331 (55.8) | 388 (83.3) | 945 (90.8) | 1664 (79.2 [77.4,80.9]) |
| Possible | 185 (31.2) | 38 (8.2) | 47 (4.5) | 270 (12.9 [11.5,14.4]) |
| Conditional | 0 (0) | 4 (0.9) | 2 (0.2) | 6 (0.3 [0.1,0.7]) |

**Table 2.** Validity assessment for the 2015 prospective cohort of ADR reports.

| | Definite | Probable | Possible | Cond | Precision % [95%CI] | Recall % [95%CI] | Node AUC % [95%CI] |
|---|---|---|---|---|---|---|---|
| Definite | 30 | 16 | 1 | 0 | 63.8 [48.5,76.9] | 83.3 [66.5,93.0] | 91.7 [84.8,98.5] |
| Probable | 4 | 792 | 117 | 32 | 83.8 [81.3,86.1] | 95.1 [93.3,96.4] | 70.7 [66.6,74.8] |
| Possible | 2 | 24 | 12 | 9 | 25.5 [14.4,40.6] | 9.2 [5.0,15.8] | 66.7 [62.0,71.3] |
| Conditional | 0 | 1 | 1 | 0 | 0.0 [0.0,80.2] | 0.0 [0.0,10.7] | 69.1 [61.3,76.9] |

# 4   Discussion

The network allowed a faster time to assessment, which has a procedural deadline of 30 days, improving daily activities in the pharmacovigilance centre. Moreover, the model was accurate on most cases, showing satisfactory results to the higher degrees of causality. On the other hand, it had a non-adequate behaviour with the two lowest degrees of causality. We believe the Bayesian network failed to learn the degree Possible because this degree is much related with the existence of concomitant diseases or conditions that could explain the ADR [20]. However, this kind of information is not collected in the ADR form as a structured field. The notifier may provide this information in a free text field (comments) or by phone. For this reason, the network does not consider any node with this question. On the other hand, the expert is aware of this information (if any) and can fully assess the case. For example, there were several reports of headache and fatigue involving new drugs used in Hepatitis C. These reports were mainly assessed by the expert as Possible, because the ADR reported could also be explained by the disease (hepatitis), contrary to the Bayesian network which assessed these cases as Probable, since these ADR are described in the SPC. We believe that this issue can be solved with the inclusion of a new question on the ADR form on a) the existence of any other eventual cause to the ADR other than the suspect drug, or b) if the drug is considered to be new in the population; consequently, new nodes in the Bayesian network are needed. The network also failed, and with greater magnitude, to learn the degree Conditional. This is a temporary degree, attributed to those cases with insufficient information and also to those cases which is expected to obtain more data. For this reason, it is a degree difficult to fit in a model. After February 2015, this interpretation has changed (which prevented a joint analysis of both cohorts together). Thus, a particular analysis will be performed to this node. Causality assessment by the expert has also some limitations [2,15]. During this activity, personal expectations and beliefs can influence the assessment. This subjectivity is hard to be replicated in a model as ours. Although our model learned data from the expert assessment, it tends to follow the causality assessment guidelines, which is not in line with this kind of subjectivity. For example, in ADR reports made by physicians the signs and symptoms are usually better described than in ADR reports made by other healthcare professionals or consumers. As a consequence, the expert (a physician) has more information about the ADR, as it is detailed by a peer, with the same language and structure. To try to solve this issue, we have included in the network the node *Notifier* which is intended to be a proxy to the manner the ADR is explained. Future developments could include learning a model with latent variables to try to capture these phenomena.

## 5  Concluding Remarks

The derived model has been used in the Northern Pharmacovigilance Centre, in Portugal, for more than two years now, for causality assessment of ADR reports. We believe that this network can be very useful to other pharmacovigilance centres, mainly to those that do not have access to a full-time expert to evaluate ADR reports. As every method for ADR causality assessment [1], the presented Bayesian network as some advantages but also some limitations. Nonetheless, the network allows to shorten the time to assessment, which is a main issue in pharmacovigilance activities, and is accurate for most of the cases. Therefore, this method does not replace the expert evaluation, but can be used to complement it. Furthermore, future work will focus on refining the model, learning a new classifier from the (now more complete) data recorded from 2015 onwards, and validate it against other alternatives.

**Acknowledgements.** This work has been developed under the scope of project NanoSTIMA [NORTE-01-0145-FEDER-000016], which was financed by the North Portugal Regional Operational Programme [NORTE 2020], under the PORTUGAL 2020 Partnership Agreement, and through the European Regional Development Fund [ERDF].

## References

1. Agbabiaka, T.B., Savović, J., Ernst, E.: Methods for causality assessment of adverse drug reactions: a systematic review. Drug Saf. **31**(1), 21–37 (2008). http://www.ncbi.nlm.nih.gov/pubmed/18095744
2. Arimone, Y., Bégaud, B., Miremont-Salamé, G., Fourrier-Réglat, A., Moore, N., Molimard, M., Haramburu, F.: Agreement of expert judgment in causality assessment of adverse drug reactions. Eur. J. Clin. Pharmacol. **61**, 169–173 (2005)
3. Arimone, Y., Miremont-Salamé, G., Haramburu, F., Molimard, M., Moore, N., Fourrier-Réglat, A., Bégaud, B.: Inter-expert agreement of seven criteria in causality assessment of adverse drug reactions. Br. J. Clin. Pharmacol. **64**(4), 482–488 (2007). http://www.pubmedcentral.nih.gov/articlerender.fcgi?artid=2048553&tool=pmcentrez&rendertype=abstract
4. Darwiche, A.: Modeling and Reasoning with Bayesian Networks. Cambridge University Press, Cambridge (2009). http://www.amazon.com/Modeling-Reasoning-Bayesian-Networks-Darwiche/dp/0521884381
5. Darwiche, A.: Bayesian networks. Commun. ACM **53**(12), 80–90 (2010). http://portal.acm.org/citation.cfm?doid=1859204.1859227
6. Højsgaard, S.: Graphical independence networks with the gRain package for R. J. Stat. Softw. **46**(10), 1–26 (2012). http://www.jstatsoft.org/v46/i10/paper
7. INFARMED: Farmacovigilância em Portugal. Technical report (2004)

8. Jones, J.: Adverse drug reactions in the community health setting: approaches to recognizing, counseling, and reporting. Fam. Community Health **5**(2), 58–67 (1982). http://www.ncbi.nlm.nih.gov/pubmed/10278126

9. Karch, F.E., Lasagna, L.: Toward the operational identification of adverse drug reactions. Clin. Pharmacol. Ther. **21**(3), 247–254 (1977). http://www.ncbi.nlm.nih.gov/pubmed/837643

10. Latoszek-Berendsen, A., Tange, H., van den Herik, H.J., Hasman, A.: From clinical practice guidelines to computer-interpretable guidelines. A literature overview. Methods Inf. Med. **49**(6), 550–570 (2010). http://www.ncbi.nlm.nih.gov/pubmed/21085744

11. Lauritzen, S.L., Spiegelhalter, D.J.: Local computations with probabilities on graphical structures and their application to expert systems. J. R. Stat. Soc. Ser. B **50**(2), 157–224 (1988). http://wrap.warwick.ac.uk/24233/

12. Lindquist, M., Staahl, M., Bate, A., Edwards, I.R., Meyboom, R.H.: A retrospective evaluation of a data mining approach to aid finding new adverse drug reaction signals in the WHO international database. Drug Saf. **23**(6), 533–542 (2000)

13. Lucas, P.: Bayesian analysis, pattern analysis, and data mining in health care. Current Opin. Crit. Care **10**(5), 399–403 (2004). http://www.ncbi.nlm.nih.gov/pubmed/15385759

14. Lucas, P.J.F., van der Gaag, L.C., Abu-Hanna, A.: Bayesian networks in biomedicine and health-care. Artif. Intell. Med. **30**(3), 201–214 (2004). http://www.ncbi.nlm.nih.gov/pubmed/15081072

15. Miremont, G., Haramburu, F., Bégaud, B., Péré, J.C., Dangoumau, J.: Adverse drug reactions: physicians' opinions versus a causality assessment method. Eur. J. Clin. Pharmacol. **46**, 285–289 (1994)

16. Naranjo, C.A., Busto, U., Sellers, E.M., Sandor, P., Ruiz, I., Roberts, E.A., Janecek, E., Domecq, C., Greenblatt, D.J.: A method for estimating the probability of adverse drug reactions. Clin. Pharmacol. Ther. **30**, 239–245 (1981)

17. R Core Team: R: A Language and environment for statistical computing (2015). http://www.r-project.org/

18. Robin, X., Turck, N., Hainard, A., Tiberti, N., Lisacek, F., Sanchez, J.C., Müller, M.: pROC: an open-source package for R and S+ to analyze and compare ROC curves. BMC Bioinf. **12**, 77 (2011)

19. Scutari, M.: Learning Bayesian networks with the bnlearn R Package. J. Stat. Softw. **35**, 22 (2010). http://arXiv.org/abs/0908.3817

20. World Health Organization. Uppsala Monitoring Centre (2017). http://www.who-umc.org/

# Bayesian Gaussian Process Classification from Event-Related Brain Potentials in Alzheimer's Disease

Wolfgang Fruehwirt[1,2], Pengfei Zhang[2], Matthias Gerstgrasser[3],
Dieter Grossegger[4], Reinhold Schmidt[5], Thomas Benke[6],
Peter Dal-Bianco[7], Gerhard Ransmayr[8], Leonard Weydemann[1],
Heinrich Garn[9], Markus Waser[9], Michael Osborne[2],
and Georg Dorffner[1(✉)]

[1] Section for AI and Decision Support,
Medical University of Vienna, Vienna, Austria
georg.dorffner@meduniwien.ac.at
[2] Department of Engineering Science, University of Oxford, Oxford, UK
[3] Department of Computer Science, University of Oxford, Oxford, UK
[4] Dr. Grossegger & Drbal GmbH, Vienna, Austria
[5] Department of Neurology, Medical University of Graz, Graz, Austria
[6] Department of Neurology, Medical University of Innsbruck,
Innsbruck, Austria
[7] Department of Neurology, Medical University of Vienna, Vienna, Austria
[8] Department of Neurology, Linz General Hospital, Linz, Austria
[9] AIT Austrian Institute of Technology GmbH, Vienna, Austria

**Abstract.** Event-related potentials (ERPs) have been shown to reflect neurodegenerative processes in Alzheimer's disease (AD) and might qualify as non-invasive and cost-effective markers to facilitate the objectivization of AD assessment in daily clinical practice. Lately, the combination of multivariate pattern analysis (MVPA) and Gaussian process classification (GPC) has gained interest in the neuroscientific community. Here, we demonstrate how a MVPA-GPC approach can be applied to electrophysiological data. Furthermore, in order to account for the temporal information of ERPs, we develop a novel method that integrates interregional synchrony of ERP time signatures. By using real-life ERP recordings of a prospective AD cohort study (PRODEM), we empirically investigate the usefulness of the proposed framework to build neurophysiological markers for single subject classification tasks. GPC outperforms the probabilistic reference method in both tasks, with the highest AUC overall (0.802) being achieved using the new spatiotemporal method in the prediction of rapid cognitive decline.

**Keywords:** Machine learning · Gaussian process classification · Event-related potentials · Alzheimer's disease · Single subject classification

© Springer International Publishing AG 2017
A. ten Teije et al. (Eds.): AIME 2017, LNAI 10259, pp. 65–75, 2017.
DOI: 10.1007/978-3-319-59758-4_7

# 1  Introduction

Due to its degenerative nature, early detection and accurate evaluation of Alzheimer's disease (AD) are crucial. However, when it comes to routine clinical practice, AD diagnosis is most commonly based on subjective clinical interpretations at a progressed stage of the disease, i.e. when symptoms are already apparent. Thus, there is a strong need to develop affordable and thereby widely available markers that facilitate the objectivization of AD assessment. Event-related potentials (ERPs), as measured by non-invasive and cost-effective electroencephalography (EEG), have been shown to reflect neurodegenerative processes in AD [1–3] and might therefore qualify as such markers.

Lately, the combination of multivariate pattern analysis (MVPA) and Gaussian process classification (GPC), a machine learning technique, has gained interest in the neuroscientific community. While MVPA incorporates interactions between multiple brain structures or function patterns, GPC allows for an easy adjustment of predictions to compensate for variable class priors (e.g. variations in diagnostic setting or disease prevalence), and - most importantly for routine practice - provides probabilistic predictions quantifying predictive uncertainty.

Only recently, a resting-state functional magnetic resonance imaging (fMRI) study showed the applicability of a MVPA-GPC approach for single subject classification tasks in AD [4]. Here, we demonstrated how this technique can be applied to electrophysiological data. Furthermore, to account for the temporal information of ERPs, we developed a novel method that integrates interregional synchrony of ERP time signatures.

To the best of our knowledge, this is the first electrophysiological ERP study to use GPC and therefore the first to use MVPA-GPC.

Utilizing real-life ERP recordings of a prospective AD cohort study, we aim to build neurophysiological markers for two crucial AD classification problems:

First, we intend to predict rapid cognitive decline. The rate of cognitive decline in AD strongly correlates with mortality and shows profound variability between individuals [5]. Hence, early prediction of individual trajectories of cognitive function is essential for treatment and care, as it allows for personalized interventions and appropriate forehanded planning of support services.

Then, we try to identify patients who test positive for the Apolipoprotein E (ApoE) ε4 allele (ε4+), the strongest genetic risk factor for AD [6]. As ApoE ε4 expression has been shown to alter ERP waveforms [7, 8] as well as functional EEG connectivity [9, 10], neurophysiological markers incorporating both aspects - as done by our framework - may allow for such estimates.

Distinct parameter settings are tested and, using the identical preselected MVPA features, GPC performance is compared to a probabilistic reference method (logistic regression classification, LRC).

# 2 Materials and Methods

## 2.1 Subjects

Sixty-three AD patients (31 with possible, 32 with probable AD diagnosis according to NINCDS-ADRDA criteria; 39 APOE ε4 carriers; 38 females; mean age 75.92 ± 8.82 standard deviation (SD); mean MMSE score 23.25 ± 3.60 SD; mean years of education 10.46 ± 2.26 SD; mean duration of illness (months) 22.89 ± 14.65 SD) were considered for this investigation. They were recruited prospectively at the tertiary-referral memory clinic of the Medical University of Innsbruck as part of the cohort study Prospective Dementia Registry Austria (PRODEM). PRODEM is a longitudinal multicenter study of AD and other dementias in a routine clinical setting by the Austrian Alzheimer Society. Ethics committee approval was obtained and patients and their caregivers gave written informed consent.

Inclusion criteria encompassed: (I) diagnosis of Alzheimer-type dementia according to NINCDS-ADRDA criteria, (II) minimum age 40 years, (III) non-institutionalization and no need for 24-hour care, (IV) availability of a caregiver who agrees to provide information on the patient's condition. Patients with comorbidities likely to preclude termination of the study were excluded.

For a maximum of 18 months patients revisited for follow-up assessments (FU) every 6 months. Prediction of rapid cognitive decline was performed for subjects who returned at least for the 12-month FU (N = 48; 22 with possible, 26 with probable AD; 18 with rapid cognitive decline; 33 APOE ε4 carriers; 29 females; mean age 75.90 ± 8.61 SD; mean MMSE score 23.71 ± 3.13 SD; mean years of education 10.83 ± 2.28 SD; mean duration of illness (months) 23.17 ± 16.02 SD).

## 2.2 Assessment of Cognitive Decline and Apolipoprotein E Genotyping

Assessment of cognitive decline was done using the Mini-Mental State Examination (MMSE, [11]). Rapid cognitive decline was defined as a decrease of 3 or more points on the MMSE between baseline and 12-month FU [12].

We used ApoE genotyping to determine ε4 expression. ApoE is the principal cholesterol carrier protein in the brain [13] and supports lipid transport and injury repair. Carriers of the ApoE ε4 allele (ε4+) are known to be of heightened risk to develop AD. Consequently, the frequency of the ε4 allele is dramatically increased in patients with AD as compared to the overall population [6].

## 2.3 Recording and Pre-processing of Event-Related Potentials

Auditory ERPs were elicited using the "oddball" paradigm, a simple discrimination task. Frequent (141) standard tones (1000 Hz) and infrequent (57) target tones (2000 Hz) appeared in a quasi-random sequence held constant across subjects. The tone duration being 100 ms, interstimulus intervals varied between 1 and 1.5 s.

Subjects were instructed to press a reaction time button, with the dominant hand, to target stimuli only. Horizontal and vertical electrooculogram (EOG) electrodes detected eye movements. The system employed was a 32-channel AlphaEEG amplifier with

NeuroSpeed software (alpha trace medical systems, Vienna, Austria). EEG electrode placement (Au-plated cups; Grass F-E5GH, Grass Technologies, West Warwick, RI, USA) was in accordance with the international 10–20 system. The electrodes were referenced to connected mastoids, the ground being positioned at FCz. The EEG amplifier had a bandpass of 0.3 to 70 Hz (3 dB points) with a 50 Hz notch filter and a sampling rate set at 256 Hz. Impedance levels were held below 10 kΩ.

After automatic horizontal and vertical regression-based EOG correction in the time domain [14] the individual sweeps to targets were visually screened for artefacts before being accepted into the average. Then, the data were bandpass filtered at 1–16 Hz using the EEGLAB toolbox [15].

## 2.4   Spatial Synchrony Measures

For multichannel covariance estimation, ERP time-courses of individual patients (1 s, starting at stimulus onset) were considered as $N$ by $T$ matrices $X_z$, $N$ being the number of electrodes, $T$ being the number of time samples, and $z$ being the patient.

$$X_z \in \mathbb{R}^{N x T} \tag{1}$$

Covariance for patient $z$ was estimated using the sample covariance matrix (SCM) $C_z$.

$$C_z = \frac{1}{T-1} XX^T \tag{2}$$

All unique elements of the SCM, i.e., the diagonal entries representing the spatial variance and one set of the off-diagonal entries representing the spatial covariance, were then combined into a feature vector $F_z$ of dimension $d = N(N+1)/2$. Using 19 electrodes $d$ equaled 190.

$$F_z \in \mathbb{R}^{N(N+1)/2} \tag{3}$$

## 2.5   Spatiotemporal Synchrony Measures

Regular covariance estimation between brain areas - as for instance used by Challis et al. [4] - comprises signal variance at each of the individual sites as well as the covariance between all site pairs. This spatial information can be embodied in the matrix form described above. ERPs, however, represent a time- and phase-locked response to a stimulus. Therefore, their unfolding in time, i.e., their time signature or pattern, contains specific temporal information putatively valuable for classification purposes.

To take these distinct ERP time signatures into account, we adopt a special type of matrix recently designed [16, 17]. Used in brain-computer interface applications to distinguish between subjects' single responses to stimuli, we adapt the construct to classify subjects themselves. We do so by utilizing averaged instead of single trial potentials. First, the averaged trial data of a given unlabeled patient is vertically concatenated with the grand-average waveforms (temporal prototypes) $\bar{X}_{(1)}$ and $\bar{X}_{(2)}$ of the two classes of patients. Holding one second of data, the respective matrix $X_z^{ST}$ has a dimensionality of $57 \times 256$.

$$X_z^{ST} = \begin{pmatrix} \bar{X}_{(1)} \\ \bar{X}_{(2)} \\ X_z \end{pmatrix} \in \mathbb{R}^{3N \times T} \tag{4}$$

Then, the covariance is estimated resulting in a $57 \times 57$ spatiotemporal SCM $C_z^{ST}$.

$$C_z^{ST} = \frac{1}{(T-1)} \left( X_z^{ST} (X_z^{ST})^T \right) = \frac{1}{(T-1)} \begin{pmatrix} \bar{X}.\bar{X}^T & (X_z\bar{X}^T)^T \\ X_z\bar{X}^T & X_zX_z^T \end{pmatrix} \in \mathbb{R}^{3N \times 3N} \tag{5}$$

$$\text{where } \bar{X}.\bar{X}^T = \frac{1}{(T-1)} \begin{pmatrix} \bar{X}_{(1)}\bar{X}_{(1)}^T & \bar{X}_{(1)}\bar{X}_{(2)}^T \\ \bar{X}_{(2)}\bar{X}_{(1)}^T & \bar{X}_{(2)}\bar{X}_{(2)}^T \end{pmatrix} \in \mathbb{R}^{2N \times 2N} \tag{6}$$

$$\text{and } X_z\bar{X}^T = \left( X_z\bar{X}_{(1)}^T, X_z\bar{X}_{(2)}^T \right) \in \mathbb{R}^{N \times 2N} \tag{7}$$

However, not all the information encoded in this special matrix is useful. The covariance block of the unlabeled patient $X_zX_z^T$ can be considered useful, as it includes spatial information of the patient's signal. $X_z\bar{X}^T$ represents the cross-covariances between the unlabeled patient $z$ and the two pattern prototypes and therefore contains the sought after temporal information (note how shuffling of the columns of $X_z$ changes $C_z^{ST}$). The covariance and cross-covariance blocks of the templates are not informative and are therefore not included in the feature vector. For this reason, we built the feature vector $F_z^{ST}$ by dividing three non-redundant blocks of $C_z^{ST}$, i.e., $X_zX_z^T$, $X_z\bar{X}_{(1)}^T$, and $X_z\bar{X}_{(2)}^T$. As $X_zX_z^T$ is symmetric, only the upper triangle was included into $F_z^{ST}$ resulting in 912 dimensions.

$$F_z^{ST} \in \mathbb{R}^{2N^2 + N(N+1)/2} \tag{8}$$

## 2.6   Machine Learning Classifiers

**Feature Selection.** Two methods for feature selection were carried out. In order to identify features with high discriminative power we computed the Kendall rank correlation coefficient versus the binary class label [4, 18]. Subsequently, only the variables with the highest absolute tau coefficients were used in the GPC or LRC model. Since this feature selection step was included in a leave-one-out cross-validation (LOOCV) algorithm, the selected features differed slightly from iteration to iteration. In line with Challis et al. [4] we examined the model performance using the highest-ranking 5, 10, 15, and 20 features. In addition we added a run with the 3 best features. Moreover, to determine implicitly the relevance of dimensions within GPC and weight feature contribution accordingly, we utilized automatic relevance determination (ARD, see Gaussian process classification).

**Gaussian Processes.** In the following, we provide a formal definition of the GPC which is a Bayesian classification method for non-linear, stochastic and complex classification problems. We start with the general definition of a 2-D Gaussian process (GP), and then define GPC.

**Definition 1.** *Gaussian process* [19, 20]*: Denote by* $f(x,t). : \mathcal{X} \mapsto \mathbb{R}$ *a stochastic process parameterized by* $\{x,t\} \in \mathcal{X}$*, where* $\mathcal{X} \in \mathbb{R}^2 \times \mathbb{R}^+$*. Then, the random function* $f(x,t)$ *is a Gaussian process if all its finite dimensional distributions are Gaussian, where for any* $m \in \mathbb{N}$*, the random variables* $(f(x_1,t_1), \cdots, f(x_m,t_m))$ *are jointly normally distributed.*

We can therefore interpret a GP as formally defined by the following class of random functions:

$$\begin{aligned} \mathcal{F} :=\{&f(\cdot) : \mathcal{X} \mapsto \mathbb{R} \\ &\text{s.t.} f(\cdot) \sim \mathcal{GP}(\mu(\cdot;\theta), \mathcal{C}(\cdot,\cdot;\Psi)), \text{with} \\ &\mu(\cdot;\theta):= \mathbb{E}[f(\cdot)] : \mathcal{X} \mapsto \mathbb{R}, \\ \mathcal{C}(\cdot,\cdot;\Psi):= &\mathbb{E}\big[(f(\cdot) - \mu(\cdot;\theta))(f(\cdot) - \mu(\cdot;\theta))^T\big] : \mathcal{X} \times \mathcal{X} \mapsto \mathbb{R}^+ \}. \end{aligned} \tag{9}$$

Before receipt of data, at each point the (prior) mean of the function is $\mu(\cdot;\theta)$ parameterized by $\theta$, and the (prior) spatial dependence between any two points is given by the covariance function (Mercer kernel) $\mathcal{C}(\cdot,\cdot;\Psi)$, parametrized by $\Psi$ (see detailed discussion in Rasmussen and Williams [19]).

**Gaussian Process Classification.** Given a model which is based on GP, we can classify an unknown field at unobserved locations $x_*$.

**Definition 2.** *Gaussian Process Classification* [19]*: A GP prior is placed over a latent function* $f(x)$*, then is "squashed" through the logistic function to obtain a prior on* $\pi(x) \triangleq p(y = +1|x) = \sigma(f(x))$*.*

To define our model and its approximate inference scheme, we make design decisions as per the following three sections.

(1) Square Exponential Kernel: The model for f(x) is specified by the kernel of the GP. One of the mostly widely used kernels in GP is the square exponential kernel (also known as the exponential quadratic, RBF, or Gaussian kernel), which specifies the covariance between any two locations in space in the following equation:

$$K(x_1, x_2) = \exp(-\frac{(x_1 - x_2)^2}{2l_1^2}) \tag{10}$$

ARD is a popular method for determining the hyperparameters in the square exponential kernels (namely $l_1$). It is implemented in the Gaussian process for machine learning (GPML) package in MATLAB [21].

(2) Expectation Propagation: For the GPC, the posterior density of a latent process $p(f_*|X, y, x_*)$ is intractable due to the fact that the likelihood $p(y|X)$ is intractable. Therefore, approximation techniques need to be proposed to get the posterior. One widely used technique is expectation propagation (EP). The EP method has been discussed in detail in Rasmussen and Williams [19] and the implementation for the study at hand is done via the GPML package [21].

(3) Prediction: The estimated class label at the unknown location is denoted by $\hat{f}_*$. The expectation propagation framework gives a Gaussian approximation to the posterior distribution. The approximate predictive mean for the latent variable $f_*$ denoted as $\mathbb{E}[f_*|X, y, x_*]$ and the uncertainty (statistical error) denoted as $\sigma_*^2|X, y, x_*$ is given by Rasmussen and Williams [19]. The approximate predictive distribution for the binary target becomes where $q[f_*|X, y, x_*]$ is the approximate latent predictive Gaussian with mean and variance given by $\mathbb{E}[f_*|X, y, x_*]$ and $\sigma_*^2|X, y, x_*$.

There are various ways of fitting parametric models for $\mu$ (defined in formula 9). In this work, we select from two mean functions. The first is a combination of a linear and a constant mean function (sum mean). The second is a constant mean function (constant mean) alone. Performance of these two functions was compared in the classification experiments.

**Classification Experiments.** To evaluate the real added value of GPC in terms of classification performance, we compared results achieved by GPC to those achieved by a probabilistic reference technique, namely, LRC.

LRC was implemented using the MATLAB package Logistic Regression for Classification (Pattern Recognition and Machine Learning Toolbox), as developed by Chen [22]. Importantly, the LRC model had the same preselected input features as the GPC model. We used LOOCV to assess the ability of the classifiers to generalize to independent data sets.

GPC and LRC were applied to two distinct single subject classification problems. Patients were classified into (I) carriers ($\varepsilon 4+$) and non-carriers ($\varepsilon 4-$) of the $\varepsilon 4$ allele, and (II) future rapid cognitive decliners (RCD) and future non-rapid cognitive decliners (nRCD).

In order to assess the degree to which unlabeled patients were identified with their correct class labels in the LOOCV, several classifier performance indices were computed (Table 1). Primary outcome measure was the area under the curve (AUC) of the receiver operating characteristic (ROC) curve, for which we calculated test statistics (null hypothesis = AUC of 0.5). A major advantage of AUC as performance measure is that it incorporates the various threshold settings inherent to the ROC curve. Remaining indices were derived from symmetric threshold values.

As opposed to conventional accuracy, balanced accuracy does not lead to an optimistic estimate when a biased classifier is tested on an imbalanced dataset [23]. Thus, it is reported additionally.

Potential model differences were evaluated by statistically comparing the entire areas under the ROC curves, while accounting for the paired nature of the data [24].

# 3   Results

## 3.1   Prediction of Rapid Cognitive Decline

Overall, the best classification in terms of AUC was achieved by spatiotemporal GPC ($GPC_{ST-RCD}$, 0.802, $p < 0.001$, best 20 features, GP function = constant mean). This model had 75.6% balanced-accuracy, 77.1% accuracy, 66.7% sensitivity, and 83.3% specificity. The best competing LRC model was of spatiotemporal nature as well ($LRC_{ST-RCD}$, best 5 features) and yielded an AUC of 0.561 ($p = 0.482$). The ROC curves of the two classifiers differed significantly at $p = 0.012$ and are depicted in Fig. 1. The best model using spatial information only was a GPC model ($GPC_{S-RCD}$, best 5 features, GP function = sum mean) and yielded an AUC of 0.607 ($p = 0.217$). Difference of $GPC_{S-RCD}$ to the superior spatiotemporal GP classifier $GPC_{ST-RCD}$ was significant ($p = 0.046$). Performance measures for the best models per feature and classifier category are given in Table 1.

## 3.2   Apolipoprotein E ε4 Classification

Again, GPC performed better than LRC in both feature categories, the corresponding models significantly rejecting the null hypothesis at $p = 0.002$ ($GPC_{S-\varepsilon4+}$, best 3 features, GP function = constant mean) and $p = 0.046$ ($GPC_{ST-\varepsilon4+}$, best 3 features, GP function = constant mean) respectively. However, the model based on spatial information ($GPC_{S-\varepsilon4+}$) ranked higher ($p = 0.017$) than the one based on spatiotemporal information ($GPC_{ST-\varepsilon4+}$). The difference between the best GPC and the best LRC ($LRC_{S-\varepsilon4+}$, best 20 features) model was significant ($p = 0.041$). Performance measures for the best models per feature and classifier category are given in Table 1.

**Table 1.** Performance measures of the best spatial and spatiotemporal GPC and LRC classifiers for cognitive decline ($N = 48$) and ApoE ε4 expression ($N = 63$)

| Model | AUC (p-value) | B. accuracy | Accuracy | Sensitivity | Specificity |
|---|---|---|---|---|---|
| $GPC_{ST-RCD}$ | 0.802 (<0.001) | 75.6% | 77.1% | 66.7% | 83.3% |
| $GPC_{S-RCD}$ | 0.607 (0.217) | 59.4% | 62.5% | 44.4% | 73.3% |
| $LRC_{ST-RCD}$ | 0.561 (0.482) | 52.8% | 45.8% | 77.8% | 26.7% |
| $LRC_{S-RCD}$ | 0.476 (0.782) | 47.2% | 43.8% | 61.1% | 33.3% |
| $GPC_{ST-\varepsilon4+}$ | 0.651 (0.046) | 64.3% | 66.7% | 76.9% | 50.0% |
| $GPC_{S-\varepsilon4+}$ | 0.735 (0.002) | 69.6% | 71.4% | 79.5% | 58.3% |
| $LRC_{ST-\varepsilon4+}$ | 0.510 (0.899) | 47.9% | 42.9% | 25.6% | 70.8% |
| $LRC_{S-\varepsilon4+}$ | 0.563 (0.404) | 57.1% | 52.4% | 38.5% | 75.0% |

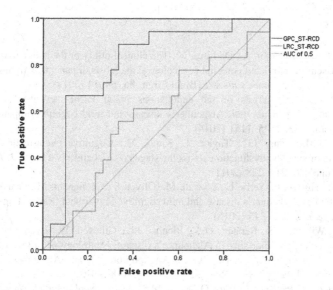

**Fig. 1.** ROC curves of the best GPC (GPC$_{\text{ST-RCD}}$) and LRC (LRC$_{\text{ST-RCD}}$) classifiers for predicting rapid cognitive decline.

## 4    Summary and Discussion

Although potential advantages of GPC for ERP analysis have already been stressed in the literature [25], this is - to the best of our knowledge - the first electrophysiological ERP study to report the use of GPC.

We demonstrated the applicability of MVPA-GPC for electrophysiological data and proposed a method that takes the synchrony of temporal ERP signatures into account. To examine potential advantages of GPC in classification performance we compared it to a probabilistic reference method (LRC). Using ERP data of a prospective cohort study we aimed at building cheap and non-invasive MVPA markers for crucial AD classification problems, i.e., the prediction of rapid cognitive decline and the distinction between carriers and non-carriers of the ApoE ε4 allele. GPC significantly outperformed LRC in both tasks, with the highest AUC overall (0.802) being achieved using the newly developed spatiotemporal method in the prediction of rapid cognitive decline.

Although the number of AD patients included in this examination is relatively large compared to other ERP studies, the modest sample size - in absolute terms - constitutes a limitation to the results. Even though single subject classification was cross-validated using a leave-one-out strategy, further studies with larger sample sizes, including extensive external validation sets, should follow.

**Acknowledgment.** The PRODEM study has been supported by the Austrian Research Promotion Agency FFG, project no. 827462, including financial contributions from Dr. Grossegger and Drbal GmbH, Vienna, Austria.

# References

1. Howe, A.S., Bani-Fatemi, A., De Luca, V.: The clinical utility of the auditory P300 latency subcomponent event-related potential in preclinical diagnosis of patients with mild cognitive impairment and Alzheimer's disease. Brain Cogn. **86**, 64–74 (2014)
2. Howe, A.S.: Meta-analysis of the endogenous N200 latency event-related potential subcomponent in patients with Alzheimer's disease and mild cognitive impairment. Clin. Neurophysiol. **125**, 1145–1151 (2014)
3. Olichney, J.M., Yang, J.C., Taylor, J., Kutas, M.: Cognitive event-related potentials: biomarkers of synaptic dysfunction across the stages of Alzheimer's disease. J. Alzheimer's Dis. **26**(Suppl. 3), 215–228 (2011)
4. Challis, E., Hurley, P., Serra, L., Bozzali, M., Oliver, S., Cercignani, M.: Gaussian process classification of Alzheimer's disease and mild cognitive impairment from resting-state fMRI. NeuroImage **112**, 232–243 (2015)
5. Hui, J.S., Wilson, R.S., Bennett, D.A., Bienias, J.L., Gilley, D.W., Evans, D.A.: Rate of cognitive decline and mortality in Alzheimer's disease. Neurology **61**, 1356–1361 (2003)
6. Liu, C.C., Kanekiyo, T., Xu, H., Bu, G.: Apolipoprotein E and Alzheimer disease: risk, mechanisms and therapy. Nat. Rev. Neurol. **9**, 106–118 (2013)
7. Rosengarten, B., Paulsen, S., Burr, O., Kaps, M.: Effect of ApoE ε4 allele on visual evoked potentials and resultant flow coupling in patients with Alzheimer. J. Geriatr. Psychiatry Neurol. **23**, 165–170 (2010)
8. Green, J., Levey, A.I.: Event-related potential changes in groups at increased risk for Alzheimer disease. Arch. Neurol. **56**, 1398–1403 (1999)
9. Lee, T.-W., Yu, Y.W.-Y., Hong, C.-J., Tsai, S.-J., Wu, H.-C., Chen, T.-J.: The influence of apolipoprotein E Epsilon4 polymorphism on qEEG profiles in healthy young females: a resting EEG study. Brain Topogr. **25**, 431–442 (2012)
10. Canuet, L., Tellado, I., Couceiro, V., Fraile, C., Fernandez-Novoa, L., Ishii, R., Takeda, M., Cacabelos, R.: Resting-state network disruption and APOE genotype in Alzheimer's disease: a lagged functional connectivity study. PLoS ONE **7**, e46289 (2012)
11. Folstein, M.F., Folstein, S.E., McHugh, P.R.: "Mini-mental state". A practical method for grading the cognitive state of patients for the clinician. J. Psychiatr. Res. **12**, 189–198 (1975)
12. Carcaillon, L., Pérès, K., Péré, J.J., Helmer, C., Orgogozo, J.M., Dartigues, J.F.: Fast cognitive decline at the time of dementia diagnosis: a major prognostic factor for survival in the community. Dement. Geriatr. Cogn. Disord. **23**, 439–445 (2007)
13. Puglielli, L., Tanzi, R.E., Kovacs, D.M.: Alzheimer's disease: the cholesterol connection. Nat. Neurosci. **6**, 345–351 (2003)
14. Anderer, P., Semlitsch, H.V., Saletu, B., Barbanoj, M.J.: Artifact processing in topographic mapping of electroencephalographic activity in neuropsychopharmacology. Psychiatry Res.: Neuroimaging **45**, 79–93 (1992)
15. Delorme, A., Makeig, S.: EEGLAB: an open source toolbox for analysis of single-trial EEG dynamics including independent component analysis. J. Neurosci. Methods **134**, 9–21 (2004)
16. Barachant, A., Congedo, M.: A Plug&Play P300 BCI Using Information Geometry. arXiv preprint arXiv:1409.0107 (2014)
17. Congedo, M., Barachant, A., Andreev, A.: A New Generation of Brain-Computer Interface Based on Riemannian Geometry arXiv:1310.8115 (2013)
18. Zeng, L.-L., Shen, H., Liu, L., Wang, L., Li, B., Fang, P., Zhou, Z., Li, Y., Hu, D.: Identifying major depression using whole-brain functional connectivity: a multivariate pattern analysis. Brain **135**, 1498–1507 (2012)

19. Rasmussen, C.E., Williams, C.K.I.: Gaussian Processes for Machine Learning. The MIT Press, Cambridge (2006)
20. Adler, R.J., Taylor, J.E.: Random Fields and Geometry. Springer, New York (2007)
21. Rasmussen, C.E., Nickisch, H.: Gaussian processes for machine learning (GPML) toolbox. J. Mach. Learn. Res. **11**, 3011–3015 (2010)
22. Chen, M.: Pattern Recognition and Machine Learning Toolbox. MATLAB Central File Exchange (2016)
23. Brodersen, K.H., Ong, C.S., Stephan, K.E., Buhmann, J.M.: The balanced accuracy and its posterior distribution. In: Proceedings of the 2010 20th International Conference on Pattern Recognition, pp. 3121–3124. IEEE Computer Society (2010)
24. Hanley, J.A., McNeil, B.J.: A method of comparing the areas under receiver operating characteristic curves derived from the same cases. Radiology **148**, 839–843 (1983)
25. Stahl, D., Pickles, A., Elsabbagh, M., Johnson, M.H., The, B.T.: Novel machine learning methods for ERP analysis: a validation from research on infants at risk for autism. Dev. Neuropsychol. **37**, 274–298 (2012)

# Data Fusion Approach for Learning Transcriptional Bayesian Networks

Elisabetta Sauta[1(✉)], Andrea Demartini[1], Francesca Vitali[2],
Alberto Riva[3], and Riccardo Bellazzi[1]

[1] Department of Electrical, Computer and Biomedical Engineering,
University of Pavia, Pavia, Italy
elisabetta.sauta01@universitadipavia.it
[2] Center of Biomedical Informatics and Biostatistics,
University of Arizona, Tucson, USA
[3] Interdisciplinary Center for Biotechnology Research,
University of Florida, Gainesville, USA

**Abstract.** The complexity of gene expression regulation relies on the synergic nature underlying the molecular interplay among its principal actors, transcription factors (TFs). Exerting a spatiotemporal control on their target genes, they define transcriptional programs across the genome, which are strongly perturbed in a disease context. In order to gain a more comprehensive picture of these complex dynamics, a data fusion approach, aimed at performing the integration of heterogeneous -omics data is fundamental.

Bayesian Networks provide a natural framework for integrating different sources of data and knowledge through the priors' use. In this work, we developed an hybrid structure-learning algorithm with the aim of exploiting TF ChIP-seq and gene expression (GE) data to investigate disease-specific transcriptional regulations in a genome-wide perspective. TF ChIP seq profiles were firstly used for structure learning and then integrated in the model as a prior probability. GE panels were employed to learn the model parameters, trying to find the best heuristic transcriptional network. We applied our approach to a specific pathological case, the chronic myeloid leukemia (CML), a myeloproliferative disorder, whose transcriptional mechanisms have not yet been deeply elucidated.

The proposed data-driven method allows to investigate transcriptional signatures, highlighting in the obtained probabilistic network a three-layered hierarchy, as a different TFs influence on gene expression cellular programs.

**Keywords:** Bayesian networks · Transcriptional regulations · -omics data integration

## 1 Introduction

A major challenge in Computational Biology is the possibility to reconstruct transcriptional regulatory networks (TRNs), investigating the cellular dynamics that guide genetic expression. Exploring the complex interactions among genes and their key regulators, transcription factors (TFs), can provide novel insight into gene expression

© Springer International Publishing AG 2017
A. ten Teije et al. (Eds.): AIME 2017, LNAI 10259, pp. 76–80, 2017.
DOI: 10.1007/978-3-319-59758-4_8

programs, especially in a pathological scenario, in which they are mainly perturbed by disease. An important characteristic of TRNs is the synergic behavior of TFs, whereby these cooperative molecules can either start and enhance, or repress the transcription of thousands of genes along the genome. The best way to capture these interactions is a genome-wide approach, obtained by the integration of different sources of -omics data. Collecting information using several technologies offers different perspectives on the considered system, but jointly analyzing such data in a single framework enables a consensus perspective to emerge. To this aim, Bayesian Networks (BNs) are an ideal formalism for heterogeneous data integration and the most effective models for decision making in uncertain knowledge reasoning [1].

In this work, we present a data fusion approach for learning Bayesian transcriptional regulatory networks, accomplished by incorporating evidence from gene expression data, and informative priors from TF ChIP-seq binding profiles.

In several studies, the use of prior knowledge has been shown to improve the fidelity of network reconstruction. Many sources of data are useful to supplement expression data, and can be incorporated at different steps of BN simulation, from prior structure definition to structure simulation and evaluation. Hartemink et al. [2] included genomic location data as model prior, forcing the search to add arcs in a specific position, and eliminating all graphs lacking these suggested edges. Perrier et al. [3] developed an algorithm that can learn BNs containing up to 50 vertices and the local search is constrained by an undirected super-structure, from which the optimal solution must be a subgraph. Kojima et al. [4] extended the same idea to BNs with a few hundred of vertices, exploiting clusters extracted from the super-structure. However, these methods add a very small number of extra edges, and clearly they cannot handle full-genome Eukaryotic TRNs. Furthermore, they do not take into account the natural directionality of these networks, from a TF to its target genes, that is instead provided by ChIP-seq experiments.

Our Bayesian learning framework exploits ChIP-seq data to determine the transcriptional relationships on a genome scale, and then it uses this information as a structural constraint. We apply this approach to a myeloproliferative disorder, the chronic myeloid leukemia (CML), in order to identify transcriptional interactions that lead aberrant gene expressions programs caused by the disease.

## 2    Methods

### 2.1    Reconstruction of CML Transcriptional Regulatory Network

A collection of TFs ChIP-seq alignment data for the K562 CML cell line was retrieved from the ENCODE database. TFs experiments were processed with an ad-hoc bioinformatics pipeline, to finally consider only the statistically significant TFs binding interactions that map the promoter of target genes. TF-gene relations were then quantitative weighted using a scoring method [5], which assigns them a *binding score*, as a strength index of the binding. The computational integration of all of these transcriptional interactions generated a genome wide TRN, conceptualized as a graph, whose nodes are TFs and target genes connected by directed and weighted edges among TFs and from TF to genes.

## 2.2  Hybrid Bayesian Network Structure Learning Algorithm

The BN procedure starts from an initial directed acyclic graph (DAG), to find the model that best fits the data, and in this disease context, the probabilistic model that best describes transcriptional regulations underlying CML. The essential steps of our framework are described below.

*A. TRN conversion and definition of its structural constraints*

In order to convert the TRN into a BN input for the algorithm, the network was firstly decomposed into two connectivity matrices, as illustrated in the Fig. 1, panel (A): a TF-TF matrix ($F_{n \times n}$) and a TF-Genes matrix ($T_{m \times m}$), where entries are 1 or 0, depending on whether or not a binding exists between two nodes. Since TRN is characterized by many regulatory feedback loops, the F matrix was used to create a subnetwork, which underwent to an iterative process for finding the initial DAG. The procedure tried to remove one arc at a time, starting from edges with lower binding score, to find the minimal connected DAG. All the excluded edges constituted an *arcs whitelist* (W), which represents the search space of the structure learning algorithm.

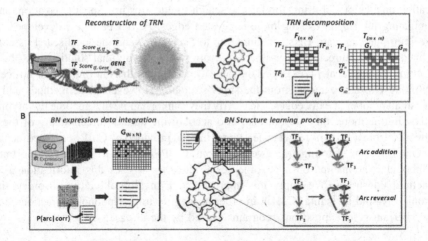

**Fig. 1.** Panel (A). TFs ChIP-seq data analysis and integration for the reconstruction of the genome wide TRN underlying CML. TRN matrix decomposition and prior constraints definition. Panel (B). GE integration of the TRN and its conversion to a BN input for our algorithm

*B. Gene expression integration and structure learning procedure*

F and T matrices were combined to create a global matrix ($G_{N \times N}$), which was firstly topologically sorted, and then integrated with gene expression (GE) profiles of CML patients, collected from GEO and Expression Atlas databases, as shown in the Fig. 1, panel (B). The underlying distribution of the resulting BN is Gaussian, assuming that each node is linearly dependent upon its continuous parents. This fully observable network represents the initialization of the heuristic search process, that tries iteratively to add or reverse a whitelisted arc. Every edge change is evaluated by the BIC scoring metric, expressed as the sum of local scores, each of which is a function only of one node and its parents in the model ($M$), described as

$$\text{BIC}(M) = n\log(\text{RSS}/n) + k\log n \qquad (1)$$

where $n$ is sample size of the GE dataset, $k$ is the number of the model parameters and RSS is the residual sum of squares from estimated model. Any local change is maintained in the network only if its score improves compared to the previous one.

GE data have further been used to calculate the correlation among TFs, which will be exploited in the initial phase of the search process. In particular, the algorithm, at each step, simultaneously tests a group of arcs (C), randomly drawn from W, using the correlation as an extraction probability associated to each edge. After this search phase, the algorithm learns in parallel the models parameters from the GE data, and computes a score for each obtained model. The BN structure is updated with the highest score solution and the process moves forward until the network change improves the BN score of the previous network.

## 3   Results and Discussion

The computational analysis and integration of 65 TF ChIP-seq experiments allowed to reconstruct a genome wide TRN, as shown in Fig. 1, panel (A), consisting of 20,876 nodes (65 TF regulators, 177 other TF target nodes, 19,427 protein coding genes and 1,207 miRNA coding genes) and 478,558 edges. This network was decomposed into two square matrices: an F matrix representing the transcriptional regulations among TFs, and a T matrix describing the TFs regulation on genes. The first one underwent to a pruning process aimed to find not only the initial DAG, but also to define the structural constraints of our method. Matrices information was then incorporated into the BN, which was integrated with 122 GE profiles of CML patients, obtained through the integration of 5 GE datasets properly normalized. The resulting BN, composed of 11,986 continuous nodes (60 TFs regulators) and 282,533 edges, is completely *data-driven* and represents the structure prior of the developed algorithm. It explores a constrained structure space to establish which transcriptional regulations are likely functional in a disease scenario. As preliminary results, our hybrid algorithm added 194 new arcs, and reverted the directionality of 39 existing edges of the initial structure. Analyzing the transcriptional regulations among TFs in the output BN, and their in-degree distribution, we identify a three-layered structure, which cannot be detectable in the starting TRN. This hierarchy, which reflects the GE impact of different classes of TFs, is composed by one *master regulator* TF, at the top, 50 *brokers* or *middle managers*, and the remaining 9 *workhorses* at the bottom, as represented in Fig. 2, panel (I). Moreover, calculating their Betweenness Centrality (BC), as an indicator of node's control exerted on the network interactions, we can highlight some TFs with higher values of BC (e.g. FOS, MAFF, ARID3A, ATF1), as illustrated in the Fig. 2, panel (II). If we try to remove these nodes one by one from the global BN, the network becomes disconnected and the information flow is not propagated to certain target genes. For example, the ARID3A removal isolates BAALC gene, that is frequently over expressed in CML patients. The ATF1 node elimination causes an interruption of the transcriptional flow toward GMFG gene, which acts as a lineage regulator for hematopoiesis.

We have presented a data fusion approach that allows to model transcriptional regulations, exploiting ChIP-seq data to build the backbone of the model, and combined to GE data, to guide the search closer to the best network. The developed algorithm follows the search and score paradigm, and its heuristic scheme permits to reduce computational time, avoiding the space investigation of all possible structures. The search phase has further been parallelized, for testing and evaluating all the extracted arcs simultaneously. We intend to investigate network variations obtained from several runs for a robustness evaluation, as faced in [6]. The proposed method will be compared to other learning strategies, such as glasso [7], even if they do not typically considerate the directionality of the transcriptional information.

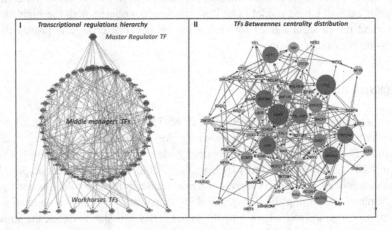

**Fig. 2.** Panel (I). Transcriptional hierarchy underlying the obtained probabilistic model. Panel (II) Betweenness centrality distribution of the TF-TF regulations subnetwork, calculated considering all the shortest paths that cross a given node.

# References

1. Jensen, F.V.: Introduction to Bayesian Networks. Springer, Secaucus (1996)
2. Hartemink, A., Gifford, D., Jaakkols, T., et al.: Combining location and expression data for principled discovery of genetic regulatory network models. PSB **7**, 437–449 (2002)
3. Perrier, E., Imoto, S., Miyano, S.: Finding optimal Bayesian network given a super-structure. JMLR **9**, 2251–2286 (2008)
4. Kojima, K., Perrier, E., Imoto, S., et al.: Optimal search on clustered structural constraint for learning Bayesian network structure. JMLR **11**, 285–310 (2010)
5. Sikora, W., Ackermann, M., Christodoulou, E., et al.: Assessing computational methods for TF target gene identification based on ChIP-seq data. PLoS Comput. Biol. **9**(11), e1003342 (2013)
6. Friedman, N., Linial, M., Nachman, I.: Bayesian networks to analyze expression data. J. Comput. Biol. **7**(3–4), 601–620 (2000)
7. Friedman, N.: Sparse inverse covariance estimation with the graphical lasso. Biostatistics **9** (3), 432–441 (2008)

# A Prognostic Model of Glioblastoma Multiforme Using Survival Bayesian Networks

Simon Rabinowicz[1], Arjen Hommersom[2,3(✉)], Raphaela Butz[2,4], and Matt Williams[5,6]

[1] Faculty of Medicine, Imperial College London, London, UK
[2] Department of Computer Science, Open University, Heerlen, The Netherlands
arjenh@cs.ru.nl
[3] Department of Software Science, Radboud University, Nijmegen, The Netherlands
[4] Institute for Computer Science, TH Köln, Cologne, Germany
[5] Department of Radiotherapy, Charing Cross Hospital, London, UK
[6] Computational Oncology Laboratory, Imperial College London, London, UK

**Abstract.** Bayesian networks are attractive for developing prognostic models in medicine, due to the possibility for modelling the multivariate relationships between variables that come into play in the care process. In practice, the development of these models is hindered due to the fact that medical data is often censored, in particular the survival time. In this paper, we propose to directly integrate Cox proportional hazards models as part of a Bayesian network. Furthermore, we show how such Bayesian network models can be learned from data, after which these models can be used for probabilistic reasoning about survival. Finally, this method is applied to develop a prognostic model for Glioblastoma Multiforme, a common malignant brain tumour.

## 1 Introduction

Glioblastoma Multiforme (GBM) is the most common malignant brain tumour in adults. It has a poor prognosis with some well recognised prognostic factors, including age and performance status [5], but there is little insight into the relationships between the various clinical factors. Providing patients with information on prognosis changes the decisions patients make about treatment [8]. Given the poor prognosis of GBM, and the treatment options available, there is a clear need for prognostic tools to guide both clinicians and patients in decision-making.

Bayesian networks (BNs) have been applied to clinical decision-making (see e.g. [2]), as they provide flexible methods for learning and reasoning about data. However, they are not suitable for dealing with time-to-event curves (e.g. survival) as variables are typically assumed to be discrete, and learning algorithms are not suited for handling right-censored data, which is the standard case in survival analysis. In line with [3,4], we propose a new method for combining discrete Bayesian networks with a Cox proportional hazard model, which we use

© Springer International Publishing AG 2017
A. ten Teije et al. (Eds.): AIME 2017, LNAI 10259, pp. 81–85, 2017.
DOI: 10.1007/978-3-319-59758-4_9

to develop a new prognostic model for GBM. The novel technical aspect of this work is that survival is modelled in continuous rather than discrete time, which we refer to as *conditional survival Bayesian networks* (CSBNs).

## 2    Conditional Survival Bayesian Networks

### 2.1    Modelling Survival in Bayesian Networks

In survival analysis, we model the survival function $S$, i.e., $S(t) = P(T > t)$, with $T$ being a random variable representing the time of death defined by a density function $p$. A popular parameterised survival model is the Cox proportional hazard model (CPHM), which estimates the *relative risk* or *hazard ratio* hr as a log-linear model:

$$\text{hr}(X) = \frac{h(t \mid X)}{h_0(t)} = \exp\left(\sum_{j=1}^{n} w_j X_j\right) \tag{1}$$

where $h_0$ is a baseline hazard function and $w = \{w_1, \ldots, w_n\}$ a set of weights. Baseline hazard can be estimated using the Kaplan-Meier estimate [1], such that $h(t \mid X)$ can be obtained.

Based on this, we define a *conditional survival Bayesian network* (CSBN) on an acyclic directed graph $G$ with nodes $V$ which represent a set of random variables. The set of nodes $V$ consists of a set of discrete nodes $D$ and (continuous) survival nodes $T$, such that $T \cap D = \varnothing$ and $T \cup D = V$. Let $\pi(V_i)$ indicate the parents of node $V_i \in V$ in graph $G$. It is assumed in CSBNs that for each $V_i \in V$, $\pi(V_i) \subseteq D$, i.e., survival nodes do not have any child nodes. This is a reasonable assumption if the survival node represents the time of death, as in most survival analysis applications.

To each CSBN, we associate a joint density function, which factorises as follows:

$$p(V) = \prod_{D_i \in D} P(D_i \mid \pi(D_i)) \prod_{T_i \in T} p(T_i \mid \pi(T_i)) \tag{2}$$

It can be shown that $p(t_i \mid \pi(T_i))$, i.e., the density function of $T_i$ given its parents, can be derived from the conditional hazard function estimated by CPHM as follows:

$$p(t_i \mid \pi(T_i)) = h(t_i \mid \pi(T_i)) \exp\left[-\int_0^{t_i} h(s \mid \pi(T_i)) \mathrm{d}s\right] \tag{3}$$

### 2.2    Learning and Inference

It is well-known that when learning from complete data, maximizing model selection criteria decomposes into finding optimal parent sets of a Bayesian network under an acyclicity constraint. Therefore, learning of the discrete and survival model can be done separately. In this paper, we used the R bnlearn package [6], selecting significant edges from a bootstrapped tabu search, using Akaike information criterion (AIC) as a model score, to estimate the structure over $D$. The CPHM, providing $\pi(T_i)$ and $p(T_i \mid \pi(T_i))$, was found by a step-wise regression that also optimises the AIC score. Subsequently, these two distributions are merged according to Eq. 2 into a CSBN.

Inference can be decomposed into a two-step process. Suppose for clarity that $|T| = 1$. Then:

$$p(t \mid d_i) = \sum_{\pi(T)} p(t \mid \pi(T)) P(\pi(T) \mid d_i) \tag{4}$$

where $P(\pi(T) \mid d_i)$ can be computed on a Bayesian network with discrete nodes only and $p(t \mid \pi(T))$ is the event rate. Arbitrary other quantities can be computed using Bayes' rule, e.g.

$$P(d_i \mid T < t) \propto P(T < t \mid d_i) P(d_i) \tag{5}$$

where $P(T < t \mid d_i)$ can be computed from the density as given in Eq. 4.

## 3 Experimental Work

### 3.1 Experimental Setup

We use data from the TCGA Research Network (http://cancergenome.nih.gov/), which contains clinical and radiotherapy data on 596 patients with GBM. These data were merged and we recoded some of the variables (based on clinical input) in order to generate clinically-relevant categories. Patients who had missing data on any of these items were removed from the dataset. The combined dataset contained key demographic variables, such as race, gender and age, data on treatment, extent of surgery, radiation dose and fractionation, and data on survival.

For investigating the quality of predictions of this CSBN, we also constructed a fully discrete Bayesian network with the same graph structure as the CSBN. In order to learn its parameters, we took the data from all the uncensored patients and discretised the time of survival (as Alive/Dead) at 12 months. Finally, we fitted the parameters of the Bayesian network on this discrete data using the same Bayesian parameter estimation method used for the CSBN.

### 3.2 Results

After preprocessing, the data consisted of 266 patients and with 12 variables, from which the Bayesian network was constructed. The median survival in the total cohort was 373 days, and 61 patients were still alive at the end of follow-up in the dataset. The resulting network is shown in Fig. 1.

We explored the four parameters which were identified by the model as affecting survival time; these were Age; Karnofsky Performance Score (KPS); Method of diagnosis and Radiation dose. Exploring these parameters, it was noted that: increasing age; lower KPS; excisional biopsy and a lower dose of radiation all conferred a poorer prognosis.

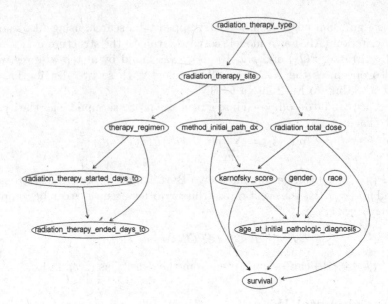

**Fig. 1.** Bayesian network structure for glioblastoma multiforme.

(Figure 2), shows that increasing age and reduced KPS were associated with worse survival. Patients who underwent a resection of the tumour had a lower median survival time of 365 days when compared to an excisional biopsy at 440 days. This is in conflict with the existing literature. However, only 16% of patients underwent resection, and thus the effect was reduced by the relatively small sample size.

We also compared the survival predictions from the continuous network to those from the discrete network, using 10-fold cross-validation and measuring the predictive error (defined as the ratio of misclassifications to the total number of

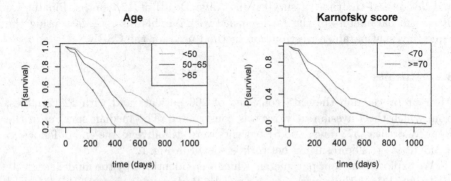

**Fig. 2.** Relationships between discrete variables and the survival variables in the Bayesian network.

cases). We find that the predictive error of the CSBN is significantly lower (37%) than the discrete BN (41%, $p < 0.01$).

# 4 Conclusions

Bayesian networks have been proposed for modelling prognosis, e.g. prognostic Bayesian networks [7]. Although they handle care process and the clinical outcome, these models are not suitable for modelling simple time-to-event data, where most of the data is of a non-temporal nature.

A translation of Cox proportional hazards model into a discrete Bayesian networks was proposed by Kraisangka and Druzdzel [3]. However, their approach requires time to be discretised, which makes reasoning about statistics such as median survival times inaccurate, and makes the numbers of parameters exponential in the number of discrete time points. For these reasons, we prefer to keep time as a continuous variable.

In this paper, we have also illustrated the new survival analysis methodology based on conditional survival Bayesian networks. We have empirically shown that, in a model based on GBM data, a continuous prediction model achieved a lower predictive error than a discretised network based on the same data. We think the combination of BN structure and probability is a useful tool for explaining the impact of prognostic factors on survival. Moreover, it provides a powerful tool for predicting survival when data is missing.

# References

1. Hilbe, J.: Survival analysis for epidemiologic and medical research. J. Stat. Softw. Book Rev. **30**(4), 1–4 (2009)
2. Hommersom, A., Lucas, P.J.: Foundations of Biomedical Knowledge Representation. Springer, Heidelberg (2015)
3. Kraisangka, J., Druzdzel, M.J.: Discrete Bayesian network interpretation of the Cox's proportional hazards model. In: Gaag, L.C., Feelders, A.J. (eds.) PGM 2014. LNCS, vol. 8754, pp. 238–253. Springer, Cham (2014). doi:10.1007/978-3-319-11433-0_16
4. Kraisangka, J., Druzdzel, M.J.: Making large Cox's proportional hazard models tractable in Bayesian networks. In: Probabilistic Graphical Models (2016)
5. Ostrom, Q.T., Gittleman, H., Farah, P., Ondracek, A., Chen, Y., Wolinsky, Y., Stroup, N.E., Kruchko, C., Barnholtz-Sloan, J.S.: CBTRUS statistical report: primary brain and central nervous system tumors diagnosed in the united states in 2006–2010. Neuro-oncology **15**(Suppl. 2), ii1–ii56 (2013)
6. Scutari, M.: Learning Bayesian networks with the bnlearn R package. J. Stat. Softw. **35**(3), 1–22 (2010)
7. Verduijn, M., Peek, N., Rosseel, P.M., de Jonge, E., de Mol, B.A.: Prognostic Bayesian networks: I: rationale, learning procedure, and clinical use. J. Biomed. Inform. **40**(6), 609–618 (2007)
8. Weeks, J., Cook, E.F., O'Day, S., Peterson, L., Wenger, N., Reding, D., Harrel, F., Kussin, P., Dawson, N., Connors, A., Lynn, J., Phillips, R.: Relationship between cancer patients' predictions of prognosis and their treatment preferences. J. Am. Med. Assoc. **279**(21), 1709–1714 (1998)

# Accurate Bayesian Prediction of Cardiovascular-Related Mortality Using Ambulatory Blood Pressure Measurements

James O'Neill[1(✉)], Michael G. Madden[1], and Eamon Dolan[2]

[1] College Engineering and Informatics, National University of Ireland, Galway, Ireland
{james.oneill,michael.madden}@nuigalway.ie
[2] Stroke and Hypertension Unit, Connolly Hospital, Dublin, Ireland
eamon.dolan026@indigo.ie

**Abstract.** Hypertension is the leading cause of cardiovascular-related mortality (CVRM), affecting approximately 1 billion people worldwide. To enable patients at significant risk of CVRM to be treated appropriately, it is essential to correctly diagnose hypertensive patients at an early stage. Our work achieves highly accurate risk scores and classification using 24-h Ambulatory Blood Pressure Monitoring (ABPM) to improve predictions. It involves two stages: (1) time series feature extraction using sliding window clustering techniques and transformations on raw ABPM signals, and (2) incorporation of these features and patient attributes into a probabilistic classifier to predict whether patients will die from cardiovascular-related illness within a median period of 8 years. When applied to a cohort of 5644 hypertensive patients, with 20% held out for testing, a K2 Bayesian network classifier (BNC) achieves 89.67% test accuracy on the final evaluation. We evaluate various BNC approaches with and without ABPM features, concluding that best performance arises from combining APBM features and clinical features in a BNC that represents multiple interactions, learned with some human knowledge in the form of arc constraints.

**Keywords:** Bayesian network · Ambulatory Blood Pressure Monitoring · Hypertension

## 1    Introduction

Cardiovascular-related mortality (CVRM) is the top cause of death worldwide, accounting for a third of all deaths worldwide, 81% of which are attributed to coronary heart disease and stroke [1]. High blood pressure plays a significant role in CVRM, untreated hypertensive patients are at the risk of stroke, heart failure, and many other health conditions that increase over time [3]. The current out-of-clinic procedure for determining hypertension involves a patient's ABP

© Springer International Publishing AG 2017
A. ten Teije et al. (Eds.): AIME 2017, LNAI 10259, pp. 86–91, 2017.
DOI: 10.1007/978-3-319-59758-4_10

being measured over a 24-h window, referred to as Ambulatory Blood Pressure Monitoring (ABPM). This work investigates how these ABPM measurements can be used; this is achieved by taking feature transformations and clusters to improve the prediction of CVRM. We also provide a comparison of Bayesian classifiers, with and without ABPM features, and also with and without the incorporation of human knowledge while learning the Bayesian network (BN).

## 2    Related Research

Dolan et al. [9] previously noted the superiority of ABPM over clinic measurements, highlighting that systolic BP in particular is the strongest predictor of CVRM among all possible ABP measurements in the form of hazard ratios, empirically highlighting the importance of nocturnal hypertension in diagnosis and treatment. However, to date there has been no attempt to incorporate features of the time series data produced during the well-established ABPM protocol; rather, only mean ABP measurements are used in diagnosis and prognosis. Bhatla and Jyoti [5] recently reviewed attempts to classify patients with cardiovascular disease by using k-nearest neighbours, Naive Bayes (NB), fuzzy logic systems, neural networks (NN), decision trees (DT). Soni et al. [4] proposed a weighted associative classifier for heart disease prediction, achieving an 81.51% test accuracy. In that case, the attributes were weighted according to the expertise of doctors to improve performance, similarly we look to incorporate human knowledge in our classification approach. A number of models re-occur in the literature that have performed well for classifying disease which are; fuzzy rule-base classifiers, boosted trees, bagging trees, logistic regression, NB and ANNs [5,6]. However, none of the above studies focus on improvements ABPM may make in both classification and risk scoring of hypertensive patients who are at possible risk of CVRM.

## 3    Experimental Methodology

***Dataset Description and Feature Extraction.*** The dataset was previously published in Dolan et al.'s [9] study where they gathered data from 5644 hypertensive patients and their respective in-clinic and out-of-clinic ABP measurements over a median period of 8 years from Beaumont Hospital in Dublin, Ireland. We split this dataset into 80% training and 20% for a held out test set. Office ABP, ABPM features, BMI and age are discretized into 10 quartiles while sex, diabetes, cardiovascular history and smoking status are predefined discrete variables. ABPM measurements were recorded for all 5644 patients. Candidate features were extracted from the ABPM time series data and scaled using z-normalization, followed by taking a cubic transformation of the normalized ABPM z-scores. A sliding window was then used over these scaled features for each 3 h 20 window with a 1 h 40 min overlap starting from 8:00 am − 11:00 pm daytime and 11:00 pm − 6:00 am during sleep. This interval length and overlap window allows for a substantial number of cubic sums to be computed to summarize the volatility of blood pressure within a 3 h period. The resulting summed values for each window are then clustered as a final step. These clusters are also considered as features for classification. Furthermore, the maximum value from local peak detection[1] are also considered, in order to account for the envelope of each signal. Feature selection is then carried out on all of the above ABPM features by carrying out a statistical test that indicates the significance of each feature by a $p$ value (small p value indicates strong evidence that the feature is statistically significant). The selected features are then tested using logistic regression as a baseline on the training dataset to determine which features should be kept and concatenated with patient attributes: diabetes mellitus, smoking status, sex, age, body mass index (BMI) office ABP and cardiovascular history for classification.

***Bayesian Classification.*** We use a BNC as human knowledge should be incorporated, the posteriors must be reliably calibrated and the model must be interpretable for informative decision making [8,10]. A hill-climbing (HC) search (score-based algorithm) is employed for learning the BN structure by sequentially adding nodes and directed arcs to an initially empty graph. Bayesian Information Criterion (BIC) is used as a scoring criterion. The BIC score for a graph G and data D is $BIC(G, D) = \sum_{\forall v \in V} log \hat{p}_{v|pa(v)}(X_{v|pa(v)} - {}^1/_2 k_{v|pa(v)})$ where $k_{(v|pa(v))}$ is the dimension of the parameter $v|pa(v)$ [7]. HC is an effective structure search strategy but it is prone to local minima, hence we introduce restarts to run a number of iterations to avoid this. Three widely used BNs include a GBN, NB and TAN, all of which are compared in this work, with and without the ABPM features. A further comparison is also carried out to assess the performance of the classifiers when arc constraints are applied. For classifiers that include ABPM features, *Nighttime SBP Peak 1*, *Nighttime SBP Peak 2*, *Nighttime SBP Peak 3*, *k-means Nighttime Systolic Cluster* and *Age* nodes are parents of the *CVRM* class node (whitelist) and blacklisted arcs exclude nodes from *CVRM* to *Smoking Status*, *Sex* and *BMI* (blacklist). The whitelisted nodes are chosen based on previous studies that highlight the importance of systolic ABP [9]. For classification without ABPM features, the *Sex* and *Age* nodes are connected to *CVRM* class node and blacklisted arcs exclude *CVRM* being a parent of *Smoking Status* and *BMI*.

## 4   Results

Table 1 presents the accuracies on the training set using a Logistic Regression model. The features above $p > 0.01$ level are selected from all tested ABPM extracted features. This includes the sum of the cubic values for each sliding window, clusters of these features ($k = 5$, determined by analyzing the clusters sum of squared distances for varying k) and the maximum value for each sliding window (not the cubic sum). The maximum peaks for systolic-ABP[2] is also

---

[1] Local peaks considers neighboring measurements when choosing which are peaks.
[2] SBP values are first z-normalized, followed by a cubic transformation.

**Table 1.** 10-CV accuracy on the training set using Logistic Regression

| Features | Accuracy | Features | Accuracy |
|---|---|---|---|
| Original mean ABPM | 66.34 % | Sliding window clusters | 70.09 % |
| Sliding window cubic sums | 71.56 % | Sliding window clusters & Max peak | 72.36 % |
| Sliding window max | 63.78 % | Sliding window clusters & Max peaks clusters | 70.82 % |

computed. The maximum peaks and sliding window clusters have shown best results with 72.36% accuracy. Using cubic transformations on z-scaled features allows to account for the variability in ABP, while the maximum SBP peaks accounts for the envelope of highest BP throughout ABPM, resulting in a 7 percentage point increase in accuracy on the training set. Based on this analysis, we incorporate these set of features into the BN classifiers and test whether they are useful for CVRM prediction by comparing them with classifiers which do not account for ABPM.

## 4.1 Bayesian Classification

This section discusses the results of performing classification on a range of BNCs: $GBN_{BIC}$, $GBN_{K2}$, NB and TAN. Constraining arcs from nighttime SBP features to the CVRM class increases the test accuracy by 3.96 % points, shown in Table 2. These experiments have shown that the incorporation of systolic ABPM

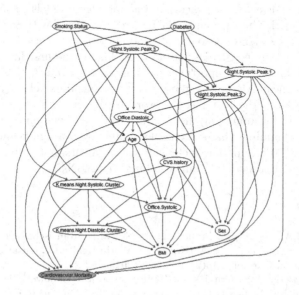

**Fig. 1.** $GBN_{K2}$ with arcs constraints

**Table 2.** Bayesian classification results - experimental results with ABPM features

| Classifiers | Including ABPM features | | | | Excluding ABPM features | | | |
|---|---|---|---|---|---|---|---|---|
| | Arc constraints | | Without arc constraints | | Arc constraints | | Without arc constraints | |
| | 10-CV | Test | 10-CV | Test | 10-CV | Test | 10-CV | Test |
| $NB$ | - | - | 75.17% | 74.45% | - | - | 75.38% | 73.98% |
| $GBN_{BIC}$ | 97.28% | 87.27% | 75.16% | 75.38% | 75.16% | 75.38% | 75.16% | 75.38% |
| $GBN_{K2}$ | 95.01% | **89.67%** | 75.16% | 75.38% | 94.95% | **88.82%** | 76.82% | 77.29% |
| $TAN$ | 84.28% | 84.26% | 82.41% | **85.78%** | 80.00% | 81.74% | 80.01% | **81.74%** |

features directly influencing the class node plays an important role in finding a different BN structure that achieves better performance. By incorporating these constraints into the network the number of all dependencies has risen, particularly for the class node. The performance improvements can be attributed to the expansion of the Markov Blanket (MB) for the classification node show in Fig. 1, allowing more interaction terms among the ABPM features in classification, forcing the network to learn a network where the class node is dependent on ABPM features has made a significant improvement.

## 5  Conclusion

The prediction of CVRM using raw ABPM time series data and patient details has shown significant improvement compared to classification without ABPM. The aforementioned BNCs have shown good performance, especially when the proposed ABPM features are incorporated into the models. The learner has recovered a network which produces the best results overall with 89.67% accuracy on a held out test set, TAN also shows good performance with 85.71% accuracy without encoding any constraints on its arcs. This is the first line of work that empirically establishes the importance of using features from ABPM data instead of merely using mean ABP measurements for the purpose of classifying patients. More generally, it is the first piece of work to report an evaluation of ABPM transformed data for improving the accuracy of CVRM outcomes. It also echoes similar empirical findings that systolic blood pressure is a significant predictor of CVRM and ABPM is superior to clinical measurements, not only in the context of applying the CPH model [9] but also for various BNs.

## References

1. Finegold, J.A., Asaria, P., Francis, D.P.: Mortality from ischaemic heart disease by country, region, and age: statistics from World Health Organisation and United Nations. Int. J. Cardiol. **168**(2), 934–945 (2013)
2. Kabir, Z., Perry, I.J., Critchley, J., O'Flaherty, M., Capewell, S., Bennett, K.: Modelling coronary heart disease mortality declines in the Republic of Ireland, 1985–2006. Int. J. Cardiol. **168**(3), 2462–2467 (2013)

3. Mozaffarian, D., et al.: Executive summary: heart disease and stroke statistics-2016 update: a report from the AmerIcan Heart Association. Circulation **133**(4), 447 (2016)
4. Soni, J., Ansari, U., Sharma, D., Soni, S.: Intelligent and effective heart disease prediction system using weighted associative classifiers. Int. J. Comput. Sci. Eng. **3**(6), 2385–2392 (2011)
5. Bhatla, N., Jyoti, K.: An analysis of heart disease prediction using different data mining techniques. Int. J. Eng. **1**(8), 1–4 (2012)
6. Austin, P.C., Tu, J.V., Ho, J.E., Levy, D., Lee, D.S.: Using methods from the data-mining and machine-learning literature for disease classification and prediction: a case study examining classification of heart failure subtypes. J. Clin. Epidemiol. **66**(4), 398–407 (2013)
7. Nagarajan, R., Scutari, M., Lèbre, S.: Bayesian Networks in R, vol. 122. Springer, Heidelberg (2013). pp. 125–127
8. Madden, M.G.: On the classification performance of TAN and general Bayesian networks. Knowl.-Based Syst. **22**(7), 489–495 (2009)
9. Dolan, E., Stanton, A., Thijs, L., Hinedi, K., Atkins, N., McClory, S., Den Hond, E., McCormack, P., Staessen, J.A., O'Brien, E.: Superiority of ambulatory over clinic blood pressure measurement in predicting mortality. Hypertension **46**(1), 156–161 (2005)
10. Freitas, A.A.: Comprehensible classification models: a position paper. ACM SIGKDD Explor. Newslett. **15**(1), 1–10 (2014)

# Temporal Methods

# Modelling Time-Series of Glucose Measurements from Diabetes Patients Using Predictive Clustering Trees

Mate Beštek[1(✉)], Dragi Kocev[2], Sašo Džeroski[2], Andrej Brodnik[3], and Rade Iljaž[4]

[1] National Institute of Public Health, Trubarjeva 2, Ljubljana, Slovenia
mate.bestek@nijz.si
[2] Jožef Stefan Institute, Jamova cesta 39, 1000 Ljubljana, Slovenia
{dragi.kocev,saso.dzeroski}@ijs.si
[3] Faculty of Computer and Information Science, Večna pot 113,
1000 Ljubljana, Slovenia
andrej.brodnik@fri.uni-lj.si
[4] Faculty of Medicine, Vrazov trg 2, 1104 Ljubljana, Slovenia
rade.iljaz@guest.arnes.si

**Abstract.** In this paper, we presented the results of data analysis of 1-year measurements from diabetes patients within the Slovenian healthcare project ECARE. We focused on looking for groups/clusters of patients with the similar time profile of the glucose values and describe those patients with their clinical status. We treated in a similar way the WONCA scores (i.e., patients' functional status). Considering the complexity of the data at hand (time series with a different number of measurements and different time intervals), we used predictive clustering trees with dynamic time warping as the distance between time series. The obtained PCTs identified several groups of patients that exhibit similar behavior. More specifically, we described groups of patients that are able to keep under control their disease, and groups that are less successful in that. Furthermore, we identified and described groups of patients that have similar functional status.

**Keywords:** eCare · Diabetes patients · Time series prediction · Predictive clustering · WONCA scores

## 1 Introduction

Predictive analytics is gaining a lot of traction lately in different industries, including healthcare. By supporting decisions about the future, backed with strong data and prediction methods, one can expect positive outcomes like more informed decisions about therapy even in case patients do not comply with the planned therapy and do not regularly input data. As feedback is one of the psychological behavior change techniques, that is also used in the field of healthcare

© Springer International Publishing AG 2017
A. ten Teije et al. (Eds.): AIME 2017, LNAI 10259, pp. 95–104, 2017.
DOI: 10.1007/978-3-319-59758-4_11

service or intervention design [1], prediction models could inform patients of the consequences of their health-related decisions.

Diabetes is a chronic condition that represents a major health care problem both on households and society [20]. Hypoglycemia prevention is one of the major challenges in diabetes research and it has been shown that predicted rather than measured continuous glucose measurements (CGM) allow a significant reduction of the number of the hypoglycemic events. This stimulates further research on the generation of preventive hypoglycemic alerts that are based on using glucose prediction methods [22]. In general, there are two approaches to blood glucose prediction: mathematical models and data-driven models that are further divided into time series based methods and one-step look-ahead prediction [8]. Data based prediction approaches can be short-term and long-term. Short-term predictions usually include CGM, and long-term predictions are more involved with long-term glucose control and the incidence of diabetic complications [9]. Due to being subject-specific modeling approaches, the burden of model tuning and data collection for CGM based model development is a major obstacle [23]. Approaches based on using the traditional empirical and theoretical mathematical models [11] for developing predictive monitoring systems (insulin pump), can be surpassed by data-driven models based approaches that even enable the usage of models between different individuals [16].

An overview of machine learning and data mining methods used for the data-based prediction is given in [12,15] where authors identify Prediction and Diagnosis as the most popular category of research (authors of [18,21] focus on predicting and diagnosing diabetes) in addition to Diabetic complications, Genetic Background, and Health Care and Management. In all the research, a wide array of machine learning algorithms were employed, including mostly supervised learning approaches and to a lesser extent unsupervised ones (association rules). Nevertheless, to the best of our knowledge, there are no previous attempts to approach this issue as a structured output/time series prediction task.

In this work, we use the data collected within the ECARE project (or EOSKRBA in Slovenian, https://eoskrba.pint.upr/) [2]. It focused on the design of new interventions [14] for healthcare which were deployed and supported by an ICT platform. Several clinical trials were performed in order to show the validity and relevance of the project. Our approach is not subject-specific. Instead, we try to predict the more long-term evolution of diabetes by identifying different groups of patients with diabetes.

We focus on a specific part of the data collected within the project: the year-long measurements of glucose levels of the patients and the answers to the questions from the WONCA (World Organisation of Family Doctors, http://www.globalfamilydoctor.com/) questionnaire. This means that the data at hand comprise of patients' clinical status, on one hand, and its yearly measurements of glucose and WONCA status on the other. Each of the time series data can contain up to 20 measurements. Moreover, each patient has a different number of measurements and they are performed at different time points. The number of measurements and the intervals depend on the patients' compliance with the prescribed care protocol. All in all, the goal is to model a time series based on the current patient data.

Considering the outlined complexities of the data at hand, we used predictive clustering trees (PCTs) [4,13] to address the modeling task at hand. PCTs are a generalization of decision trees able to predict structured output data types including time series data. Moreover, considering the fact that each patient has a different number of measurements taken at different time intervals, we coupled the PCTs with the dynamic time warping (DTW) [17] to obtain the models. The result of this study is the discovery of groups/clusters of patients that exhibit similar dynamics of the glucose measurements and the WONCA scores and the description of these clusters with the patients' clinical status.

The remainder of this paper is organized as follows. Section 2 presents the predictive clustering trees methodology. Next, Sect. 3 describes the data and the experimental design. Section 4 discusses the results of the study and Sect. 5 concludes and provides directions for further work.

## 2 Predictive Clustering Trees for Time Series Modelling

The general overview of our methodology consists of obtaining and processing clinical trial data which we input to the CLUS system in order to learn Predictive Clustering Trees (PCTs) [4]. We experimented with different settings regarding size and depth of PCTs and present two trees – one predicting glucose time series and the other predicting questionnaire results. We also describe clinical impacts of the result.

PCTs generalize decision trees [5] and can be used for a variety of learning tasks, including different types of prediction and clustering. The PCT framework views a decision tree as a hierarchy of clusters (see Figs. 3 and 4): the top node of a PCT corresponds to one cluster (group) containing all data, which is recursively partitioned into smaller clusters while moving down the tree. The leaves represent the clusters at the lowest level of the hierarchy and each leaf is labeled with its cluster's prototype (prediction). PCTs can be learned by the system CLUS available at http://clus.sourceforge.net.

PCTs are built with a greedy recursive top-down induction (TDI) algorithm, similar to that of C4.5 or CART [5]. The heuristic used in this algorithm for selecting the attribute tests in the internal nodes is intra-cluster variation (ICV Pairwise Distances) summed over the subsets induced by the test. Lower intra-subset variance results in more accurate predictions. The cluster variance is calculated as the sum of the squared pairwise distances between the cluster elements, i.e.,

$$Var(C) = \frac{1}{2|C|^2} \sum_{X \in C} \sum_{Y \in C} d^2(X, Y). \tag{1}$$

where $C$ is the cluster, $X$ and $Y$ are examples from $C$ and $d$ is the distance measure. Note that, no cluster prototypes are required for the computation of variance in this case. The prototype $c$ of a cluster of time series $C$ is then calculated as

$$c = argmin_q \sum_{X \in C} d^2(X, q). \tag{2}$$

After building a tree (and a PCT), it is typical to prune it, in order to deal with noise and other types of imperfection in the data. We employ *MaxDepth* and *MaxSize* pruning to increase the interpretability of PCTs while maintaining (or increasing) their predictive performance (on unseen cases).

The predictive clustering trees approach has a number of desirable properties. No prior assumptions are made on the probability distributions of the dependent and the independent variables. PCTs can handle discrete or continuous independent variables, as well as missing values. In addition, they are tolerant to redundant variables and noise. Furthermore, they are computationally inexpensive and are easily interpretable. Also, from a clustering point of view, the PCTs are unique in the sense that they provide cluster descriptions while constructing the clusters. All in all, PCTs are robust, efficient and interpretable models with satisfactory predictive performance.

Considering the guidelines given in [7], we used an extension of PCTs for modeling time series data described in detail in [6,10,19]. More specifically, we used PCTs that use dynamic time warping (DTW) [17] distance between time series to calculate the variance function needed for split selection. DTW can capture non-linear distortion along the time axis. It accomplishes this by assigning multiple values of one of the time series to a single value of the other. As a result, DTW is suitable to use if the time series are not properly synchronized, e.g., if one is delayed, or if the two time series are not of the same length.

$d_{\text{DTW}}(X, Y)$ with $X = \alpha_1, \alpha_2, \ldots, \alpha_I$, $Y = \beta_1, \beta_2, \ldots, \beta_J$ is defined based on the notion of a warping path between $X$ and $Y$. A warping path is a sequence of grid points $F = f_1, f_2, \ldots, f_K$ on the $I \times J$ plane. Let the distance between two values $\alpha_{i_k}$ and $\beta_{j_k}$ be $d(f_k) = |\alpha_{i_k} - \beta_{j_k}|$, then an evaluation function $\Delta(F)$ is given by $\Delta(F) = 1/(I + J) \sum_{k=1}^{K} d(f_k) w_k$. The weights $w_k$ are as follows: $w_k = (i_k - i_{k-1}) + (j_k - j_{k-1}), i_0 = j_0 = 0$. The smaller the value of $\Delta(F)$, the more similar $X$ and $Y$ are. In order to prevent excessive distortion, we assume an adjustment window ($|i_k - j_k| \leq r$). $d_{\text{DTW}}(X, Y)$ is the minimum of $\Delta(F)$.

DTW takes into account differences in scale and baseline of the values of the time series. If a given time series is identical to a second time series, but scaled by a certain factor or offset by some constant, then the two time series will be distant. For many applications, these differences are, however, not important and only the shape of the time series matters.

## 3    Data Description and Experimental Design

The datasets consist of measurements of 120 subjects (diabetes patients). Each of the patients is described with three types of data [3]:

- **Clinical status:** described with 41 variables (simple numeric or nominal variables) such as age, gender, frequency and amount of alcohol consumption, typical frequency of measurement of blood pressure, glucose, and body weight.
- **Glucose measurements:** Yearly measurements of glucose that patients take at home (time series variable).
- **WONCA scores:** Assessment of one's own functional status throughout a year. The scores represent a measuring instrument for physicians providing them with social and psychometric data about a patient. There are 8 categories (time series variables) describing different aspects of everyday life including physical fitness, feelings, daily activities, social activities, change in health, overall health, feeling of pain and overall weighted summary.

In Figs. 1 and 2, we show some of the properties of the data at hand. More specifically, in Fig. 1, we give histograms of the age by gender and the glucose values as measured before entering the study and, in Fig. 2, we show histograms of additional measurements taken before entering the study, namely the overall cholesterol and also both types of cholesterol. Higher values of LDL and low levels of HDL increase your chance of heart disease while high levels of HDL decrease your chances of heart disease. HDL should be higher than 1 for male and 1.2 for female, and LDL should be lower than 3. These graphs show that the majority of patients are over 50 years old, have increased glucose levels (due to having diabetes), mostly have somewhat acceptable cholesterol levels or slightly increased with HDL distributed around the normal value of 1 and LDL mostly in acceptable levels. We can say that these patients are not at a high risk and seem to manage to keep their diabetes in control.

The clinical status of the patients was assessed in a physicians' office (laboratory examinations) before the patients were admitted to the study: yearly measurements of glucose and the answering the WONCA questionnaires. After their examination, the patients were released to go home and were required to complete different tasks on their own. These tasks included measuring the blood glucose and filling out the WONCA questionnaires. The patients were suggested to perform these measurements 1 per month, but not all of them complied to this. The mean value of the number of measurements was 4.5 and 3 for glucose and WONCA, respectfully. We constructed two types of datasets representing the two tasks addressed here:

- *Glucose time series*: Consisting of the patients' clinical status and the glucose measurements.

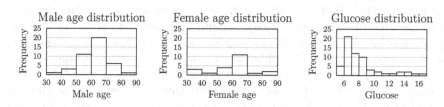

**Fig. 1.** Male age, female age and glucose distribution

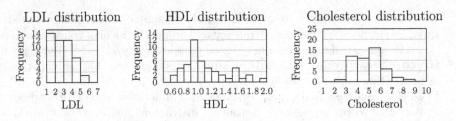

**Fig. 2.** LDL, HDL and overall cholesterol

- *WONCA time series*: We constructed 8 datasets, each one targeting the different WONCA categories outlined above.

For analyzing all of the datasets, we used the CLUS system for learning predictive clustering trees. More specifically, we used the extension of PCTs that is able to predict time series [19]. Next, we selected dynamic time warping (DTW) as a distance measure between the time series. We were bound to make this selection because of the properties of the measurements at hand: time series with a different number of measurements taken at different time intervals. Furthermore, we experimented with the two pruning algorithms described above: *MaxDepth* was set to 3 and *MaxSize* was set to 20.

## 4    Results

**Prediction of Yearly Glucose Measurements.** In Fig. 3, we show the obtained PCT with *MaxDepth* set to 3 on the glucose measurements. The model identified eight different clusters considering the time series profiles of the patients belonging to the leafs. The first attribute, *labalthcp*, indicates levels of Alanine Aminotransferase (ALT) enzyme in the blood (can indicate liver damage or disease). The next splits are at frequency of self-measurement of blood pressure (*pogostostsamomeritevkrvnitlak*) where the values indicate the frequency of at least 2 to 3 times a month, and diabetes status (*stanjesladkornebolezni*) where diabetes without complications is indicated. In the third level split, sex, frequency of measuring glucose levels (*pogostostsamomeritevkrvnisladkor*), and alcohol consumption frequency (*alcoholfrequencyhcp*) are used. The cluster C1 indicates male patients that measure their blood pressure at least 2 to 3 times a month and have their ALT value higher than 0.37. It shows glucose measurements that fluctuate slightly on a few occasions but otherwise seem steady at around 7 indicating a patient with a managed diabetes (no relapses and regular measurements). Similar conclusions can be drawn from the cluster C3 where even more and lower (6.5) measurements are shown indicating an even better-managed diabetes. Comparing patients in C1 and C3 we can see that patients in C3 focus more on measuring their blood glucose levels as those in C1 that focus more on measuring their blood pressure. Both C1 and C3 present patients that are well aware of their health status and try their best in order to

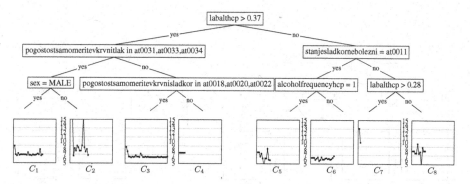

**Fig. 3.** Predictive clustering tree for modelling the glucose measurements obtained with *MaxDepth* set to 3. The Intra-cluster variance (calculated via pairwise distances) is 0.716.

be healthy with those in C3 with slightly better results. In addition to C1 and C3, the C4 cluster shows a flat line which indicates a managed glucose level but low compliance to the therapy in terms of the number of glucose measurements taken: these patients do not measure often their blood pressure or glucose. In cluster C2 we can note that the measurements fluctuate even into very high glucose levels of more than 15. This indicates patients that do not have well-managed diabetes. These fluctuations are a sign off quickly changing glucose levels which cause health deterioration in the long term and sooner on-come of diabetes complications. The patients in this group are female and often measure their blood glucose. Continuing with the analysis of the right side, the tree is first split on whether diabetes is without complications. For those patients that fall into this category, the frequency of alcohol consumption is considered next. Patients that consume one glass of alcohol per month or fewer show fluctuations in their glucose levels but the overall levels are never too high as is shown in C5. The cluster C6 exhibits similar behavior as cluster C5 but at even lower glucose levels which do indicate a managed diabetes. Finally, clusters C7 and C8 indicate patients that have diabetes with complications. C7 describes patients with only two measurements both of high values up to more than 13, while cluster C8 shows relapses and a poorly managed diabetes.

**Prediction of the WONCA Scores.** We have constructed PCTs for each score separately and here we illustrate, in Fig. 4, the obtained PCT with *MaxDepth* set to 3 on the overall WONCA scores. The tree identifies 8 clusters of patients with several of them showing only straight lines. This suggests little change through time, but we have different WONCA sum values at which such change is not occurring - four clusters show values at around 15 namely C1, C3, C4, and C6, while cluster C7 shows values around 12 and C5 shows values around 7. Cluster C2 is the only one showing fluctuations indicating change through time.

The tree is first split based on the *age*. For patients older than 42, the next attribute used to split the tree is the institution type. For the general

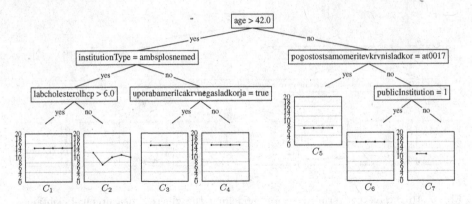

**Fig. 4.** WONCA sum prediction tree with MaxDepth set to 3. The Intra-cluster variance (calculated via pairwise distances) is 2.381.

practitioners (*ambsplosnemed*) the cholesterol level is used for further splitting the tree producing clusters C1 and C2, and for other types of institution, the attribute about usage of blood glucose measuring devices (*uporabamerilcakrvnegasladkorja*) is used for splitting the tree further, producing clusters C3 and C4. Patients that are younger than 42 and do not measure their blood glucose (*pogostostsamomeritevkrvnisladkor* = at0017) have a rather bad WONCA sum of less than 8 (cluster C5). For the patients taking measurements and visiting public healthcare providers (*publicInstitution* = 1), have a rather high WONCA sum of 16 (cluster C6) in comparison to less than 12 for cluster C7 that indicates patients who visit private healthcare providers.

Finally, for example, we observe that younger patients that do not have a high cholesterol, tend to fluctuate more with their WONCA values. This suggests that these patients are not facing any complications in terms of a high cholesterol and are thus not well managed in terms of the WONCA results as shown in C2. We can also see that older people should measure their blood pressure levels otherwise, their WONCA values will be very low which represents a basis for further complications of diabetes.

## 5    Conclusions

In this paper, we presented the results of data analysis of measurements from diabetes patients. The data used here were obtained from the Slovenian healthcare project ECARE that monitored the status of several patients throughout a period of 1 year. The main idea behind the work presented here is to provide stronger decision support for doctors that will handle future patients. More specifically, we focused on modeling yearly glucose measurements and WONCA scores using the patients' clinical status. In other words, we opted to search for patients with the similar time profile of the glucose values and describe those patients with their clinical status. We had the same goal for the functional status (WONCA scores).

Considering the complexity of the data at hand (time series with a different number of measurements and different time intervals), we selected PCTs as modeling framework. The extension of the PCTs employed here uses dynamic time warping as the distance between time series. The obtained PCTs identified several groups of patients that exhibit similar behavior. More specifically, we described groups of patients that are able to keep their disease under control and groups that are less successful. Furthermore, we identified and described groups of patients that have similar functional status.

We want to emphasize the potential weaknesses of supervised methods that often rely on a proxy for a gold standard e.g. diagnosis. These are overcome here by using time-course of glucose measurements as the predictive variable.

We plan to extend this work along several directions. To begin with, we plan to improve the prognostic power of the methods by using ensembles of PCTs. Next, we will use feature ranking for structured data to elucidate the most important features that need to be monitored. Furthermore, we can analyze this data in the one-look-ahead approach: predict the next glucose value based on the clinical status and the previous measurements of glucose. Finally, we will use the knowledge from the models to design a decision support system for helping the medical practitioners.

# References

1. Abraham, C., Michie, S.: A taxonomy of behavior change techniques used in interventions. Health Psychol.: Off. J. Div. Health Psychol. Am. Psychol. Assoc. **27**(3), 379–387 (2008)
2. Beštek, M., Brodnik, A.: Interoperability and mHealth precondition for successful eCare. In: Adibi, S. (ed.) Mobile Health (mHeath) The Technology Road Map. Springer Series in Bio-/Neuroinformatics. Springer, Switzerland (2014). doi:10. 1007/978-3-319-12817-7_16
3. Beštek, M., Brodnik, A.: Preconditions for successful eCare. Inform. Med. Slov. **20**(1–2), 17–29 (2015)
4. Blockeel, H.: Top-down induction of first order logical decision trees. Ph.D. thesis, Katholieke Universiteit Leuven, Leuven, Belgium (1998)
5. Breiman, L., Friedman, J., Olshen, R., Stone, C.J.: Classification and Regression Trees. Chapman & Hall/CRC, Boca Raton (1984)
6. Debeljak, M., Squire, G.R., Kocev, D., Hawes, C., Young, M.W., Džeroski, S.: Analysis of time series data on agroecosystem vegetation using predictive clustering trees. Ecol. Model. **222**(14), 2524–2529 (2011)
7. Džeroski, S.: Introduction: the challenges for data mining. In: 5th International Workshop Knowledge Discovery in Inductive Databases, KDID 2006, pp. 259–300 (2007)
8. Eljil, K.A.A.S.: Predicting hypoglycemia. In: Diabetic Patients Using Machine Learning Techniques, pp. 1–92. Faculty of the American University of Sharjah College of Engineering, UAE (2014)
9. Georga, E., Protopappas, V.C.: Short-term vs. long-term analysis of diabetes data: application of machine learning and data mining techniques. In: IEEE 13th International Conference on Bioinformatics and Bioengineering (BIBE) (2013)

10. Gjorgjioski, V.: Distance-based learning from structured data. Ph.D. thesis, International postgraduate school Jožef Stefan, Ljubljana, Slovenia (2015)

11. Karpel'ev, V.A., Filippov, Y., Tarasov, Y., Boyarsky, M.D., Mayorov, A., Shestakova, M.V., Dedov, I.I.: Mathematical modeling of the blood glucose regulation system in diabetes mellitus patients. Vestn. Ross. Akad. Med. Nauk **70**(5), 549–560 (2015). https://www.scopus.com/inward/record.uri?eid =2-s2.0-84948981003&doi=10.15690%2Fvramn.v70.i5.1441&partnerID=40&md5 =b578cbe1711a9bee957a656eb681c1c2

12. Kavakiotis, I., Tsave, O., Salifoglou, A., Maglaveras, N., Vlahavas, I., Chouvarda, I.: Machine learning and data mining methods in diabetes research. Comput. Struct. Biotechnol. J. **15**, 104–116 (2017). http://dx.doi.org/10.1016/j.csbj.2016.12.005

13. Kocev, D., Vens, C., Struyf, J., Džeroski, S.: Tree ensembles for predicting structured outputs. Pattern Recognit. **46**(3), 817–833 (2013)

14. Lenert, L., Norman, G.J., Mailhot, M., Patrick, K.: A framework for modeling health behavior protocols and their linkage to behavioral theory. J. Biomed. Inform. **38**(4), 270–280 (2005)

15. Marinov, M., Mosa, A.S.M., Yoo, I., Boren, S.A.: Data-mining technologies for diabetes: a systematic review. J. Diabetes Sci. Technol. **5**(6), 1549–1556 (2011). http://dst.sagepub.com/lookup/doi/10.1177/193229681100500631

16. Reifman, J., Rajaraman, S., Gribok, A., Ward, W.K.: Predictive monitoring for improved management of glucose levels. J. Diabetes Sci. Technol. **1**(4), 478–486 (2007). https://www.scopus.com/inward/record.uri?eid=2-s2.0-52449101078 &partnerID=40&md5=07e50cc16ed4d23ac38ba713efed205a

17. Sakoe, H., Chiba, S.: Dynamic programming algorithm optimization for spoken-word recognition. IEEE Trans. Acoust. Speech Signal Process. **ASSP–26**, 43–49 (1978)

18. Shivakumar, B.L.: A survey on data-mining technologies for prediction and diagnosis of diabetes (2014)

19. Slavkov, I., Gjorgjioski, V., Struyf, J., Džeroski, S.: Finding explained groups of time-course gene expression profiles with predictive clustering trees. Mol. BioSyst. **6**(4), 729–740 (2010)

20. Sowjanya, K., Singhal, A., Choudhary, C.: MobDBTest: a machine learning based system for predicting diabetes risk using mobile devices. In: 2015 IEEE International Advance Computing Conference (IACC), pp. 397–402 (2015)

21. Sumalatha, G., Muniraj, N.J.R.: Survey on medical diagnosis using data mining techniques. In: 2013 International Conference on Optical Imaging Sensor and Security (ICOSS), pp. 1–8 (2013). http://ieeexplore.ieee.org/lpdocs/epic03/wrapper. htm?arnumber=6678433

22. Zecchin, C., Facchinetti, A., Sparacino, G., Cobelli, C.: Reduction of number and duration of hypoglycemic events by glucose prediction methods: a proof-of-concept in silico study. Diabetes Technol. Therapeutics **15**(1), 66–77 (2013)

23. Zhao, C., Yu, C.: Rapid model identification for online subcutaneous glucose concentration prediction for new subjects with type i diabetes. IEEE Trans. Biomed. Eng. **62**(5), 1333–1344 (2015). https://www.scopus.com/inward/record.uri?eid =2-s2.0-84929075114&doi=10.1109%2FTBME.2014.2387293&partnerID=40&md5 =e23db851493020e4f51367ce89f66dcb

# Estimation of Sleep Quality by Using Microstructure Profiles

Zuzana Rošťáková[1,2(✉)], Georg Dorffner[2], Önder Aydemir[2,3],
and Roman Rosipal[1]

[1] Institute of Measurement Science, Slovak Academy of Sciences,
Bratislava, Slovakia
zuzana.rostakova@savba.sk
[2] Section for Artificial Intelligence and Decision Support,
Center for Medical Statistics, Informatics and Intelligent Systems,
Medical University of Vienna, Vienna, Austria
[3] Department of Electrical and Electronics Engineering,
Karadeniz Technical University, Trabzon, Turkey

**Abstract.** Polysomnograhy is the standard method for objectively measuring sleep, both in patient diagnostics in the sleep laboratory and in clinical research. However, the correspondence between this objective measurement and a person's subjective assessment of the sleep quality is surprisingly small, if existent. Considering standard sleep characteristics based on the Rechtschaffen and Kales sleep models and the Self-rating Sleep and Awakening Quality scale (SSA), the observed correlations are at most 0.35. An alternative way of sleep modelling - the probabilistic sleep model (PSM) characterises sleep with probability values of standard sleep stages Wake, S1, S2, slow wave sleep (SWS) and REM operating on three second long time segments. We designed sleep features based on the PSM which correspond to the standard sleep characteristics or reflect the dynamical behaviour of probabilistic sleep curves. The main goal of this work is to show whether the continuous sleep representation includes more information about the subjectively experienced quality of sleep than the traditional hypnogram. Using a linear combination of sleep features an improvement in correlation with the subjective sleep quality scores was observed in comparison to the case when a single sleep feature was considered.

**Keywords:** Probabilistic sleep model · Hypnogram · Self–rating Sleep and Awakening Quality scale · Sleep features

## 1 Introduction

Polysomnography (PSG) is the standard method for objectively measuring sleep, both in patient diagnostics in the sleep laboratory and in clinical research. Besides revealing important events pointing towards sleep disorders, such an objective biomarker can also be expected to reflect the quality of sleep in

© Springer International Publishing AG 2017
A. ten Teije et al. (Eds.): AIME 2017, LNAI 10259, pp. 105–115, 2017.
DOI: 10.1007/978-3-319-59758-4_12

terms of how rested the subject feels in the morning. Yet, if one looks at the correspondence between this objective measurement and a person's subjective assessment of one's sleep quality then it is found to be surprisingly small, if existent. In [1] the authors cluster patients by values of the Pittsburgh Sleep Quality Index (PSQI) and the Epworth Sleepiness Scale (ESS), both subjective measures of long-term (typically a month) sleep and daytime wakefulness qualities. They found no significant differences in any PSG sleep variable between those clusters, indicating that those subjective variables measure something distinct from the objective sleep recording.

In [2] the authors perform a multi–variable regression predicting different psychomotor performance results from subjective and PSG–based sleep variables. The highest correlation was between a set of variables containing total sleep time (TST), sleep efficiency (EFF), wake after sleep onset (WASO) and sleep onset latency (SLAT), with $R^2 = 0.21$ predicting the performance in a simple reaction time test, but only within a group of normal sleepers.

Our own previous results [3,4] show that correlations between any PSG sleep variable and subjective assessments on the same night (in particular Saletu's Subjective Sleep and Awakening Scale, or SSA, [5]) are poor, at best, with Pearson or Spearman correlation coefficients hardly above 0.4. The same work, however, showed that the novel probabilistic model of sleep, representing the microstructure of sleep as compared to standard hypnograms, can lead to variables with significantly higher correlation coefficients, pointing to the fact that standard sleep scoring does not extract the maximum information about sleep from the electrophysiological signals, in particular, electroencephalography (EEG).

In this paper we investigate whether linear combinations of several variables can achieve a higher correlation with subjective sleep quality than single variables. We do this for both probabilistic and traditional stage–based sleep profile to also investigate whether in a multi–variable setting the former can also outperform the latter.

## 2   Data Set Description

In this study, the electroencephalographic (EEG) data from 540 polysomnographic sleep recordings from the SIESTA database [6] were used. The 540 PSG nights were recorded from 270 subjects in two consecutive nights spent in the sleep laboratory.

The microstructure of each sleep recording was calculated by using the probabilistic sleep model (PSM) introduced in [4]. In the PSM method three seconds long time segments are used to calculate probability values to be in a certain stage (Wake, S1, S2, slow wave sleep (SWS) and REM). Figure 1 shows an example of the microstructure of sleep by depicting the probabilities for each sleep stage over time. The standard Rechtschaffen and Kales scores[1] obtained by the automatic scoring system Somnolyser 24 × 7 [9] are plotted as well.

---

[1] Nowadays, the American Academy of Sleep Medicine (AASM) sleep model is preferred in the clinical praxis, but we do not expect significant changes in results when using the AASM scores instead of the Rechtschaffen and Kales sleep model.

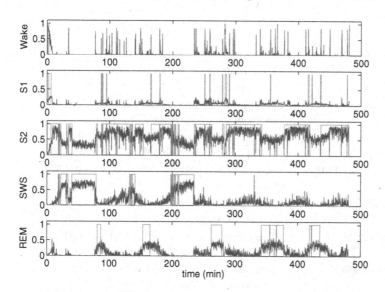

**Fig. 1.** An example of the microstructure of sleep for an all-night recording (blue) and corresponding Rechtschaffen and Kales scores (red). (Color figure online)

After the subjects woke up, they were asked to fill in the Self–rating Sleep And Awakening Quality (SSA) questionnaire [5]. The scale consist of 7, 8 and 5 questions on sleep quality, awakening quality and somatic complaints, leading to a total score with a value between 20 (best quality) and 80 (worst quality).

## 3  Sleep Features

### 3.1  Hypnogram Features

A set of 25 standard sleep variables derived from a hypnogram were calculated (Table 1). The descriptions and abbreviations of the features are given in the second and third column of Table 1, respectively.

### 3.2  PSM Based Sleep Features

For a continuous probabilistic sleep profile $X$ observed over a time interval $T$ we aimed to design variables that have a correspondence to the standard sleep measures and to include variables that optimally exploit potentially characteristics found in the continuous profiles.

**Band Power.** The band power (BP) was computed by the following formula

$$BP|_{f_1}^{f_2} = \sum_k \|F_x(k)|_{f_1}^{f_2}\|^2; \tag{1}$$

**Table 1.** The hypnogram features

| Hypnogram feature | Description | Abbreviation |
|---|---|---|
| Time in bed (min) | Time from lights out to the end of the recording | TIB |
| Total sleep period (min) | Time from the first to the last epoch in any sleep stage | TSP |
| Total sleep time (min) | Sum of epochs in one of the sleep stages S1, S2, SWS, REM | TST |
| Wake within TSP (min) | Sum of wake-epochs within the total sleep period | WTSP |
| Wake after final awakening (min) | Time from final awakening to the end of the recording | WAFA |
| Lights out to S1 | Time from lights out to the first occurrence of stage 1 | LS1 |
| Lights out to S2 | Time from lights out to the first occurrence of stage 2 | LS2 |
| Sleep latency | Time from lights out to the first occurrence of three consecutive epochs in stage S1 or to the first occurrence of stage S2 | SLAT |
| Sleep efficiency (%) | $\frac{TST}{TIB} \times 100$ | EFF |
| Stage (min) | Time spent in a given sleep stage | S1, S2, S3, S4, SWS, REM |
| Stage (%) | $\frac{\text{Time spent in a given sleep stage}}{TST} \times 100$ | S1p, S2p, S3p, S4p, SWSp, REMp |
| Frequency of awakening | Number of awakenings within the total sleep period | FW |
| Awakening-index | Number of awakenings within the total sleep period per hour sleep | FWTST |
| Frequency of stage shifts | Number of stage shifts within the total sleep period | FS |
| Stage shift-index | Number of stage shifts within the total sleep period per hour sleep | FSTST |

where $F_x(k)|_{f_1}^{f_2}$ denotes the coefficients of the Fast Fourier Transform of $x$ between a lower cut-off frequency $f_1 = 0$ and an upper cut-off frequency $f_2 = 0.001$.

**Entropy.** Entropy characterises the level of uncertainty of a signal. For each microstructure of sleep stages it was computed as follows

$$ent = -\int_T \frac{X(t)}{\int_T X(s)ds} \log \frac{X(t)}{\int_T X(s)ds} dt, \qquad \text{where } \log 0 = 0 \text{ by definition.} \quad (2)$$

**Log-Spaced Power Spectral Density.** The power spectral density of the microstructure of sleep EEG was estimated through the modified covariance method. This method fits a $p^{th}$ order autoregressive (AR) model to a signal, which is assumed to be the output of an AR system driven by white noise and minimising the forward and backward prediction errors. The order of the AR model was empirically selected as 13.

To characterise the output vector $X_{PSD}$ of the normalised estimate of the AR system parameters with only one number we chose the common logarithm of its mean value

$$psd = \log_{10} \overline{X_{PSD}}. \tag{3}$$

**Moving Window Features.** Two statistical features were extracted by using a moving window through the probabilistic profile of sleep EEG. The moving window has two kinds of parameters – a height $h$ and a length $l$ (the extension of the window). Each of the non–overlapping windows of length $l$ was represented by the number of probability values which were higher than the height parameter. This procedure results in a sequence MWS (*moving window sequence*) of length $L$ that resemble a smoothed version of the entire profile.

The moving window features were calculated as the arithmetic mean (Am) and skewness (Sm) of MWS [2,3]

$$Am = \sum_{m=1}^{L} \frac{MWS(m)}{L} = \overline{MWS}, \tag{4}$$

$$Sm = \frac{\frac{1}{L}\sum_{m=1}^{L}\left(MWS(m) - \overline{MWS}\right)^3}{\left(\sqrt{\frac{1}{L}\sum_{m=1}^{L}\left(MWS(m) - \overline{MWS}\right)^2}\right)^3}. \tag{5}$$

The optimal window parameters were set as $l = 140$ and $h = 0.22$.

**Arithmetic Mean (AM), Geometric Mean (GM), Median (Med)** were considered as features of the discrete observation of a probabilistic profile $X$.

**Area Under a Curve.** In the case of PSM the area under a probabilistic sleep profile $X$ forms an analogy to the time spent in a given sleep stage

$$AUC(X) = \int_T X(t)dt. \tag{6}$$

**Moments of a Curve.** The first moment $mom_1$ of a curve $X$ characterises the expected value of the curve according to time [8]

$$mom_1 = \int_T t\frac{X(t)}{AUC(X)}dt. \tag{7}$$

Higher order central moments

$$mom_k = \int_T (t - \mu)^k \, \frac{X(t)}{AUC(X)} dt, k = 2, 3, \ldots. \qquad (8)$$

describe the variability in $X$. In this study only the second order central moment $mom_2$ was considered.

**Moments of a Feature Function of a Curve.** A feature function is a strictly positive transformation of a curve which highlights a set of curves features [8], for example

$$I_{max}(t) = c \left( X(t) - \min_{t \in T} X(t) \right)^r, \; r \in \mathbb{R}, \qquad (9)$$

which concentrates its weight to the local maxima of the curve $X$ or

$$I_m(t) = c|X^{(m)}(t)| \, , X^{(m)} \text{ is the } m^{th} \text{ derivative of } X, \; m = 1, 2, \qquad (10)$$

which highlights global characteristics. The constant $c$ guarantees that the area under a feature function is equal to 1. For all three feature functions the first order moment $mom_1$ was computed.

**Curve Length.** The curve length characterises changes of a curves profile over a time interval $T$

$$cl = \int_T \sqrt{1 + (X'(t))^2} dt. \qquad (11)$$

## 4  Methodology

To relate objective and subjective measures of sleep we performed a linear regression for modelling the total score from the SSA scale by a linear combination of variables describing the sleep architecture.

The whole dataset was divided into two parts. The first part served for modelling the SSA scores as a linear combination of sleep features. Because of presence of either redundant or irrelevant sleep features, right before models fitting the feature selection procedure was performed in order to simplify the model. More details about the procedure are given in the next section.

The second part (testing dataset) was used for checking the models' quality by computing the Spearman's correlation coefficient between the real and predicted total SSA scores for the testing dataset.

To avoid problems caused by randomly splitting the dataset into two parts, a 10-fold cross-validation was considered.

The procedure was performed separately for the variables from the PSM and hypnogram as well as for joined datasets of sleep features. Furthermore, this was done for the sleep recordings of the first night and second night separately.

Finally, to detect whether the differences in the correlations estimated using different sets of sleep features are significant, the Student t–test and the Wilcoxon rank–sum test were considered.

## 4.1  Feature Selection

High number of features considered for the sleep models (25 standard sleep features and 75 features for PSM) may cause inaccuracies in estimation of parameters in the linear regression model. Moreover, some of the features are redundant – they are either highly correlated with other features or include only small information about the sleep process.

Dimensionality reduction is important in machine learning. It leads not only to a decrease in computational time, but may also increase the comprehensibility and the performance of the model. It includes two main approaches – feature extraction methods and feature selection techniques.

Feature extraction methods transform high–dimensional data into a vector space with lower dimension by designing new variables expressed as a linear combination of the original ones. Principal Component Analysis, Factor Analysis and other techniques are typical representatives.

On the other hand, feature selection methods work only with original features and they aimed to find the smallest subset of features with the most informative features.

For the Rechtschaffen and Kales sleep model there is a standard set of sleep features used in the majority of sleep studies. We aimed to design similar set of features for PSM and therefore the feature selection approach is more appropriate for our case.

In this study, we performed sequential feature selection procedure which executes a sequential search among each candidate feature subset in order to find out the smallest subset of features which is able to predict the SSA scores in the best way. This algorithm is implemented for example in the function *sequentialfs* in the MATLAB environment [10].

Similarly to the previous case the 20–fold cross–validation was used. In each of 20 trials, one fold served as a validation dataset, the remaining part of 19 folds formed a training dataset. A linear regression model was fitted to the training dataset using a candidate subset of features. The mean squared error (MSE) between the real and predicted SSA scores for the validation dataset measured the quality of the model. The algorithm started with an empty feature set and then candidate feature subsets were created by sequentially adding each of the features not yet selected until there was no improvement in prediction.

The process resulted into 20 possibly different linear regression models. The model with the lowest MSE in all was used in the further analysis.

## 5  Results and Discussion

In Table 2 average Spearman's correlation coefficients between original and predicted total SSA scores are listed. Regarding the first night, the highest correlations were obtained by considering joined datasets of sleep features for the PSM and standard hypnogram ($\approx$0.38). Using the standard sleep features only the average correlation ($\approx$0.37) was higher than in the case of the PSM sleep features ($\approx 0.34$). On other hand, because of high standard deviations the differences in results between all three cases were not significant.

Comparing results for the hypnogram features and PSM based sleep variables separately or results of joined datasets of features (Table 2) for the second night the correlations are higher for the first two cases. However, the differences between results were still not significant.

Similar values of mean squared error (Table 3) obtained in the last iteration of the feature selection step confirmed similar performance of the standard hypnogram and PSM.

These results contradict our expectations, that PSM includes more information about the sleep process. A possible reason for no or only slight improvement in the correlations may be that the designed sleep features for the PSM model do not describe the dynamical behaviour of the probabilistic sleep curves properly. The features extracted from the PSM (Sect. 3.2) were chosen so that they are natural counterparts of the standard hypnogram features or they highlight specific properties of a sleep probabilistic curve. However, it is difficult to say whether our features are the most appropriate. The major task for future research is to design new features which would improve the results. Another idea is to use the whole probabilistic sleep curves instead of their one–dimensional characteristics for modelling the results of the subjectively scored sleep and awakening quality.

**Table 2.** Average Spearman's correlation coefficients and their standard deviations for the PSM based sleep measures, the standard hypnogram sleep features as well as for joined datasets of sleep measures.

|  | 1. Night | 2. Night |
| --- | --- | --- |
| PSM |  |  |
| Wake stage | $0.3551 \pm 0.1732$ | $0.2224 \pm 0,2500$ |
| S1 stage | $0.2024 \pm 0.2134$ | $0.2886 \pm 0.1778$ |
| S2 stage | $0.3145 \pm 0.2202$ | $0.1921 \pm 0.2053$ |
| SWS stage | $0.2390 \pm 0.2158$ | $0.1653 \pm 0.1892$ |
| REM stage | $0.1358 \pm 0.2025$ | $-0.0048 \pm 0.1836$ |
| All stages together | $0.3396 \pm 0.1800$ | $0.2954 \pm 0.1626$ |
| Standard features | $0.3679 \pm 0.2230$ | $0.2762 \pm 0.1349$ |
| PSM + standard features | $0.3809 \pm 0.2496$ | $0.2688 \pm 0.2238$ |

On the other hand, considering a single sleep feature for predicting the SSA scores in the testing dataset, the correlations were significantly lower or approximately equal to the case when a linear combination of features was used. This was true for both sleep models as well as for joined datasets of sleep features.

In addition, we were interested in sleep features which were selected in the majority of trials. Regarding to the standard sleep features, only the time spent in the S1 stage (total or relative) was selected in more than 7 trials (Table 4). In the case of PSM especially sleep variables related to the Wake and S1 stages

**Table 3.** An example of the mean squared error between real and predicted values of the Self–rating questionnaire for sleep and awakening quality and sequences of chosen sleep features in the feature selection step for the first night. For both sleep models first three and the last iteration of a randomly chosen trail are described.

| Model | 1. Iteration | 2. Iteration | 3. Iteration | Last iteration |
|---|---|---|---|---|
| Standard features | 76.64 | 67.20 | 62.75 | 61.86 |
| | - | S1p | eff, S1p | eff, S1, S1p, S4p |
| PSM | 76.11 | 66.5 | 64.17 | 58.47 |
| | - | Am (Wake) | Am (Wake), psd (Wake) | Am (Wake), psd (Wake), psd (S1), $mom_2$ (Wake), $mom_1$ (REM), ... |

dominated. Moreover, using the PSM sleep variables related only to the Wake stage the third highest average correlation was observed for the first night.

Considering the joined datasets of sleep features, approximately the same sleep variables related to wakefulness or light sleep were selected.

Finally we observed that the average correlations for the first night are higher in comparison to the results of the second night. We hypothesise this is caused by the "first night effect" – subjects usually sleep poor in a new environment and feel tired in the morning. This is reflected by higher variability in values of sleep features related to the Wake or S1 stage which were the most selected sleep variables in the feature selection process as well as by possibly increased variability in values of the SSA scores. Because of visible changes in sleep features and SSA scores which are typical for the majority of the first nights data, the observed correlations are expected to be higher than in the case of the second night where this effect is diminished.

**Table 4.** List of features selected in at least 7 of 10 trials for each sleep model and each night separately. In the case of joined datasets of the hypnogram and PSM based sleep features the first mentioned are accentuated with italics.

| | | |
|---|---|---|
| 1. Night | PSM | $mom_2$ (Wake), AM (Wake), Am (Wake), Sm (S1) |
| | Standard features | S1, S1p |
| | PSM + standard features | $mom_2$ (Wake), Sm (S1), psd (SWS), $mom_1$ (REM), *S2, FS* |
| 2. Night | PSM | Sm (S1), AM (Wake), Med (Wake), ent (REM) |
| | Standard features | S1p |
| | PSM + standard features | AM (Wake), Sm (S1), *S1p* |

# 6   Conclusion

In this study we compared the quality of prediction of subjective sleep quality scores using sleep features extracted from the standard hypnogram, the probabilistic sleep model or both sleep models together.

PSM based sleep features led to approximately equal or only slightly higher average correlation coefficients in comparison to the case when a subset of standard hypnogram sleep features was considered. Because of high standard deviations the Student t-test and Wilcoxon test were not able to reject the hypothesis that the correlations are equal. The sleep features from both sleep models together did not lead to significant improvement in prediction of the SSA scores. From this point of view the standard hypnogram and PSM seem to be of equivalent quality.

On the other hand, correlation coefficients estimated when the SSA scores were modelled by a single sleep feature were either significantly lower or approximately equal to the correlations between SSA scores and a linear combination of sleep features.

In the feature selection step, sleep features representing the wakefulness or light sleep were selected in the majority of trials for the hypnogram or PSM related features separately as well as joined datasets of sleep variables. This indicates that the subjectively scored sleep quality is influenced mainly by the amount of the time spent awake or in the light sleep during the night.

**Acknowledgements.** This research was supported by the Ernst Mach Stipendien der Aktion Österreich–Slowakei ICM-2016-03516, the Slovak Research and Development Agency (grant number APVV–0668–12), the Ministry of Health of the Slovak Republic (grant number MZ 2012/56–SAV–6) and by the VEGA 2/0011/16 grant. Dr. Aydemir's contribution was supported by a scholarship from The Scientific and Technological Research Council of Turkey (TUBITAK).

# References

1. Buysse, D.J., et al.: Relationships between the Pittsburgh Sleep Quality Index (PSQI), Epworth Sleepiness Scale (ESS), and clinical/polysomnographic measures in a community sample. J. Clin. Sleep Med. **4**(6), 563–571 (2008)
2. Edinger, J.D., et al.: Psychomotor performance deficits and their relation to prior nights' sleep among individuals with primary insomnia. Sleep **31**(5), 599–607 (2008)
3. Rosipal, R., Lewandowski, A., Dorffner, G.: In search of objective components for sleep quality indexing in normal sleep. Biol. Psychol. **94**(1), 210–220 (2013)
4. Lewandowski, A., Rosipal, R., Dorffner, G.: Extracting more information from EEG recordings for a better description of sleep. Comput. Methods Programs Biomed. **108**(3), 961–972 (2012)
5. Saletu, B., et al.: Short-term sleep laboratory studies with cinolazepam in situational Insomnia induced by traffic noise. Int. J. Clin. Pharmacol. Res. **7**(5), 407–418 (1987)
6. Klosch, G., et al.: The SIESTA project polygraphic and clinical database. IEEE Eng. Med. Biol. Mag. **20**(3), 51–57 (2001)
7. Jain, A., Zongker, D.: Feature selection: evaluation, application, and small sample performance. IEEE Trans. Pattern Anal. Mach. Intell. **19**(2), 153–158 (1997)
8. James, G.M.: Curve alignment by moments. Ann. Appl. Stat. **1**(2), 480–501 (2007)

9. Anderer, G., Gruber, S., Parapatics, M., Woertz, T., Miazhynskaia, G., Klösch, B., Saletu, J., Zeitlhofer, M., Barbanoj, H., Danker-Hopfe, S., Himanen, B., Kemp, T., Penzel, M., Grőzinger, D., Kunz, P., Rappelsberger, A., Schlögl, G., Dorffner, G.: An E-health solution for automatic sleep classification according to Rechtschaffen and Kales: validation study of the Somnolyzer 24 × 7 utilizing the SIESTA database. Neurophysiology **51**, 115–133 (2005)

10. MATLAB, version 8.3.0 (R2014a), The MathWorks Inc., Natick, Massachusetts (2014)

# Combining Multitask Learning and Short Time Series Analysis in Parkinson's Disease Patients Stratification

Anita Valmarska[1,2(✉)], Dragana Miljkovic[1], Spiros Konitsiotis[3], Dimitris Gatsios[4], Nada Lavrač[1,2], and Marko Robnik-Šikonja[5]

[1] Jožef Stefan Institute, Ljubljana, Slovenia
{anita.valmarska,dragana.miljkovic,nada.lavrac}@ijs.si
[2] Jožef Stefan International Postgraduate School, Ljubljana, Slovenia
[3] Department of Neurology, Medical School, University of Ioannina,
Ioannina, Greece
skonitso@uoi.gr
[4] Department of Biomedical Research, University of Ioannina, Ioannina, Greece
dgatsios@cc.uoi.gr
[5] Faculty of Computer and Information Science, University of Ljubljana,
Ljubljana, Slovenia
marko.robnik@fri.uni-lj.si

**Abstract.** Quality of life of patients with Parkinson's disease degrades significantly with disease progression. This paper presents a step towards personalized medicine management of Parkinson's disease patients, based on discovering groups of similar patients. Similarity is based on patients' medical conditions and changes in the prescribed therapy when the medical conditions change. The presented methodology combines multitask learning using predictive clustering trees and short time series analysis to better understand when a change in medications is required. The experiments on PPMI (Parkinson Progression Markers Initiative) data demonstrate that using the proposed methodology we can identify some clinically confirmed patients' symptoms suggesting medications change.

## 1 Introduction

Parkinson's disease (PD) is the second most common neurodegenerative disease (after Alzheimer's disease) that affects many people worldwide. Due to the death of nigral neurons, patients experience both motor and non-motor symptoms, affecting their quality of life. The reasons for the cell death are still poorly understood, and there is currently no cure for Parkinson's disease. Physicians try to manage patients' symptoms by introducing medications therapies, consisting of antiparkinson medications. Physicians need to be careful with the prescribed medications therapies since the prolonged intake in particular of higher dosages of antiparkinson medications can have side-effects.

Data mining algorithms have been successfully used to learn predictive models and to discover insightful patterns in the data. Both approaches, predictive

© Springer International Publishing AG 2017
A. ten Teije et al. (Eds.): AIME 2017, LNAI 10259, pp. 116–125, 2017.
DOI: 10.1007/978-3-319-59758-4_13

and descriptive data mining, have been successfully used also in medical data analysis. The use of data mining methods may improve diagnostics, disease treatment and detection of disease causes. In personalized health [7] data mining can be used to improve drug recommendations and medical decision support, leading to reduced costs of medical solutions. The discovered patterns can provide the clinicians with new insights regarding the status of the treated patients and can support decision making regarding patients' therapy recommendations.

To the best of our knowledge, data mining techniques have not been used before to analyze clinicians' decisions of changing drug prescription as a reaction to the change of patients' symptoms when using antiparkinson medications through prolonged periods of time. Physicians follow certain guidelines for therapy prescriptions, and the response of patients to medications is usually recorded in clinical studies using simple statistical methods. This paper uses multitask learning with predictive clustering trees on short time series data—describing patients' status at multiple time points—in order to determine symptoms which trigger physicians' decisions to modify the medications therapy. We consider trigger symptoms to be those symptoms that a patient cannot tolerate and the physician is pressed to change the medications therapy in order to control them. The proposed methodology addresses the task of determining subgroups of patients with similar symptoms and therapy. As each patient potentially receives several groups of medications, predicting their changes with multitask learning is effective and improves control over drug interactions.

After presenting the background and motivation in Sect. 2, Sect. 3 describes the Parkinson's Progression Markers Initiative (PPMI) symptoms data set [11], together with the medications used for symptoms control. In Sect. 4 we propose a methodology for analyzing Parkinson's disease symptoms by learning predictive clustering trees from short time series data. We also describe how changes in symptoms-based clustering of patients are connected to the changes in medications therapies with the goal to find symptoms which trigger therapy modifications. Section 5 presents the results of data analysis. Finally, Sect. 6 presents the conclusions and plans for further work.

## 2   Background and Motivation

Data mining research in the field of Parkinson's disease can be divided into three groups: classification of Parkinson's disease patients, detection of Parkinson's disease symptoms, and detection of subtypes of Parkinson's disease patients.

Due to the overlap of Parkinson's disease symptoms with other diseases, only 75% of clinical diagnoses of Parkinson's disease are confirmed to be idiopathic Parkinson disease at autopsy [8]. Classification techniques offer decision support to specialists by increasing the accuracy and reliability of diagnosis and reducing possible errors. Gil and Johnson [5] use Artificial Neural Networks (ANN) and Support Vector Machines (SVM) to distinguish Parkinson's disease patients from healthy subjects. Ramani and Sivagami [14] compare effectiveness of different data mining algorithms in diagnosis of Parkinson's disease patients. The authors use data set consisting of 31 people (23 are Parkinson's disease patients).

Tremor is a symptom strongly associated with Parkinson's disease. Several approaches to computational assessment of tremor have been proposed. Methods such as time domain analysis [18], spectral analysis [16] and non-linear analysis [16] have addressed tremor detection and quantification. Many recent works are based on body fixed sensors (BFS) for long-term monitoring of patients [13].

Parkinson's disease (PD) is a heterogeneous neurodegenerative condition with different clinical phenotypes, genetics, pathology, brain imaging characteristics and disease duration [3]. This variability indicates the existence of disease subtypes. Using k-means clustering, Ma et al. [10] identify four groups of Parkinson's disease patients which is consistent with the conclusions from [9,15].

Classification and clustering models usually focus on diagnosing new patients. None of the listed methods follow the progression of the disease, and to the best of our knowledge, no data mining research in the field of Parkinson's disease analyzed the development of the disease in combination with the medications that the patients receive. Identification of groups of patients based on how they react to a certain therapy can be helpful in the assignment of personalized therapies and more adequate patient treatment. For that purpose, we propose a methodology for determination of trigger symptoms, which influence the physician's decision about therapy modification. In addition, the methodology aims to uncover the side-effects of the modified therapy.

The basic data mining technique we use is induction of predictive clustering trees (PCTs). PCTs [1] adapt the basic top-down induction of decision trees method towards clustering and allows for multitask learning. In multitask learning (MTL) are learned simultaneously. This can improve generalization and prevent overfitting [2]. In our research we use multitask learning algorithms implemented in the CLUS data mining framework [1] to obtain multi-target decision trees, simultaneously predicting three target variables: change of levodopa dosage, change of dopamine agonists dosage, and change of MAO B inhibitors dosage.

## 3    The Parkinson's Disease Data Set

In this paper we use the PPMI data collection [11] gathered in the observational clinical study to verify progression markers in Parkinson's disease. Below we present the selection of PPMI data used in the experiments.

### 3.1    PPMI Symptoms Data Sets

The medical condition and the quality of life of a patient suffering from Parkinson's disease is determined using the Movement Disorder Society (MDS)-sponsored revision of the Unified Parkinson's Disease Rating Scale (MDS-UPDRS) [6]. This is a four part questionnaire consisting of 65 questions concerning the development of the disease symptoms. Each answer is given on a five point Likert scale, where $0 =$ normal (patient's condition is normal, symptom is not present), and $4 =$ severe (symptom is present and severely affects the independent functioning of the patient).

The cognitive state of the patient is determined using the Montreal Cognitive Assessment (MoCA). It is a questionnaire consisting of 11 questions (maximum 30 points), assessing different cognitive domains. In addition to the MoCA data, physicians also use the Questionnaire for Impulsive-Compulsive Disorders (QUIP) to address four major and three minor impulsive compulsive disorders.

In their everyday practice, physicians use a vector of chosen symptoms to follow the development of the disease and decide when to intervene with medications treatment modifications. They focus their attention on both *motor* and *non-motor* aspects of patients' quality of life. Physicians evaluate the motor aspect of patient's quality of life using the following symptoms: *bradykinesia, tremor, gait, dyskinesia,* and *ON/OFF fluctuations.* The *non-motor* aspect of patient's quality of life is determined using *daytime sleepiness, impulsivity, depression, hallucinations,* and *cognitive disorder.* In addition to *motor* and *non-motor* symptoms, physicians also consider epidemiological symptoms which include *age, employment, living alone,* and *disease duration.* According to the collaborating clinicians, physicians are more inclined to change the therapy of younger patients (younger than 65[1]), who are still active, who live alone, and for the patients diagnosed with Parkinson's disease for a shorter time (less than 8 years), where physicians will try more changes to the therapy in order to find the most suitable therapy, rather than therapy prolongation with increased medications dosage.

In modifying the patient's medications treatment based on numerical evaluation of symptoms, the physicians decide whether the symptom is *problematic* and needs their immediate attention or not. Table 1 presents the *motor* and *non-motor* attributes influencing the physicians' decisions for medications modifications, the data sets they are part of, and the intervals of values that are considered *normal* or *problematic* for Parkinson's disease patients. For example, the value of *tremor* is defined as the mean value of all questions concerning tremor from MDS-UPDRS Part II and Part III. Intervals of *normal* and *problematic* values are determined by the clinical expert. For all UPDRS items value 0 is normal and value 1 is a slight or minor symptom. Value 2 is mild, 3 is moderate and 4 is severe. Thus, in most cases and given the progressive nature of PD, for most symptoms 0 and 1 are not problematic and are baring for the patients, but become annoying and hampering when they progress in the range 2–4, which leads to distinguishing values *normal* and *problematic* [6].

## 3.2 PPMI Concomitant Medications Log

The PPMI data collection offers information about all of the concomitant medications that the patients used during their involvement in the study. We concentrate on whether a patient receives a therapy with antiparkinson medications and which combination of antiparkinson medications she received between consecutive time points when the MDS-UPDRS and MoCA test were administered. The three main families of drugs used for treating motor symptoms are levodopa, dopamine agonists and MAO-B inhibitors [12].

---

[1] Retirement age for men (https://en.wikipedia.org/wiki/Retirement_age).

**Table 1.** Description of *motor* (upper part) and *non-motor* (lower part) symptoms used by Parkinson's disease physicians in everyday practice to estimate patient's quality of live. The values intervals (*normal* and *problematic*) are defined by the clinician.

| Symptom | Data set | Question | Normal values interval | Problematic values interval |
|---|---|---|---|---|
| *bradykinesia* | MDS-UPDRS Part III | 3.14 | 0–1 | 2–4 |
| *tremor* | MDS-UPDRS Part II and III | Mean value | 0 | 1–4 |
| *gait* | MDS-UPDRS Part III | 3.10 | 0–1 | 2–4 |
| *dyskinesia* | MDS-UPDRS Part IV | 4.3 | 0–1 | 2–4 |
| *ON/OFF fluctuations* | MDS-UPDRS Part IV | 4.5 | 0 | 1–4 |
| *daytime sleepiness* | MDS-UPDRS Part I | 1.8 | 0–1 | 2–4 |
| *impulsivity* | QUIP | SUM | 0–1 | $\geq 2$ |
| *depression* | MDS-UPDRS Part I | 1.3 | 0–1 | 2–4 |
| *hallucinations* | MDS-UPDRS Part I | 1.2 | 0–1 | 2–4 |
| *cognitive disorder* | MoCA | SUM | 26–30 | <26 |

### 3.3 Experimental Data Set

In our work we consider the above mentioned *motor*, *non-motor*, and *epidimilogical* symptoms, with the exception of *employment*, *living alone*, and *cognitive disorder*. The PPMI data collection does not have data about patients' employment and living arrangements. We decided to omit the *cognitive disorder* attribute due to its present values, which were either *normal* or missing.

For each patient in the data set, the *motor* and *non-motor* symptoms data were obtained and updated periodically (on each patient's visit to the clinician's), providing the clinicians with the opportunity to follow the development of the disease. The data set contains 897 instances, containing information about 368 PPMI patients. Most of the considered patients have data records about two or three visits to the clinician. The maximum number of considered visits is 4.

## 4    Methodology

Our goal is to support physicians in their decisions regarding the patients' therapies. The physicians have several groups of medications at their disposal with which they try to preserve good quality of patient's life. They use and switch between different groups of drugs and their dosages to treat different symptoms (i.e. levodopa is used for *motor* symptoms), and also to prevent overuse of any specific drug in order to reduce side-effects and undesired drug interactions. Our multitask approach based on Predictive Clustering Trees (PCTs) [1] allows for modeling all medication groups simultaneously. By simultaneously predicting several target variables, the physicians are able to observe the interactions between different groups of medications, which is not possible with univariate models. We use predictive clustering trees on time-stamped symptoms and medications data. Figure 1 outlines the proposed five-step methodology, which uses symptoms data over time (i.e. over several patient's visits) and respective changes

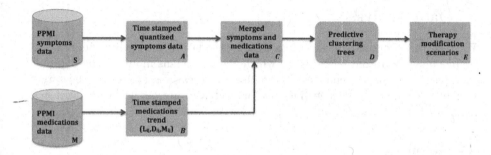

**Fig. 1.** Outline of the methodology for determining medications change scenarios in PPMI data using predictive clustering trees.

in medications therapies. Our goal is to identify symptoms scenarios for which physicians are to consider modifications of therapies.

The input to the methodology are PPMI data sets of patient symptoms (described in Sect. 3.1) and the PPMI medications log data set (described in Sect. 3.2). The output of the methodology are models of patients' symptoms for which particular changes of medications were made.

In step $A$ we construct a time stamped symptoms data set consisting of the symptoms attributes described in Sect. 3.1. This data set consists of patient-visit pairs $(p_i, v_{ij})$ and is based on the patients and their visits to the clinician.

In step $B$ we construct a data set of medications changes which are represented with $(p_i, m_{ij}, m_{ij+1})$ tuples, where $m_{ij}$ and $m_{ij+1}$ are medication therapies of patient $p_i$ in two consecutive visits, $v_{ij}$ and $v_{ij+1}$. A patient receives a therapy which is any combination of levodopa, dopamine agonists, and MAO B inhibitors. For each of the three medications groups, we determine whether its dosage in the time of visit $v_{ij+1}$ has changed (*increased* or *decreased*) or remained unchanged with respect to the dosage at visit $v_{ij}$. The output of step $B$ is a data set of medications changes, presented as tuples $(L_{ij}, D_{ij}, M_{ij})$, indicating whether between visits $v_{ij}$ and $v_{ij+1}$ a change of dosage in levodopa, dopamine agonist, or MAO B inhibitors took place.

In step $C$ we concatenate the data sets obtained in steps $A$ and $B$ into a merged data set of symptoms and medications data. We use patient-visit pairs $(p_i, v_{ij})$ describing patient's symptoms at visit $v_{ij}$ and the changes of medications in the same visit with respect to next visit $v_{ij+1}$. This data set consists of eleven attributes (Sect. 3.3), describing the condition of the patient, and three attributes (levodopa, dopamine agonists, and MAO B) indicating the changes in their dosage, respectively.

The merged data set is used in step $D$ to determine scenarios for medications changes. The three medications groups are used as multi-target variables (multiple classes) in the predictive clustering trees learning approach. We want to determine which symptoms influence decisions of physicians to modify the therapies that the patients receive. The discovered therapy modifications scenarios are analyzed by the physician in step $E$.

## 5    Data Analysis

We test the proposed methodology using the data set described in Sect. 3.3 by modeling symptom patterns for which the physicians modified the patients' therapies. We use the changes of the three medications groups as the target variables. Changes in dosage (increase or decrease) are labeled with label *yes*, while unchanged drug dosages are labeled with label *no*.

### 5.1    Results

A pruned multi-target PCT-based decision tree model for modification of medications based on the patient's status is shown in Fig. 2. The proposed multi-task approach models the dosage changes of all three antiparkinson medication groups simultaneously, allowing for the detection of drug interactions based on the patient's status.

The leaves of the predictive clustering tree hold information about the proposed therapy modifications. The components of the lists presented in each tree leaf predict dosage changes of levodopa, dopamine agonists, and MAO B inhibitors, respectively. For example, the list presented in the first leaf from left to right (Path 1), [yes, yes, yes], indicates that the dosages of levodopa, dopamine agonists, and MAO B changed. The instances (patient-visit pairs) influencing

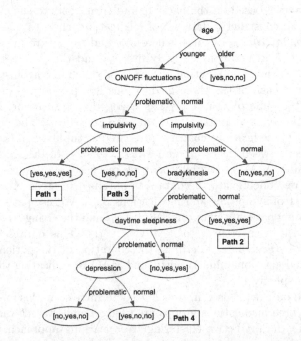

**Fig. 2.** Pruned predictive clustering tree modeling dosage changes for three groups of medications. The tree should be read in the direction from the *leaf* to the *root* of the tree. Medication dosage changes are modeled by patients' symptoms.

this change are presented along the path from the tree root to the respective leaf. In this example, these are the patients who are younger than 65, have problems with ON/OFF fluctuations, and have problems with their impulsivity. Leaf [yes, no, no] on the right (Path 3) suggests that physicians shall only consider changes in levodopa. These dosage changes can be justified by patients' symptoms, i.e. patients have problems with ON/OFF fluctuations and no problems with their impulsivity. Moreover, in younger patients with other problematic symptoms: impulsivity, bradykinesia and daytime sleepiness, the physicians will also change only levodopa dosages (Path 4 in Fig. 2). This might also reflect the current clinical practice—many patients want treatment of their motor symptoms first.

Path 2 in Fig. 2 shows that for younger patients with problematic impulsivity and even without problems with ON/OFF fluctuations and bradykinesia, physicians will change the dosages of all three medication groups. These results are in accordance with the literature on Parkinson's disease [19] and were confirmed by the clinical expert.

The testing results reveal that if the patient experiences ON/OFF fluctuations problems (left subtree in Fig. 2), physicians will react with the change of dosage of her levodopa medications [4]. If the patients experience non-motor symptoms (e.g., impulsivity, depression), physicians will react by modifying the dosages of dopamine agonists [17]. This is in accordance with the literature on Parkinson's disease and was confirmed by the expert. Increased dosages of dopamine agonists can produce non-motor related side-effects. Physicians will react by lowering the dosage of dopamine agonists (consequently increasing the dosage of levodopa). This was revealed in our post analysis, where we followed the actual changes of levodopa and dopamine agonists.

While prediction is not the ultimate goal of the developed methodology, reasonably high classification accuracy increases the clinician's trust in using the model. Table 2 presents the classification accuracy of the default model, the accuracy of the multitask classification model, and the classification accuracy of single-task models for each of the medications groups. These results are obtained by 10-fold cross validation. The default classification accuracy is obtained by always predicting the majority class. For the multitask learning classifier we present the results for each learning task separately. Single-task models are constructed and tested for each medications group separately.

**Table 2.** Comparison of the classification accuracies obtained by the default model, the pruned multitask model and the pruned single-task models.

| Medications group | Default model | Multitask model | Single-task model |
|---|---|---|---|
| Levodopa | 0.637 | 0.685 | 0.686 |
| Dopamine agonists | 0.435 | 0.642 | 0.642 |
| MAO B | 0.518 | 0.602 | 0.586 |

The multitask learning PCT approach produces models with comparable accuracy to the single-task PCT approach. The advantage of using the multitask approach is the ability to observe the interaction between the targets simultaneously.

## 6    Conclusions

This paper presents the methodology developed to detect trigger symptoms for change of medications therapy of Parkinson's disease patients. We consider trigger symptoms to be the ones which press the physicians to make modifications of the treatment for their patients. We tested the methodology on a chosen subset of time-stamped PPMI data. The data set offers an insight into the patients' symptoms progression through time, as well as the response of physicians following problematic states of either motor or non-motor symptoms. We identify clinically confirmed patients' symptoms indicating the need for medication changes.

The proposed approach allows the identification of patient subgroups for which certain medications modifications have either a positive or a negative effect. By post analysis of the patients who respond well to the medications modification and those who do not, and the underlying characteristics of each group, we may be able to assist the physicians with the therapy modifications for a given patient by narrowing the number of possible scenarios. In future work we will focus on describing these subgroups of patients and present therapy change scenarios which have a positive influence on the control of symptoms and also on therapy scenarios which are more likely to lead to side effects.

**Acknowledgements.** This work was supported by the PD_manager project, funded within the EU Framework Programme for Research and Innovation Horizon 2020 grant 643706. We acknowledge the financial support from the Slovenian Research Agency (research core fundings No. P2-0209 and P2-0103). This research has received funding from the European Unions Horizon 2020 research and innovation programme under grant agreement No. 720270 (HBP SGA1). The data used in the preparation of this article were obtained from the Parkinson's Progression Markers Initiative (PPMI) database (www.ppmi-info.org/data). For up-to-date information on the study, visit www.ppmi-info.org. PPMI—a public-private partnership—is funded by the Michael J. Fox Foundation for Parkinson's Research and funding partners. List of funding partners can be found at www.ppmi-info.org/fundingpartners.

## References

1. Blockeel, H., Raedt, L.D., Ramon, J.: Top-down induction of clustering trees. In: Proceedings of the Fifteenth International Conference on Machine Learning (ICML), pp. 55–63 (1998)
2. Caruana, R.: Multitask learning. Mach. Learn. **28**(1), 41–75 (1997)
3. Foltynie, T., Brayne, C., Barker, R.A.: The heterogeneity of idiopathic Parkinson's disease. J. Neurol. **249**(2), 138–145 (2002)
4. Fox, S.H., Katzenschlager, R., Lim, S.-Y., Ravina, B., Seppi, K., Coelho, M., Poewe, W., Rascol, O., Goetz, C.G., Sampaio, C.: The movement disorder society evidence-based medicine review update: treatments for the motor symptoms of Parkinson's disease. Mov. Disord. **26**(S3), S2–S41 (2011)

5. Gil, D., Johnson, M.: Diagnosing Parkinson by using artificial neural networks and support vector machines. Glob. J. Comput. Sci. Technol. **9**(4), 63–71 (2009)
6. Goetz, C.G., Tilley, B.C., Shaftman, S.R., Stebbins, G.T., Fahn, S., Martinez-Martin, P., Poewe, W., Sampaio, C., Stern, M.B., Dodel, R., et al.: Movement disorder society-sponsored revision of the unified Parkinson's disease rating scale (MDS-UPDRS): scale presentation and clinimetric testing results. Mov. Disord. **23**(15), 2129–2170 (2008)
7. Holzinger, A.: Trends in interactive knowledge discovery for personalized medicine: cognitive science meets machine learning. IEEE Intell. Inform. Bull. **15**(1), 6–14 (2014)
8. Hughes, A.J., Daniel, S.E., Kilford, L., Lees, A.J.: Accuracy of clinical diagnosis of idiopathic Parkinson's disease: a clinico-pathological study of 100 cases. J. Neurol. Neurosurg. Psychiatry **55**(3), 181–184 (1992)
9. Lewis, S., Foltynie, T., Blackwell, A., Robbins, T., Owen, A., Barker, R.: Heterogeneity of Parkinson disease in the early clinical stages using a data driven approach. J. Neurol. Neurosurg. Psychiatry **76**(3), 343–348 (2005)
10. Ma, L.-Y., Chan, P., Gu, Z.-Q., Li, F.-F., Feng, T.: Heterogeneity among patients with Parkinson's disease: cluster analysis and genetic association. J. Neurol. Sci. **351**(1), 41–45 (2015)
11. Marek, K., Jennings, D., Lasch, S., Siderowf, A., Tanner, C., Simuni, T., Coffey, C., Kieburtz, K., Flagg, E., Chowdhury, S., et al.: The Parkinson's progression markers initiative (PPMI). Prog. Neurobiol. **95**(4), 629–635 (2011)
12. National Collaborating Centre for Chronic Conditions. Parkinson's Disease: National Clinical Guideline for Diagnosis and Management in Primary and Secondary Care. Royal College of Physicians, London (2006)
13. Patel, S., Lorincz, K., Hughes, R., Huggins, N., Growdon, J., Standaert, D., Akay, M., Dy, J., Welsh, M., Bonato, P.: Monitoring motor fluctuations in patients with Parkinson's disease using wearable sensors. IEEE Trans. Inf Technol. Biomed. **13**(6), 864–873 (2009)
14. Ramani, R.G., Sivagami, G.: Parkinson disease classification using data mining algorithms. Int. J. Comput. Appl. **32**(9), 17–22 (2011)
15. Reijnders, J., Ehrt, U., Lousberg, R., Aarsland, D., Leentjens, A.: The association between motor subtypes and psychopathology in Parkinson's disease. Parkinsonism Relat. Disord. **15**(5), 379–382 (2009)
16. Riviere, C.N., Reich, S.G., Thakor, N.V.: Adaptive Fourier modeling for quantification of tremor. J. Neurosci. Methods **74**(1), 77–87 (1997)
17. Seppi, K., Weintraub, D., Coelho, M., Perez-Lloret, S., Fox, S.H., Katzenschlager, R., Hametner, E.-M., Poewe, W., Rascol, O., Goetz, C.G., et al.: The movement disorder society evidence-based medicine review update: treatments for the non-motor symptoms of Parkinson's disease. Mov. Disord. **26**(Suppl. S3), S42–S80 (2011)
18. Timmer, J., Gantert, C., Deuschl, G., Honerkamp, J.: Characteristics of hand tremor time series. Biol. Cybern. **70**(1), 75–80 (1993)
19. Weintraub, D., Koester, J., Potenza, M.N., Siderowf, A.D., Stacy, M., Voon, V., Whetteckey, J., Wunderlich, G.R., Lang, A.E.: Impulse control disorders in Parkinson disease: a cross-sectional study of 3090 patients. Arch. Neurol. **67**(5), 589–595 (2010)

# Change-Point Detection Method for Clinical Decision Support System Rule Monitoring

Siqi Liu[1](✉), Adam Wright[2], and Milos Hauskrecht[1]

[1] Department of Computer Science, University of Pittsburgh, Pittsburgh, USA
siqiliu@cs.pitt.edu, milos@pitt.edu
[2] Brigham and Women's Hospital and Harvard Medical School, Boston, USA
awright@bwh.harvard.edu

**Abstract.** A clinical decision support system (CDSS) and its components can malfunction due to various reasons. Monitoring the system and detecting its malfunctions can help one to avoid any potential mistakes and associated costs. In this paper, we investigate the problem of detecting changes in the CDSS operation, in particular its monitoring and alerting subsystem, by monitoring its rule firing counts. The detection should be performed online, that is whenever a new datum arrives, we want to have a score indicating how likely there is a change in the system. We develop a new method based on Seasonal-Trend decomposition and likelihood ratio statistics to detect the changes. Experiments on real and simulated data show that our method has a lower delay in detection compared with existing change-point detection methods.

## 1 Introduction

A clinical decision support system (CDSS) is a complex computer-based system aimed to assist clinicians in patient management [10]. It consists of multiple interconnected components. The monitoring and alerting component of the CDSS is used to encode and execute expert defined rules that monitor the patient related information. If the rule condition is satisfied, an alert or a reminder is raised and presented to a physician either via email or a pop-window. Examples are alerts on pneumococcal vaccination for the patient at risk, regular yearly checkups, or an alert for a occurrence of some adverse event. In general, the monitoring and alerting component of the CDSS provides a safety net for many clinical and patient conditions that may be missed in the regular clinical workflow.

The monitoring and alerting component is integrated in the CDSS, and it draws information from both the patient's electronic health record (EHR) and other CDSS components. As a result any changes in the information stored in the EHR (variable coding or terminology changes) or system updates made to other components of the system may affect its intended function [22]. Moreover, the rules in the CDSS are regularly reviewed and updated, and any mistake in the rule logic may change the rule. Hence it is critical to assure that the alerting system and its rules continue to function as intended. Our objective is to develop methods that are able to monitor and detect changes in alert rule behaviors so that any serious misbehavior or an error can be quickly identified and corrected.

© Springer International Publishing AG 2017
A. ten Teije et al. (Eds.): AIME 2017, LNAI 10259, pp. 126–135, 2017.
DOI: 10.1007/978-3-319-59758-4_14

To accurately detect the changes in the alerting component of the CDSS, it would be ideal to have measurements on many different aspects of the system, but in reality, it is not feasible to collect such data. In our case, all we have is the daily firing counts of different rules in the CDSS. A typical solution to detect the changes in the mean of the time series is to apply change-point detection methods [3]. However, rule firing count data may be subject to many different sources of variation that influence the data readings and consequently the performance of the change-point methods. Some of these sources may be identifiable. An example is weekly signal variation where the different days of the week influence the rule firing counts. Additionally, many possible sources are unknown or not accounted for in observed data. For example, sudden changes in the population of patients screened by the rules may cause an increase or a decrease in the numbers of alerts. As a result, it is challenging to develop change-point methods that can distinguish real changes from noise or a natural signal variation given such limited information.

Another challenge for designing an accurate change-point detector for CDSS rule monitoring is that it should run in an online (or sequential), instead of offline (or retrospective), mode. In retrospective analysis, the detector has access to the whole collection of data, and tries to find all changes that occurred in the past. In this case, it usually has enough data both before and after the point to decide whether it is a change-point. In contrast, in online detection, there is a time limit for the detector to make a decision on whether there is a change. For example, detecting a malfunction in the system after 6 months is not very helpful. Given this constraint, only limited amount of recent/past data are available when the decision is made, making the detection even more difficult.

To address the above challenges, we develop and test a new change-point detection method based on Seasonal-Trend (STL) decomposition [4] and likelihood ratio statistics. Like existing change-point detection methods, our method gives a score at each time point, indicating how likely a change has occurred. The major advantages of our method are that it accounts for periodic (or seasonal) variation in the data before calculating the statistics, and that the statistical models for calculating the scores are robust to the remaining noise. Therefore, it is able to calculate more accurate scores that are more likely to indicate true changes in the time series.

## 2    Method

The firing counts of each CDSS rule form a univariate time series. Our goal is to detect (in real time) changes in the behavior of the time series, in particular, its mean. Our detection framework is based on a sliding window, that is, at each time point, it looks back a constant amount of time, referred to as a window. All analysis is done only on the data within the window. We use the sliding window to restrict our attention to recent data, because the time series are noisy and may drift over time, that is, they exhibit a nonstationary behavior. In such a case, old data can add bias to the inference on recent data. The sliding window not only deals with nonstationary behaviors, but also reduces the computational cost of the algorithm, so that it is suitable for online detection. For the data within the window, we perform several steps to get the final output, the score that reflects how significant the change in time series behavior is. They are described in the following sections.

## 2.1   Data Transformation to Stabilize Variance

Our data are counts and show heteroscedasticity, that is, the variance changes with the mean. Therefore, we apply the square-root transformation ($f(x) = \sqrt{x + 0.5}$) to stabilize the variance [1], which is commonly used for Poisson distribution.

## 2.2   Seasonal-Trend Decomposition

The rule firing time-series often show strong seasonal variation reflecting typical workflow and types of patients screened by the CDSS on different days of the week. For example, between every Friday and Saturday, the mean of the rule firing counts typically drops, but this is normal, and can be simply explained by a fewer number of patient visits on weekends. However, this weekly variation may negatively affect the change-point inferences. Our solution is to account for and systematically remove this variation as much as possible prior to change-point inference.

Seasonal-Trend (STL) decomposition [4] assumes that a time series is a sum of three component signals: seasonal (periodic signal), trend (long-term low-frequency signal), and remainder (noise). It can separate out these components from the original time series by nonparametric smoothing with locally weighted regression, or LOESS [5,6]. See Fig. 1 for an example of STL. An advantage of

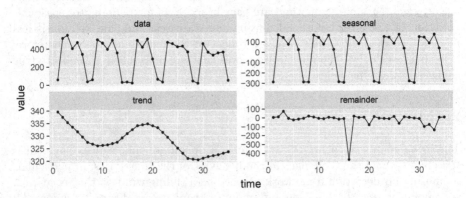

**Fig. 1.** Seasonal-Trend (STL) decomposition of a time series. The top-right graph shows the original data, which has a strong (weekly) seasonality. The following graphs show the seasonal, trend, and remainder signals decomposed from the original time series. Notice the point at time 16 is an outlier.

STL is that its seasonal and trend components are robust to outliers, which are isolated data points with aberrant values that do not follow the pattern of the majority of the data. We use STL to remove the seasonal signal and use the sum of the trend and remainder for further analysis.

## 2.3   Likelihood Ratio Statistics

Given a set of data points $x = \{x_1, x_2, \ldots, x_n\}$ within a window at time $t$ with the seasonal signal removed, we want to derive a score indicating how likely a change in the mean has occurred in the time span $[t - n + 1, t]$. Ultimately, we want to know if a change has occurred or not (1 or 0), and a score is a continuous quantity representing our belief that a change has occurred. By applying a threshold to the scores, we can convert them to binary labels indicating changes.

To calculate the scores, we formulate the following hypothesis test for each possible change-point $c, 1 < c \leq n$.

$$H_0 : x_i \sim F(\mu_0), 1 \leq i \leq n. \quad H_1 : x_i \sim F(\mu_1), x_j \sim F(\mu_n), 1 \leq i < c \leq j \leq n. \tag{1}$$

$F$ is a distribution family with a parameter for the mean. If we fix $c$, the change-point, then the likelihood ratio statistic would be

$$r_c = \log L_{H_c} - \log L_{H_0}, \tag{2}$$

where $L_{H_0}$ is the maximum likelihood of the sample $x$ under the null hypothesis, and $L_{H_c}$ is under the alternative hypothesis with known $c$. Since we do not know $c$, and instead want to detect whether there is a change at *any* point, the score for the sample $x$ is

$$r_* = \max_{1 < c \leq n} r_c, \tag{3}$$

and the corresponding maximizer is the suspected change-point.

The statistics depend on the distribution family $F$. Because the data can be quite noisy and contain outliers, we use Student's t-distribution to model the data. Specifically, the probability density function (PDF) is

$$p(x|\nu, \mu, \sigma^2) = \frac{\Gamma(\frac{\nu+1}{2})}{\Gamma(\frac{\nu}{2})\sqrt{\pi\nu\sigma^2}} \left(1 + \frac{(x-\mu)^2}{\nu\sigma^2}\right)^{-\frac{\nu+1}{2}}, \tag{4}$$

where $\nu$ is the degrees of freedom, $\mu$ is the location, and $\sigma^2$ is the scale.

We consider $\nu$ as given and only estimate $\mu$ and $\sigma^2$. For t-distributions, the maximum likelihood estimators (MLEs) do not have a closed-form solution, so we follow [15] and develop an EM algorithm for estimating the parameters under either the null or the alternative hypothesis. The EM algorithm is based on an equivalent form of the distribution as an infinite mixture of Gaussians, which includes an additional hidden variable $\tau$:

$$\tau \sim \text{Gamma}(\nu/2, \nu/2), \quad x \sim N(\mu, \sigma^2/\tau), \tag{5}$$

where the parameters of the Gamma distribution are shape and rate. The marginal distribution of $x$ in Eq. 5 is the t-distribution in Eq. 4.

Based on the above, for the null hypothesis, the EM algorithm is as follows. The E-step is

$$w_i = \mathrm{E}[\tau_i | x_i, \nu, \mu_0, \sigma^2] = \frac{\nu + 1}{\frac{(x_i - \mu_0)^2}{\sigma^2} + \nu}. \tag{6}$$

The M-step is

$$\mu_0 = \frac{\sum_i w_i x_i}{\sum_i w_i}, \quad \sigma^2 = \frac{\sum_i w_i (x_i - \mu)^2}{n}. \tag{7}$$

We alternate between the E-step and M-step till convergence, use the final values of $\mu_0$ and $\sigma^2$ as the MLEs.

For the alternative hypothesis, the E-step is almost the same as Eq. 6, except that $\mu_0$ is replaced by $\mu_1$ or $\mu_n$ depending on whether $i < c$ or not. The M-step is

$$\mu_1 = \frac{\sum_{i<c} w_i x_i}{\sum_{i<c} w_i}, \quad \mu_n = \frac{\sum_{i \geq c} w_i x_i}{\sum_{i \geq c} w_i},$$

$$\sigma^2 = \frac{\sum_{i<c} w_i (x_i - \mu_1)^2 + \sum_{i \geq c} w_i (x_i - \mu_n)^2}{n}. \tag{8}$$

## 2.4  Further Improvements

The data are counts, and even with transformation, low counts are problematic, because the variance is too low. To improve the performance, we add a small noise to the data. Specifically, for every point $x$ in the time series after transformation, we add a noise as

$$x' = x + \epsilon, \quad \epsilon = u - 0.5, \quad u \sim Beta(a, a). \tag{9}$$

We use a (symmetric) beta-distribution, so the size of the noise is within control, $\epsilon \in [-0.5, 0.5]$, and the mean of $\epsilon$ is 0.

The second improvement is based on the following observation. When calculating the likelihood ratio statistics, if say $c = 2$ or $n$, only one point is used for estimating $\mu_1$ or $\mu_n$, so the sample size is small. But our data contain outliers, which can bias the inference especially when the sample size is small. We can make sure the sample size is always greater than $l$ by restricting $l < c \leq n - l + 1$, but an obvious drawback is that the expected delay of the detection would increase. However, noticing that new data always come from the right of the sliding window, and usually the change can be detected quickly, we restrict $l < c \leq n$ instead, so the sample size for estimating $\mu_1$ is at least $l$, while the delay of the detection is not affected at all, if without the restriction it would be detected within $n - l$ observations after the change.

# 3  Experiments

## 3.1  Experiment Design

We test our framework and compare it to alternative methods on rule firing counts from a large teaching hospital collected over a period of approximately 5 years [22]. We run and evaluate the methods by considering both (1) known and (2) simulated changes in their time series.

In the first part of the experiment we use 14 CDSS rules with a total of 22 labeled change-points. These reflect known changes in the rule logic, or confirmed changes in the firing rates due to various issues. In the second part we simulate changes on the existing rule firing counts to help us analyze the sensitivity of the methods to the magnitude of the changes. We use the firing counts of 4 CDSS rules with no known change-points, and simulate change-points on these data by randomly sampling 10 segments of length 240 per rule and simulating a change in the middle of these segments. We simulate the change at time $c$ in time series $x$ by changing the values as $x'_i = \lambda x_i, i \geq c$. In different experiments, we set $\lambda$ to 2/1, 3/2, 6/5, 1/2, 2/3, and 5/6 respectively, to cover both increasing and decreasing changes in different sizes. The final values $x_i$ are rounded, so they are still nonnegative integers consistent with counts. We use multiplicative instead of additive changes, because the data are counts and have heteroscedasticity.

We use AMOC curves [7] to evaluate the performance of the methods. In general, a change-point can be detected within an acceptable delay. Meanwhile, normal points can be falsely detected, resulting in false positives. In an AMOC curve, the delay of a detection is plotted against the false positive rate (FPR) by varying the threshold on the scores. If a change is not detected at all, a penalty is used as the delay. In our experiments, the maximum delay is 13, which is related to the sliding window size explained later, and the penalty is 14. The first 140 points in each time series are used as a warm-up, and no scores are produced.

We compare the following methods:

- RND: a baseline that gives uniformly sampled scores.
- SCP: single change-point detection method for normal distribution [13].
- MW: a method based on Mann-Whitney nonparametric statistics [18].
- Pois: a method based on Poisson likelihood ratios [3].
- NDT1: our method without restricting $l < c$.
- NDT2: our method with restricting $l < c$, where $l = 7$.

A window of 14 is used for change detection, while a window of 140 is for STL. The square-root transformation is also used for SCP.

We use the robust STL implemented in R [19] and set the period to 7 (a week) and $s.window = 7$ (even smaller values are not recommended [4]). Default values are used for other parameters. $\nu = 3$ for Eqs. 4 and 5. $a = 1$ for Eq. 9.

## 3.2  Results on Data with Known Change-Points

AMOC curves on the real data are shown in Fig. 2a. Notice that our methods dominate all the other methods almost everywhere, meaning for almost any given FPR, our methods have a lower delay in detection compared with the others. Comparing NDT1 and NDT2, we notice NDT2 is better, especially when the FPR is low, showing the effectiveness of the second improvement. We also did experiments for NDT without any improvements in Sect. 2.4, and the performance (not listed) is worse than NDT1.

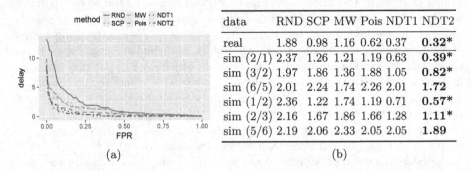

| data | RND | SCP | MW | Pois | NDT1 | NDT2 |
|------|-----|-----|-----|------|------|------|
| real | 1.88 | 0.98 | 1.16 | 0.62 | 0.37 | **0.32*** |
| sim (2/1) | 2.37 | 1.26 | 1.21 | 1.19 | 0.63 | **0.39*** |
| sim (3/2) | 1.97 | 1.86 | 1.36 | 1.88 | 1.05 | **0.82*** |
| sim (6/5) | 2.01 | 2.24 | 1.74 | 2.26 | 2.01 | **1.72** |
| sim (1/2) | 2.36 | 1.22 | 1.74 | 1.19 | 0.71 | **0.57*** |
| sim (2/3) | 2.16 | 1.67 | 1.86 | 1.66 | 1.28 | **1.11*** |
| sim (5/6) | 2.19 | 2.06 | 2.33 | 2.05 | 2.05 | **1.89** |

(a)                                                    (b)

**Fig. 2.** (a) AMOC curves on real data averaged over all change-points. (b) The mean AUC-AMOC averaged over all change-points. *Wilcoxon tests show that NDT2 significantly (.05) outperforms other methods.

The means of the areas under the AMOC curves (AUC-AMOC) are in Fig. 2b (row 1). These are calculated by treating the data around each change-point as a single example. For each example, the AUC summarizes the AMOC curve by integrating the delay w.r.t. the FPR. These results show that our methods are the best in terms of the overall performance.

## 3.3  Results on Data with Simulated Change-Points

AMOC curves on the simulated data are shown in Fig. 3. They are grouped by experiment settings, that is the fold of the simulated changes (the value of $\lambda$). Each subgraph corresponds to a different fold, shown in the label on the top. A general trend in these graphs is that, as the change gets smaller, all the curves get closer to the random baseline (RND). This reflects that the smaller the change, the harder to detect it (in time). But except when the change is at the smallest setting (the last column of the graphs), our methods dominate the other methods almost everywhere by a noticeable margin.

Figure 2b (row 2–7) shows the mean AUC-AMOC for different folds of changes ($\lambda$). NDT2 performs the best in all cases, although when $\lambda = 6/5, 5/6$, the difference is not significant.

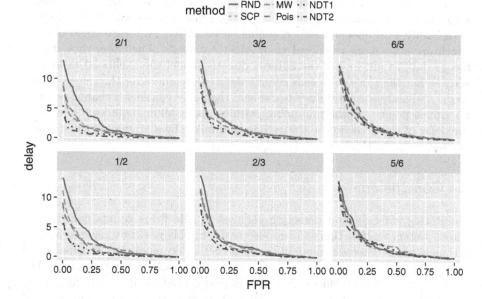

**Fig. 3.** AMOC curves on simulated data averaged over all change-points. The label on top of each subgraph indicates the fold of the changes ($\lambda$).

# 4  Related Work and Discussion

Although change-point detection has been studied by statisticians for a long time, most work focused on offline detection. For example, Sen and Srivastava [20] study likelihood ratio test for detecting changes in mean for normal distributions. Pettitt [18] proposes a nonparametric statistics for detecting changes. Killick et al. [14] and Frizlewicz [9] improve efficiency in detecting multiple change-points. An exception is work in quality control (e.g., CUSUM [17]), where tests are performed in a sequential manner to detect errors in real time. However, these methods usually need a reference value, so it is hard to apply them to nonstationary time-series data like ours. Furthermore, all the methods assume the data are independent and identically distributed. The assumption does not hold for our data.

Research in time-series outlier detection is also related. Fox [8] defines and studies two types of outliers. Tsay [21] extends them to four types. These concepts are defined in terms of ARIMA (autoregressive integrated moving average) models. Chen and Liu [2] improve upon the previous work by jointly estimating model parameters and outlier effects. Although some types of outliers correspond to change-points, their work has some limitations. First, they assume the model generating the time series is ARIMA. As soon as the time series do not follow ARIMA, their theoretical justifications are gone, and some algorithms may not even work properly without modifications. Second, they all deal with offline detection, so the algorithms are usually inefficient for online detection.

In the data mining community, there is some work addressing online change-point detection. Yamanishi and Takeuchi [23] propose a framework for detecting both additive outliers and change-points based on AR (autoregressive) models, which are even more restricted than ARIMA models. Therefore, they do not fit our data at all. In [12] the authors directly model the likelihood ratio with kernels, but their method needs enough data before and after the change-point, so they actually solve a different problem: they only consider a change-point to be a fixed point (say the mid-point) within a large sliding window. Therefore, the delay of the detection is always bounded below by a large number, which is not preferable in practice.

## 5    Conclusion

Monitoring a CDSS and detecting changes in rule firing counts can help us detect system malfunctions and reduce costs. In this work, we have developed a change-point detection method based on STL decomposition and likelihood ratio statistics, and two improvements to further boost the performance. The method can be applied efficiently to detect changes in real time. Experiments on real data with both known and simulated changes have shown that our method outperforms traditional change-point detection methods in terms of false positive rate and detection delay.

In the future we plan to test the methodology on hundreds of CDSS rules and study the feasibility of the method in detecting rule firing changes in terms of precision-alert-rate (PAR) curves [11]. In terms of the methodology, our detection methods currently work only with the time-series of rule counts and ignore context information other than the day of the week (accounted for by STL). An interesting open problem is how to add additional covariates into the change-point models that can account for other types of variations similarly to spike detection work in [16].

**Acknowledgement.** This research was supported by grants R01-LM011966 and R01-GM088224 from the NIH. The content of this paper is solely the responsibility of the authors and does not necessarily represent the official views of the NIH.

## References

1. Bartlett, M.S.: The use of transformations. Biometrics **3**(1), 39–52 (1947)
2. Chen, C., Liu, L.M.: Joint estimation of model parameters and outlier effects in time series. J. Am. Stat. Assoc. **88**(421), 284–297 (1993)
3. Chen, J., Gupta, A.K.: Parametric Statistical Change Point Analysis. Birkhäuser Boston, Boston (2012)
4. Cleveland, R.B., Cleveland, W.S., McRae, J.E., Terpenning, I.: STL: a seasonal-trend decomposition procedure based on loess. J. Off. Stat. **6**(1), 3–73 (1990)
5. Cleveland, W.S., Cleveland, W.S.: Robust locally weighted regression and smoothing scatterplots. J. Am. Stat. Assoc. **74**(368), 829–836 (1979)

6. Cleveland, W.S., Devlin, S.J.: Locally weighted regression: an approach to regression analysis by local fitting. J. Am. Stat. Assoc. **83**(403), 596–610 (1988)

7. Fawcett, T., Provost, F.: Activity monitoring: noticing interesting changes in behavior. In: ACM SIGKDD International Conference on Knowledge Discovery and Data Mining, vol. 1, pp. 53–62 (1999)

8. Fox, A.J.: Outliers in time series. J. Roy. Stat. Soc.: Ser. B (Methodol.) **34**(3), 350–363 (1972)

9. Fryzlewicz, P.: Wild binary segmentation for multiple change-point detection. Ann. Stat. **42**(6), 2243–2281 (2014)

10. Garg, A.X., Adhikari, N.K.J., McDonald, H., Rosas-Arellano, M.P., Devereaux, P.J., Beyene, J., Sam, J., Haynes, R.B.: Effects of computerized clinical decision support systems on practitioner performance and patient outcomes: a systematic review. JAMA **293**(10), 1223–1238 (2005)

11. Hauskrecht, M., Batal, I., Hong, C., Nguyen, Q., Cooper, G.F., Visweswaran, S., Clermont, G.: Outlier-based detection of unusual patient-management actions: an ICU study. J. Biomed. Inform. **64**, 211–221 (2016)

12. Kawahara, Y., Sugiyama, M.: Change-point detection in time-series data by direct density-ratio estimation. In: SIAM International Conference on Data Mining, pp. 389–400. Society for Industrial and Applied Mathematics, April 2009

13. Killick, R., Eckley, I.: changepoint: an R package for changepoint analysis. J. Stat. Softw. **58**(3), 1–19 (2014)

14. Killick, R., Fearnhead, P., Eckley, I.A.: Optimal detection of changepoints with a linear computational cost. J. Am. Stat. Assoc. **107**(500), 1590–1598 (2012)

15. Liu, C., Rubin, D.B.: ML estimation of the t distribution using EM and its extensions. ECM ECME. Stat. Sin. **5**, 19–39 (1995)

16. Liu, S., Wright, A., Hauskrecht, M.: Online conditional outlier detection in non-stationary time series. In: FLAIRS Conference (2017)

17. Page, E.S.: Continuous inspection schemes. Biometrika **41**(1/2), 100–115 (1954)

18. Pettitt, A.N.: A non-parametric approach to the change-point problem. J. Roy. Stat. Soc.: Ser. C (Appl. Stat.) **28**(2), 126–135 (1979)

19. R Core Team: R: A Language and Environment for Statistical Computing. R Foundation for Statistical Computing, Vienna (2016)

20. Sen, A., Srivastava, M.S.: On tests for detecting change in mean. Ann. Stat. **3**(1), 98–108 (1975)

21. Tsay, R.S.: Outliers, level shifts, and variance changes in time series. J. Forecast. **7**(May 1987), 1–20 (1988)

22. Wright, A., Hickman, T.T.T., McEvoy, D., Aaron, S., Ai, A., Andersen, J.M., Hussain, S., Ramoni, R., Fiskio, J., Sittig, D.F., Bates, D.W.: Analysis of clinical decision support system malfunctions: a case series and survey. J. Am. Med. Inform. Assoc. **23**(6), 1068–1076 (2016)

23. Yamanishi, K., Takeuchi, J.: A unifying framework for detecting outliers and change points from non-stationary time series data. In: ACM SIGKDD International Conference on Knowledge Discovery and Data Mining, pp. 676–681. ACM (2002)

# Discovering Discriminative and Interpretable Patterns for Surgical Motion Analysis

Germain Forestier[1,2]([⊠]), François Petitjean[2], Pavel Senin[3],
Fabien Despinoy[4], and Pierre Jannin[4]

[1] MIPS, University of Haute-Alsace, Mulhouse, France
**germain.forestier@uha.fr**
[2] Faculty of Information Technology, Monash University, Melbourne, Australia
[3] Los Alamos National Laboratory, Los Alamos, NM 87545, USA
[4] INSERM MediCIS, Unit U1099 LTSI, University of Rennes 1, Rennes, France

**Abstract.** The analysis of surgical motion has received a growing interest with the development of devices allowing their automatic capture.
In this context, the use of advanced surgical training systems make an
automated assessment of surgical trainee possible. Automatic and quantitative evaluation of surgical skills is a very important step in improving
surgical patient care. In this paper, we present a novel approach for the
discovery and ranking of discriminative and interpretable patterns of surgical practice from recordings of surgical motions. A pattern is defined
as a series of actions or events in the kinematic data that together are
distinctive of a specific gesture or skill level. Our approach is based on
the discretization of the continuous kinematic data into strings which
are then processed to form bags of words. This step allows us to apply
discriminative pattern mining technique based on the word occurrence
frequency. We show that the patterns identified by the proposed technique can be used to accurately classify individual gestures and skill
levels. We also present how the patterns provide a detailed feedback on
the trainee skill assessment. Experimental evaluation performed on the
publicly available JIGSAWS dataset shows that the proposed approach
successfully classifies gestures and skill levels.

**Keywords:** Surgical motion analysis · Skill assessment · Pattern mining

## 1 Introduction

In recent years, analysis of surgical motion has received a growing interest following the development of devices enabling automated capture of surgeon motions
such as tracking, robotic and training systems. Surgical training programs now
often include surgical simulators which are equipped with sensors for automatic
surgical motions recording [1,2]. The ability to collect surgical motion data brings
unprecedented opportunities for automated objective analysis and assessment of
surgical trainees progression. The main goal of this effort is to support surgeons
in technical skills acquisition, as these are shown to correlate with a reduction

© Springer International Publishing AG 2017
A. ten Teije et al. (Eds.): AIME 2017, LNAI 10259, pp. 136–145, 2017.
DOI: 10.1007/978-3-319-59758-4_15

of patient complications [3]. Hence, automated evaluation of surgical skill level is an important step in surgical patient care improvement.

This article tackles the issue of identifying discriminative and interpretable patterns of surgical practice from recordings of surgical motions. We define a *pattern* a series of actions or events in the kinematic data that together are distinctive of a specific gesture or a skill level. We show, that by using these patterns, we can reach beyond the simple classification of observed surgeons into categories (*e.g.*, expert, novice) by providing a quantitative evidence-supported feedback to the trainee as per where he or she can improve. The proposed approach, based on SAX-VSM algorithm [4], considers surgical motion as continuous multi-dimensional time-series and starts by discretizing them into sequence of letters (*i.e.*, strings) using Symbolic Aggregate approXimation (SAX) [5]. In turn, SAX sequences are decomposed into subsequences of few consecutive letters via sliding window. The relative frequencies of these subsequences, *i.e.*, the number of times they appear in a given sequence or in a set of sequences, are then used to identify discriminative patterns that characterize specific surgical motion. To discover the patterns, we rely on the Vector Space Model (VSM) [6] which has been originally proposed as an algebraic model for representing collection of text documents. The identified discriminative patterns are then used to perform classification by identifying them in to-be-classified recordings. Furthermore, by highlighting discriminative patterns in the visualization of original motion data, we are able to provide an intuitive visual explanation about *why* a specific skill assessment is provided. We evaluated our method on the kinematic data from the JHU-ISI Gesture and Skill Assessment Dataset (JIGSAWS) [7] that is currently the largest publicly accessible database for surgical gesture analysis. Our experiments have shown that the proposed method accurately classifies gestures and skill levels. The main contributions of this paper are:

- A framework for identifying discriminative and interpretable patterns in surgical activity motion based on SAX [5] and VSM [4].
- Experimental evaluation highlighting the relevance of the proposed method for gestures classification and skill assessment.
- A visualization technique enabling self-assessment of trainee skills.

## 2    Background

Surgical motion analysis is mainly based on kinematic data recorded by surgical robot [8,9] and video data [10–12]. Kinematic data usually include multiple attributes such as the position of robot's tools, rotations, and velocities. From such data, significant amount of work has been devoted to the segmentation of surgical tasks into more detailed gestures [13–15]. Segmenting surgical motion into gestures makes it possible to obtain a finer description of surgical task leading to more detailed feedback on skill assessment [16,17]. Previous work concerned with gesture segmentation using kinematic and video data uses Hidden Markov Models [18,19], Conditional Random Fields [20] and Linear Dynamical Systems [11]. Main drawback of these approaches is the difficulty for the trainee to understand the output and to use it as a feedback to improve performance. In contrast, our approach seeks not only to identify that a surgical motion has been performed by a novice surgeon, but also to explain *why* it has been classified as such. This step is critical in justifying the reasons why the trainee is still considered as a novice and to help him or her to focus on the specific steps that require improvement.

# 3   Method

## 3.1   Symbolic Aggregate ApproXimation (SAX)

We propose to use Symbolic Aggregate approXimation (SAX) [5] to discretize the input time series [21]. For time series $T$ of length $n$, SAX obtains a lower-dimensional representation by first performing a z-normalization then dividing the time series into $s$ equal-sized segments. Next, for each segment, SAX computes a mean value and maps it to a symbol according to a pre-defined set of breakpoints dividing the data space into $\alpha$ equiprobable regions, where $\alpha$ is the user specified alphabet size. While dimensionality reduction is a desirable feature for exploring global patterns, the high compression ratio $(n/s)$ significantly affects performance in cases where localized phenomena are of interest. Thus, for the local pattern discovery, SAX is typically applied to a set of subsequences that represent local features – a technique called subsequence discretization [22] which is implemented via a sliding window. Note that other time-series discretization approaches could have been used at this step [23].

## 3.2   Bag of Words Representation of Kinematic Data

Following the approach proposed in [4], a sliding window technique is used to convert a time series $T$ of length $n$ into the set of $m$ SAX words, where $m = (n - l_s) + 1$ and $l_s$ the sliding window length. A sliding window of length $l_s$ is applied across the time series $T$ and the overlapping extracted subsequences are converted into SAX words and then put in a collection. This collection is a *bag of words* representation of the original time series $T$.

In the case of kinematic data, this process is performed independently for each dimension of the data (*e.g.*, $x$ coordinate, $y$ coordinate, etc.). All features are normalized on a per-trial per-feature basis. Each word extracted in each dimension of the data is postfixed with the name of the dimension (*e.g.* $x$, $y$, etc.). We assume that depending of the gesture or the skill level to classify, different kinematic features can be relevant. Note, that this methodology can be used regardless of the available kinematic data (*e.g.* number of features, etc.). Figure 1 illustrates the conversion of kinematic data for one trial into a bag of words using SAX.

## 3.3   Vector Space Model (VSM)

We rely on the original definition of vector space model as it is known in Information Retrieval (IR) [4,6]. The *tf\*idf* weight for a term $t$ is defined as a product

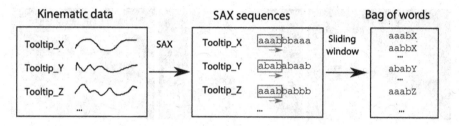

**Fig. 1.** Conversion of kinematic data for one trial into a bag of words using SAX [5] and a sliding window of size 4 (in red). (Color figure online)

of two factors: term frequency (*tf*) and inverse document frequency (*idf*). The first factor corresponds to logarithmically scaled term frequency [24].

$$\text{tf}_{t,d} = \begin{cases} \log(1 + f_{t,d}), & \text{if } f_{f,d} > 0 \\ 0, & \text{otherwise} \end{cases} \tag{1}$$

where $t$ is the term, $d$ is a bag of words (a document in IR terms), and $f_{t,d}$ is the frequency of $t$ in $d$. The inverse document frequency [24] is defined as

$$\text{idf}_{t,D} = \log\frac{|D|}{|d \in D : t \in d|} = \log\frac{N}{\text{df}_t} \tag{2}$$

where $N$ is the cardinality of a corpus $D$ (the total number of classes) and the denominator $\text{df}_t$ is the number of bags where the term $t$ appears. Then, *tf\*idf* weight value for a term $t$ in the bag $d$ of a corpus $D$ is defined as

$$\text{tf*idf}(t, d, D) = \text{tf}_{t,d} \times \text{idf}_{t,D} = \log(1 + f_{t,d}) \times \log\frac{N}{\text{df}_t} \tag{3}$$

for all cases where $f_{t,d} > 0$ and $\text{df}_t > 0$, or zero otherwise.

Once all frequencies are computed, the term frequency matrix becomes the term weight matrix, whose columns are used as *class term weight* vectors to perform classification using Cosine similarity. For two vectors **a** and **b**, the Cosine similarity is based on their inner product and defined as

$$\text{similarity}(\mathbf{a}, \mathbf{b}) = cos(\theta) = \frac{\mathbf{a} \cdot \mathbf{b}}{||a|| \cdot ||b||} \tag{4}$$

### 3.4   Training and Classifying Kinematic Data

The training step starts by transforming the kinematic data into SAX representation using two parameters: the size of the sliding window $l_s$, and the size of the alphabet $\alpha$. Then, the algorithm builds a corpus of $N$ bags corresponding to the subsequences extracted from the $N$ classes of kinematic data, i.e. same skill level or same gesture depending on the application. The *tf\*idf* weighting is then applied to create $N$ real-valued weight vectors of equal length, representing the different class of kinematic data.

**Fig. 2.** Snapshots of the three surgical tasks in the JIGSAWS dataset (from left to right): suturing, knot-tying, needle-passing [7].

In order to classify an unlabeled kinematic data, the method transforms it into a terms frequency vector using exactly the same sliding window and SAX parameters used for the training part. It computes the cosine similarity measure (Eq. 4) between this term frequency vector and the $N$ *tf\*idf* weight vectors representing the training classes. The unlabeled kinematic data is assigned to the class whose vector yields the maximal cosine similarity value.

## 4    Experimental Evaluation

The JIGSAWS dataset [7] includes 8 subjects with 3 different skill levels (novice, intermediate and expert) performing 3–5 trials of three tasks (suturing, knot tying, and needle passing). The Fig. 2 illustrates the three tasks. Each trial lasts about 2 min and is represented by the kinematic data of both master and slave manipulators of the da Vinci robotic surgical system recorded at a constant rate of 30 Hz. Kinematic data consists of 76 motion variables including positions and velocities of both master and slave manipulators. All trials in the JIGSAWS dataset were manually segmented into 15 surgical gestures. Video of the trials are also available and are synchronized with the kinematic data. A detailed description of the dataset is available in [25].

Our training step first transforms the kinematic data time series into SAX representation configured by two parameters: the sliding window length ($l_s$) and SAX alphabet size ($\alpha$). The number of segments per window was kept equal to the length of the window which means that every point of the time series was transformed into a letter. This choice was made to allow us to map back the patterns on the original time series. Parameters $l_s$ and $\alpha$ were optimized using cross-validation on the training data. As they can differ for each specific classification problem, their values are provided along with the experimental results.

### 4.1    Gesture Classification

We considered the gesture boundaries to be known and we used the kinematic data alone. We present results for two cross-validation configurations provided

**Table 1.** Gesture classification performance, assuming known boundaries and using kinematic data only.

| Method | Metric | Leave-one-supertrial-out | | | Leave-one-user-out | | |
|---|---|---|---|---|---|---|---|
| | | Suturing | Needle passing | Knot tying | Suturing | Needle passing | Knot tying |
| | $(l_s,\alpha)$ | (8,19) | (13,18) | (15,7) | (8,19) | (14,18) | (10,12) |
| *Proposed* | Micro | **93.69** | **81.08** | **92.45** | **88.27** | **75.29** | **89.76** |
| | Macro | **79.95** | **74.67** | **89.78** | **68.77** | **67.54** | **82.29** |
| LDS [25] | Micro | 84.61 | 59.76 | 81.67 | 73.64 | 47.96 | 71.42 |
| LDS [25] | Macro | 63.87 | 46.55 | 74.51 | 51.75 | 32.59 | 63.99 |
| HMM [25] | Micro | 92.56 | 75.68 | 89.76 | 80.83 | 66.22 | 78.44 |
| HMM [25] | Macro | 79.66 | 72.36 | 87.29 | 65.03 | 62.70 | 72.68 |

with the JIGSAWS data [7]. In the first configuration – leave one supertrial out (LOSO) – for each iteration of cross-validation (five in total), one trial of each subject was left out for the test and the remaining trials were used for training. In the second configuration – leave one user out (LOUO) – for each iteration of the cross-validation (eight in total), all the trials belonging to a particular subject were left out for the test. These are the standard benchmark configurations provided in [7]. We report micro (average of total correct predictions across all classes) and macro (average of true positive rates for each class) performance results as defined in [25].

Table 1 presents the results for gesture classification assuming known boundaries and using kinematic data only. For comparison purposes, we also report state-of-the-art results for Linear Dynamical Systems (LDS) and Hidden Markov Models (HMM) from [25]. The proposed method outperforms both LDS and HMM methods in terms of micro and macro performances for the three tasks and the two cross-validation configurations. These results show that our method accurately identifies patterns that are specific to a gesture motion. One of the interesting features of the proposed method is the ability to use different kinematic data depending of the gesture. As our method computes the frequencies for each component of the kinematic data for each gesture independently, the most discriminative attributes of a given gesture naturally stand out. Furthermore, the *tf\*idf* regularization discards the motion patterns that are common to every gesture (*i.e.*, irrelevant for classification as not distinctive of any class).

The LOUO configuration is known to be particularly challenging, because we attempt to classify gestures of a subject without having any of his or her other attempts. The good performance of our approach can be explained by its ability to identify highly discriminative patterns that are the most distinctive of each gesture. These results also indicate that our method generalizes well, as shown by the fact that it can accurately classify gestures from unobserved trainees.

**Table 2.** Skill classification performance per trial using kinematic data only.

| Method | Metric | Leave-one-supertrial-out | | |
|--------|--------|----------|----------------|-----------|
| | | Suturing | Needle passing | Knot tying |
| | $(l_s,\alpha)$ | (10,9) | (12,13) | (5,14) |
| *Proposed* | Micro | 89.74 | **96.30** | 61.11 |
| | Macro | 86.67 | 95.83 | 53.33 |
| SHMM [19] | Micro | **97.40** | 96.20 | **94.40** |

## 4.2 Skills Classification

For skill classification, we performed experiments to identify the skill level (novice, intermediate or expert) at the trial level. In this experiment, we used the leave one trial out (LOSO) cross-validation configuration. Table 2 presents the results for the three tasks and reports micro and macro performances. The results are better for Suturing and Needle Passing tasks than for Knot Tying task. The poor performance on the Knot Tying task can be explained by the minor difference between the Expert and Intermediate subjects for this task (mean GRS is 17.7 and 17.1 for expert and intermediate respectively). We also report the state-of-the-art results from [19] for the Suturing task. The SHMM approach gives better results for the per trial classification configuration as it uses global temporal information, whereas our method is focusing on the local patterns regardless of their location within larger time series. Furthermore, the SHMM approach [19] uses gestures boundaries to learn the temporal model while our method is not using this information.

## 4.3 Interpretable Patterns Visualization

Our approach outputs a set of discriminative patterns weighted by the class specificity for each of the input class. These lists of ranked patterns can be studied to better understand what makes each class distinctive. As the use of *tf\*idf* (Eq. (3)) discards patterns that are common to all classes, only patterns having discriminative power remain.

The list of weighted discriminative patterns can be used to visualize, on a given trial, where are the areas that are specific to the current skill level of the trial. We propose to use a heat map-like visualization technique that provides immediate insight into the layout of the "important" class-characteristic patterns (as described in [4]). Figure 3 shows, for the Suturing task, the two individual $5^{th}$ trials of subjects B (Novice) and E (Expert), using $(x, y, z)$ coordinates for the right hand. In this figure, we used respectively the *tf\*idf* weights vectors of the $5^{th}$ fold for the Novice on subject B and for the Expert on subject E. The red areas correspond to specific motions that are correlated with a skill level. For Subject B (Fig. 3a), these areas correspond to motions that were only observed among the novices. By contrast, green areas correspond to motions that are common to all subjects regardless of their skill. This visualization provides

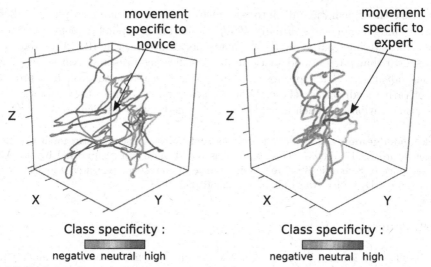

(a) Trial 5 of Suturing task of subject B (novice) using Novice class *tf*\**idf* vector weights of 5th fold (best viewed in color).

(b) Trial 5 of Suturing task of subject E (expert) using Expert class *tf*\**idf* vector weights of 5th fold (best view in color).

**Fig. 3.** Example of interpretable feedback using a heat-map visualization of subsequence importance to a class identification. The value corresponds to the combination of the *tf-idf* weights of all patterns which cover the point.

a rich information about what makes a specific skill level distinctive and can also be used to provide individual and personalized feedback. As the videos of the trials are also available, this result has to be displayed side-by-side with the videos in order to show to the trainee the movements that are specific. Note that a more detailed analysis could be performed by observing which kinematic data features are specific in these areas or by performing the analysis on a per gesture basis. Visualization (like Fig. 3) for all the subject trials for the Suturing task are available on the companion webpage[1].

Note that as the *tf*\**idf* weight vectors are computed prior to the classification step, it is possible to display this heat-map visualization in real-time during the trial. We provide a video on the companion webpage that shows the real-time computation of this visualization while a trainee performs a suturing task. We believe that this tool is an interesting addition to existing learning tools for surgery as it provides a way to obtain a feedback on which parts of an exercise have been used to classify the attempt.

## 5    Conclusion

In this paper, we presented a new method for discovery of discriminative and interpretable patterns in surgical activity motion. Our method uses SAX to

---

[1] http://germain-forestier.info/src/aime2017/.

discretize the kinematic data into sequence of letters. A sliding window is then used to build bag of words. Finally, *tf\*idf* framework is applied to identify motion class-characteristic patterns. Experiments performed on the JIGSAWS dataset has shown that our method successfully classifies gestures and skill levels. The strong advantage of the proposed technique is the ability to provide a precise quantitative feedback for the classification results. Of course, the evaluation of our visualization approach needs to be performed within curriculum.

**Acknowledgement.** This work was supported by the Australian Research Council under award DE170100037. This material is based upon work supported by the Air Force Office of Scientific Research, Asian Office of Aerospace Research and Development (AOARD) under award number FA2386-16-1-4023.

# References

1. Tsuda, S., Scott, D., Doyle, J., Jones, D.B.: Surgical skills training and simulation. Curr. Probl. Surg. **46**(4), 271–370 (2009)
2. Forestier, G., Petitjean, F., Riffaud, L., Jannin, P.: Optimal sub-sequence matching for the automatic prediction of surgical tasks. In: Holmes, J.H., Bellazzi, R., Sacchi, L., Peek, N. (eds.) AIME 2015. LNCS (LNAI), vol. 9105, pp. 123–132. Springer, Cham (2015). doi:10.1007/978-3-319-19551-3_15
3. Dlouhy, B.J., Rao, R.C.: Surgical skill and complication rates after bariatric surgery. N. Engl. J. Med. **370**(3), 285 (2014)
4. Senin, P., Malinchik, S.: SAX-VSM: interpretable time series classification using SAX and vector space model. In: IEEE International Conference on Data Mining, pp. 1175–1180 (2013)
5. Lin, J., Keogh, E., Wei, L., Lonardi, S.: Experiencing SAX: a novel symbolic representation of time series. Data Min. Knowl. Discov. **15**(2), 107 (2007)
6. Salton, G., Wong, A., Yang, C.S.: A vector space model for automatic indexing. Commun. ACM **18**(11), 613–620 (1975)
7. Gao, Y., Vedula, S.S., Reiley, C.E., Ahmidi, N., Varadarajan, B., Lin, H.C., Tao, L., Zappella, L., Béjar, B., Yuh, D.D., et al.: JHU-ISI gesture and skill assessment working set (JIGSAWS): a surgical activity dataset for human motion modeling. In: Modeling and Monitoring of Computer Assisted Interventions (M2CAI)-MICCAI Workshop, pp. 1–10 (2014)
8. Reiley, C.E., Hager, G.D.: Decomposition of robotic surgical tasks: an analysis of subtasks and their correlation to skill. In: Modeling and Monitoring of Computer Assisted Interventions (M2CAI) – MICCAI Workshop (2009)
9. Reiley, C.E., Plaku, E., Hager, G.D.: Motion generation of robotic surgical tasks: learning from expert demonstrations. In: IEEE International Conference on Engineering in Medicine and Biology Society, pp. 967–970 (2010)
10. Béjar Haro, B., Zappella, L., Vidal, R.: Surgical gesture classification from video data. In: Ayache, N., Delingette, H., Golland, P., Mori, K. (eds.) MICCAI 2012. LNCS, vol. 7510, pp. 34–41. Springer, Heidelberg (2012). doi:10.1007/978-3-642-33415-3_5
11. Zappella, L., Béjar, B., Hager, G., Vidal, R.: Surgical gesture classification from video and kinematic data. Med. Image Anal. **17**(7), 732–745 (2013)

12. Zia, A., Sharma, Y., Bettadapura, V., Sarin, E.L., Ploetz, T., Clements, M.A., Essa, I.: Automated video-based assessment of surgical skills for training and evaluation in medical schools. Int. J. Comput. Assist. Radiol. Surg. **11**(9), 1623–1636 (2016)

13. Reiley, C.E., Hager, G.D.: Task versus subtask surgical skill evaluation of robotic minimally invasive surgery. In: Yang, G.-Z., Hawkes, D., Rueckert, D., Noble, A., Taylor, C. (eds.) MICCAI 2009. LNCS, vol. 5761, pp. 435–442. Springer, Heidelberg (2009). doi:10.1007/978-3-642-04268-3_54

14. Despinoy, F., Bouget, D., Forestier, G., Penet, C., Zemiti, N., Poignet, P., Jannin, P.: Unsupervised trajectory segmentation for surgical gesture recognition in robotic training. IEEE Trans. Biomed. Eng. **63**(6), 1280–1291 (2015)

15. Gao, Y., Vedula, S.S., Lee, G.I., Lee, M.R., Khudanpur, S., Hager, G.D.: Unsupervised surgical data alignment with application to automatic activity annotation. In: IEEE International Conference on Robotics and Automation, pp. 4158–4163 (2016)

16. Zhou, Y., Ioannou, I., Wijewickrema, S., Bailey, J., Kennedy, G., O'Leary, S.: Automated segmentation of surgical motion for performance analysis and Feedback. In: Navab, N., Hornegger, J., Wells, W.M., Frangi, A.F. (eds.) MICCAI 2015 Part I. LNCS, vol. 9349, pp. 379–386. Springer, Cham (2015). doi:10.1007/978-3-319-24553-9_47

17. Kowalewski, T.M., White, L.W., Lendvay, T.S., Jiang, I.S., Sweet, R., Wright, A., Hannaford, B., Sinanan, M.N.: Beyond task time: automated measurement augments fundamentals of laparoscopic skills methodology. J. Surg. Res. **192**(2), 329–338 (2014)

18. Reiley, C.E., Lin, H.C., Varadarajan, B., Vagvolgyi, B., Khudanpur, S., Yuh, D., Hager, G.: Automatic recognition of surgical motions using statistical modeling for capturing variability. Stud. Health Technol. Inform. **132**, 396 (2008)

19. Tao, L., Elhamifar, E., Khudanpur, S., Hager, G.D., Vidal, R.: Sparse hidden Markov models for surgical gesture classification and skill evaluation. In: Abolmaesumi, P., Joskowicz, L., Navab, N., Jannin, P. (eds.) IPCAI 2012. LNCS, vol. 7330, pp. 167–177. Springer, Heidelberg (2012). doi:10.1007/978-3-642-30618-1_17

20. Tao, L., Zappella, L., Hager, G.D., Vidal, R.: Surgical gesture segmentation and recognition. In: Mori, K., Sakuma, I., Sato, Y., Barillot, C., Navab, N. (eds.) MICCAI 2013. LNCS, vol. 8151, pp. 339–346. Springer, Heidelberg (2013). doi:10.1007/978-3-642-40760-4_43

21. Höppner, F.: Time series abstraction methods-a survey. In: GI Jahrestagung, pp. 777–786 (2002)

22. Patel, P., Keogh, E., Lin, J., Lonardi, S.: Mining motifs in massive time series databases. In: IEEE International Conference on Data Mining, pp. 370–377 (2002)

23. Moskovitch, R., Shahar, Y.: Classification-driven temporal discretization of multivariate time series. Data Min. Knowl. Disc. **29**(4), 871–913 (2015)

24. Manning, C.D., Raghavan, P., Schütze, H., et al.: Introduction to Information Retrieval, vol. 1. Cambridge University Press, Cambridge (2008)

25. Ahmidi, N., Tao, L., Sefati, S., Gao, Y., Lea, C., Bejar, B., Zappella, L., Khudanpur, S., Vidal, R., Hager, G.D.: A dataset and benchmarks for segmentation and recognition of gestures in robotic surgery. IEEE Trans. Biomed. Eng. (2017)

# Natural Language Processing

# Automatic Classification of Radiological Reports for Clinical Care

Alfonso E. Gerevini[1], Alberto Lavelli[2], Alessandro Maffi[1], Roberto Maroldi[1,3], Anne-Lyse Minard[1,2(✉)], Ivan Serina[1], and Guido Squassina[3]

[1] Università degli Studi di Brescia, Brescia, Italy
[2] Fondazione Bruno Kessler, Trento, Italy
minard@fbk.eu
[3] Spedali Civili di Brescia, Brescia, Italy

**Abstract.** Radiological reporting generates a large amount of free-text clinical narrative, a potentially valuable source of information for improving clinical care and supporting research. The use of automatic techniques to analyze such reports is necessary to make their content effectively available to radiologists in an aggregated form. In this paper we focus on the classification of chest computed tomography reports according to a classification schema proposed by radiologists of the Italian hospital *ASST Spedali Civili di Brescia*. At the time of writing, 346 reports have been annotated by a radiologist. Each report is classified according to the schema developed by radiologists and textual evidences are marked in the report. The annotations are then used to train different machine learning based classifiers. We present in this paper a method based on a cascade of classifiers which make use of a set of syntactic and semantic features. By testing the classifiers in cross-validation on manually annotated reports, we obtained a range of accuracy of 81–96%.

## 1 Introduction

The use of electronic health record (EHR) in the last years has allowed hospitals to collect a large amount of digital contents (both structured data and narrative text). Such contents have generated new challenges and opportunities in the medical domain since, for example, they can be used to improve the clinical workflows, the efficacy and quality of patient care and can also be used in research in medicine. In particular, natural language processing (NLP) techniques are fundamental and efficient for the automatic extraction of information and allow an effective use of unstructured clinical narratives of the EHR, including radiological reports.

In our context, the use of automatic techniques to analyze clinical narratives (e.g., radiological reports) is fundamental in order to make their content effectively available to radiologists in a structured form; in fact, around 5,500 reports of chest computed tomography are generated every year by the radiology department involved in this project and all these unstructured data cannot be easily summarized and evaluated by humans.

© Springer International Publishing AG 2017
A. ten Teije et al. (Eds.): AIME 2017, LNAI 10259, pp. 149–159, 2017.
DOI: 10.1007/978-3-319-59758-4_16

In this paper we focus on the classification of chest computed tomography reports according to a classification schema proposed by radiologists of the Italian hospital *ASST Spedali Civili di Brescia*. Being able of correctly classifying old and new reports according to such a classification schema can have many benefits. First of all, it would simplify epidemiologic studies, by allowing physicians to automatically discard irrelevant reports or, equally, to find relevant ones; in a similar way, it would allow physicians to retrieve relevant cases for teaching purposes. Other benefits concern quality assessment of radiologic practice and logistic management of the hospital. For example, the automatic classification could be used to compare the number of follow-ups requested by different physicians; it could also enable automatic comparison between the provisional diagnosis and the result of the examination.

Our goal is to build a system that can be used to automatically classify all the reports generated until now. Moreover, it could be integrated in the software used by radiologists for writing the reports; this would allow to obtain a "real time" classification of a report (as soon as the radiologist has written it), which should then be confirmed (or modified, if needed) by radiologists. This would produce a twofold effect: (i) the manual validation of the automatic results of our system would help to build a more accurate classifier; (ii) if accurate enough, this would reduce the classification effort required by the physician.

We organize the paper as follows. In Sect. 2 we define some background concepts and review related work; then in Sect. 3 we describe our classification schema and our dataset. In Sect. 4 we describe the machine learning techniques adopted for the automatic annotation and classification of reports and we give an overview of our classification system and in Sect. 5 we present the evaluation of our approach. Finally we conclude the paper with future work in Sect. 6.

## 2    Background and Related Work

A radiology report is the formal product of a diagnostic imaging referral, used for communication and documentation purposes. In general there are different guidelines for effective reporting of diagnostic imaging, although essentially a report consists of free text, possibly organized in a number of standard sections. Medical reports and clinical narratives are characterized by non-standard language: they contain abbreviations, ungrammatical language, acronyms and typing errors; this is due to the fact that reports are often written in haste or dictated to speech recognition software. In addition, abbreviations and acronyms are sometimes idiosyncratic to the specific hospital or department.

Natural language processing techniques are needed to convert the unstructured text of these reports into a structured form, therefore enabling automatic identification of information. NLP applications rely on a sequence of steps that extract structured textual features from the radiology report. The basic steps are *segmentation* (i.e. splitting the reports into their sections), *sentence splitting, tokenization, stemming* or *lemmatization, part-of-speech tagging, chunking*, etc. These steps enable to perform semantic analysis, i.e. assigning meaning to the words and phrases by linking them to semantic types and concepts (*concept recognition*). A further step is *negation detection*, i.e. checking whether concepts or relations in the text are negated. The final result of these steps is a set of features that can be used for the actual task, for instance, text classification. The features can be processed by a machine learning based classifier or by a set of rules hand-crafted by experts. Hybrid approaches are also possible.

Pons et al. [12] present a systematic review of NLP applications for radiology developed until 2016, both in operational use or not. They point out five main categories of study: (a) diagnostic surveillance, (b) cohort building for epidemiologic studies, (c) query-based case retrieval, (d) quality assessment of radiologic practice, (e) clinical support services. All the applications studied in [12] are developed for English. We can in particular mention the work presented in [14] about recognition of recommendations in radiology reports and a work on the classification of radiology reports in two classes: whether a report contains a "cancer alert" or not [5]. The latter reported a F-measure of 0.77 on the binary classification. As for Italian, some works which use supervised learning in order to extract information from radiology reports are [1,3]. In both [3,14], a corpus of manually annotated reports was created to be used as a training set. In [3], segments of text were annotated with tags representing concepts of interest in the radiological domain. In [1] the aim is to find relations among biomedical entities: a very large set of reports was automatically annotated with NER (Named Entity Recognition) tools to be used as a training set. To automatically extract medical entities, standard taxonomies (e.g. Snomed-CT, ICD9) can be used; in some cases, entities can then be mapped to their unique UMLS CUI (Concept Unique Identifiers).

Our work consists in the classification of radiology reports following a multi-level schema, whereas the previously mentioned works focused mainly on a unique level. For this reason we perform the classification through a cascade of classifiers which are used to annotate a report at the phrase-level and to classify it in different classes. The differences between our task and the ones mentioned above make difficult to compare their performance.

## 3    Data Representation and Annotation

The system for reports classification proposed is based on a classification schema defined by the radiologists of *Spedali Civili di Brescia*. The schema consists of five high level classes that may assume two or more different values. Our approach relies on supervised machine learning methods, so it was necessary to perform a manual classification of a set of reports.

In this section we describe first the classification schema, then the data used and their manual annotation.

### 3.1    Classification Schema

In Fig. 1 we show the classification schema proposed by the radiologists and adapted after discussion. The classification is composed of five levels:

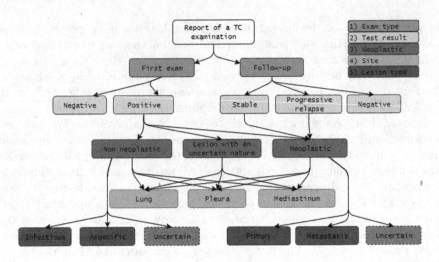

**Fig. 1.** Classification schema

1. examination type (*first examination* or *follow-up*);
2. result of the examination (*positive* or *negative*, *stable* or *progressive relapse*);
3. neoplastic nature of the lesion (*neoplastic*, *non-neoplastic* or *uncertain*);
4. site of the lesion (*lung*, *pleura* or *mediastinum*);
5. type of the lesion (*infectious* or *aspecific*, *primary* or *metastasis*).

### 3.2   Data

We used CT (computed tomography) reports from one of the radiology departments of *Spedali Civili di Brescia*. The reports were extracted from the database of the hospital and anonymized by removing patient names as well as medical staff names. The reports at our disposal are composed of text only, no CT images are associated to the text. The reports usually contain three parts: *quesito clinico*, i.e. the reason why the examination is requested by the doctors (often contains a provisional diagnosis made by the doctors using the results of the previous analysis); *quadro clinico*, i.e. the case history of the patient; *referto*, i.e. the report written by the doctor when analyzing the CT. We had at our disposal around 10,000 unlabeled reports of chest CT done in 2015 and 2016.

**Manual Annotation.** The manual annotation of a set of reports is necessary in order to develop supervised machine learning methods and to evaluate them. We adapted an interface based on the MT-EQuAl tool [6] to perform the manual annotation. It is a web interface based on PHP and MySQL. For each report the annotator performs two tasks: classification of the report through a form (according to the schema described in Sect. 3.1), and annotation of evidences of the classification in the text (using five tags corresponding to the five levels of the annotation schema). For example, "posterior basal segment" is an evidence for the "Site" *lung*. The guidelines provided to the annotators are very simple and consist mainly in asking to annotate the maximum extent of text that is an evidence of a level of the classification.

The interface enables the annotator to duplicate the report (i.e. to annotate twice the same report) if more than one classification is possible. Currently, the classification schema forces to choose only one test result for each report, to associate it to either neoplastic lesion or non-neoplastic lesion or uncertain lesion, and to indicate only one type of lesion. However, reports can describe the evolution of lesions that can be of different nature. For example in the lungs, nodules can be identified and some can be suspected to be metastasis and other to be neoplastic primitive. In this case the report can be duplicated and classified twice with two different lesion types.

The task has been performed by an expert (a specialist in radiology). Periodically, after he has annotated a certain amount of reports, the classification/annotation is discussed within the project team and in case of inconsistencies or missing evidences the classification is corrected with the specialist.

**Data Statistics.** In total the medical expert annotated 346 reports selected manually out of the 10,000 unlabeled reports in order to represent adequately all the classes of our schema, which have been used to train and test our approach. Reports are composed on average of 174 tokens. Some statistics about class distribution are presented in Table 1. From these figures we can observe that the distribution of values in some classes is unbalanced, i.e. one value is much more frequent than the others. For example, there are very few reports associated to the "Site" *pleura* compared to the "Site" *lung*.

**Table 1.** Statistics about the annotated data.

| | Reports of a chest CT examination: 346 | | | | |
|---|---|---|---|---|---|
| Exam type | First exam | | Follow-up | | |
| | 136 | | 210 | | |
| Result | Negative | Positive | Stable | Progressive relapse | Negative |
| | 16 | 120 | 79 | 54 | 77 |
| Site | Lung | | Pleura | Mediastinum | |
| | 188 | | 19 | 40 | |
| Neoplastic | Non neoplastic | | Uncertain | Neoplastic | |
| | 54 | | 29 | 161 | |
| Lesion type | Infectious | Aspecific | | Metastasis | Neoplastic primitive |
| | 25 | 28 | | 53 | 28 |

# 4 Report Classification

## 4.1 Text Processing

Reports are preprocessed using the TextPro suite [11] in order to extract features from text. TextPro performs sentence splitting, tokenization, different linguistic analysis which gives us morphological analysis, lemmatization, Part-of-Speech tagging, identification of syntactic phrases and time expression detection. Making use of these linguistic analysis and of some external resources, a tool developed specifically for this project identifies prefixes and suffixes contained in the words (e.g. *-tomia* [-tomy]), negation cues (e.g. *non* [not]) and the presence of numbers and measurements. In addition, if a word is derived from another (e.g. *riduzione* [reduction] is derived from *ridurre* [reduce]) it is associated to its derived term and if it has synonyms a unique (preferred) term is added.[1]

## 4.2 Automatic Annotation

The classification can be improved by automatically identifying sequences of words which are evidences of the different classes. These sequences will be then used as additional features for the classification. In order to annotate the relevant sequences of words, we implemented a supervised machine learning based system. We used tinySVM algorithm through the Yamcha toolkit [10]. For each level of the classification schema a model is built. Several features are used: surface features (token types), syntactic features (Part-of-Speech, chunk phrases, etc.) and semantic features (prefixes, suffixes, numerical expressions, etc.).

The automatic annotation step cannot be evaluated independently from the classification task as the manual annotation of the reports is partial, i.e. not all the phrases related to one class (e.g. "Site", "Lesion type") have been annotated by the expert, but only those that are evidences of the classification he chose.

## 4.3 Classification

We experimented two different ways to build our classification system. The former uses a combination of both information extraction and text classification techniques. The latter is based on information extraction techniques only. In the following, for each method we describe how the final system classifies a new (unseen) report and how it is trained (using the corpus of annotated reports).

Both methods try to distinguish between *first examination* and *follow-up* (first level of the classification schema) in the same way, i.e. searching for typical expressions (patterns) associated to follow-ups in the whole text. The patterns are automatically generated from the training set, using the manual annotations. Then the text is divided into sentences, which are independently classified according to the remaining levels of the classification schema. The final classification of a report is obtained by merging together all the sentence classifications, according to the following rules:

---

[1] The list of synonyms and their preferred term has been built manually.

- if the report is a follow-up and there is at least one sentence classified as *progressive relapse*, this classification prevails on *stable*;
- if at least one sentence is classified as *neoplastic*, the report is classified as *neoplastic* (*neoplastic* prevails on *uncertain* and *uncertain* prevails on *non-neoplastic*);
- the sites are collected from all positive sentences.

The two methods differ in the way the automatic annotation of the sentences is performed.

**Method 1.** Given the first level of the report classification, this method follows two steps:

Step 1. The report is automatically annotated (by our tool described in Sect. 4.2) tagging significant words or phrases with four different tags. The tags are identical to the ones used by the annotator: each one is associated with a level of classification.

Step 2. The tagged words and phrases are classified into a specific class value in order to perform sentence classification (for example, an item tagged with tag "Site" must be classified into *lung*, *pleura* or *mediastinum*). Eight different text classifiers are trained for this purpose, using the annotated sections of the reports.

**Method 2.** This method relies on an enhanced version of the automatic annotation tool, which tags the text using one tag for each class value (e.g. for the "Site" there will three tags, one for *lung*, one for *pleura* and one for *mediastinum*), instead of using four generic tags.

### 4.4 Machine Learning Algorithms

For the implementation and evaluation of our methods, we used five different machine learning techniques: Naive Bayes classifier [4], decision trees [13], random decision forests [9], neural networks [8] and support vector machines [2].

More specifically, the Naive Bayes classifier assumes that all the attributes in the dataset are *independent* of each other; although this is a strong assumption, it usually allows to build simple models that work surprisingly well. Support vector machines are based on the resolution of quadratic programming problems and they can work efficiently also with a low number of observations across many predictor variables. A key advantage of decision trees with regard to other approaches is that the corresponding final model generated can be easily understood by humans; on the other hand, their effectiveness is sometimes limited by the reduced space of hypotheses that they can represent. Random decision forests construct a number of decision trees at training time; they can provide better results than a single decision tree, although the final classification model is more difficult to understand. Finally, neural networks are an approach based on a potentially large collection of neural units, and they have a proven track record of success for different problem domains.

## 5   Evaluation and Discussion

We experimented different learning algorithms: the Naive Bayes classifier (NB); the Sequential Minimal Optimization algorithm (SMO) for training the support vector machines; the J48 implementation for the decision trees; Random Forests (RF); the MultiLayer Perceptron model (MLP) for training the neural networks. To implement the different classifiers, we used the Weka open source Java data mining library [7], with the default configurations for each classifier.

The performance was evaluated using the following measures:

- accuracy (Acc), which is the number of correct predictions divided by the total number of predictions.
- macro-averaged F-measure (FM), which is based on the commonly used F-measure metric; in particular the F-measure is computed locally over each category first and then the average over all categories is considered.

**Table 2.** Accuracy and macro-averaged F-measure using ten-fold cross-validation on the labeled data set considering different machine learning algorithms.

|                 | NB       |      | SMO      |      | J48      |      | RF   |      | MLP      |      |
|-----------------|----------|------|----------|------|----------|------|------|------|----------|------|
|                 | Acc      | FM   | Acc      | FM   | Acc      | FM   | Acc  | FM   | Acc      | FM   |
| Result          | **97,2** | 96,3 | 96,9     | 95,9 | **97,2** | 96,2 | 96,7 | 95,5 | 96,4     | 95,2 |
| Result f-up     | 87,3     | 86,9 | **91,1** | 90,9 | 87,3     | 86,7 | 89,2 | 88,9 | 87,2     | 87,0 |
| Neoplastic      | 82,6     | 75,7 | **83,8** | 73,0 | 81,0     | 71,5 | 83,4 | 74,5 | 83,4     | 73,2 |
| Site lung       | 87,4     | 84,5 | **93,0** | 90,5 | 90,4     | 87,6 | 92,8 | 90,1 | 91,3     | 88,0 |
| Site med.       | **97,2** | 86,5 | **97,2** | 86,5 | **97,2** | 86,5 | 97,2 | 86,5 | 97,0     | 85,7 |
| Site pleura     | 94,3     | 88,6 | **96,0** | 92,2 | 92,8     | 85,2 | 95,7 | 91,9 | **96,0** | 92,9 |
| L.t. neopl.     | 91,2     | 90,3 | **92,5** | 92,1 | 80,0     | 77,6 | 91,2 | 90,6 | **92,5** | 92,0 |
| L.t. non neopl. | **87,7** | 87,5 | 86,0     | 85,4 | 68,4     | 68,3 | 82,5 | 81,8 | 86,0     | 85,7 |

In Table 2, we show the efficacy of machine learning algorithms using ten-fold cross-validation on the labeled dataset (using only the phrases of the reports manually annotated by the experts). In particular, we used two text classifiers for the "Result" level (Result and Result f-up), which respectively distinguish between *positive* or *negative* and *stable* or *progressive relapse*. Then we have one classifier for the "Neoplastic" level (Neoplastic) which distinguishes among *neoplastic*, *non neoplastic* or *uncertain*. Three classifiers for the "Site" level (Site lung, Site pleura and Site med.) and finally two classifiers for the "Lesion type" level, one for neoplastic reports (L.t. neopl.) which distinguishes between *primitive* or *metastasis* and one for non neoplastic reports (L.t. non neopl.) which distinguishes between *infectious* or *aspecific*. We can observe the general good performance of the support vector machines with accuracy and F-measures close to or greater than 85%.

In Table 3, we analyze the results of the classification on a test set (consisting of about the 20% of the 346 reports) considering the two classification methods

**Table 3.** Evaluation of the two classification methods on the test set in terms of accuracy and macro-averaged F-measure. The last column reports the number of documents for the different classes in the test set.

| | Method 1 | | | | | | | | | | Method 2 | | Number of |
| | NB | | SMO | | J48 | | RF | | MLP | | | | |
| | Acc | FM | Acc | FM | Acc | FM | Acc | FM | Acc | FM | Acc | FM | reports |
|---|---|---|---|---|---|---|---|---|---|---|---|---|---|
| Exam type | 95,6 | 94,9 | 95,6 | 94,9 | 95,6 | 94,9 | 95,6 | 94,9 | 95,6 | 94,9 | 95,6 | 94,9 | 68 |
| Result | 77,9 | 67,1 | 79,4 | 70,1 | 79,4 | 70,1 | **82,4** | 75,5 | **82,4** | 75,5 | 80,9 | 71,5 | 68 |
| Result f-up | 54,8 | 67,1 | 67,7 | 70,1 | 67,7 | 70,1 | 67,7 | 75,5 | 64,5 | 75,5 | **74,2** | 71,5 | 31 |
| Neoplastic | 70,8 | 56,1 | **77,1** | 66,0 | 60,4 | 45,3 | 75,0 | 57,0 | 70,8 | 56,3 | 62,5 | 32,8 | 48 |
| Site lung | 70,8 | 52,1 | 72,9 | 53,5 | 72,9 | 53,5 | **75,0** | 55,0 | 70,8 | 52,1 | 70,8 | 55,8 | 48 |
| Site pleura | 87,5 | 46,7 | 87,5 | 59,1 | **89,6** | 61,5 | **89,6** | 47,3 | **89,6** | 47,3 | 87,5 | 46,7 | 48 |
| Site med. | 72,9 | 67,9 | 72,9 | 67,9 | 70,8 | 66,1 | 72,9 | 67,9 | 66,7 | 63,6 | **81,3** | 70,5 | 48 |
| L. t. neopl. | **66,7** | 64,3 | 63,6 | 60,8 | 63,6 | 60,8 | 63,6 | 60,8 | 63,6 | 60,8 | 60,6 | 43,6 | 33 |
| L.t.nonneopl. | **70,0** | 69,4 | 60,0 | 57,8 | 60,0 | 57,8 | 60,0 | 57,8 | 60,0 | 57,8 | 30,0 | 24,4 | 10 |

described in Sect. 4.3 and different supervised machine learning algorithms. With regards to Table 2, in Table 3 we have an additional line Exam type that provides the results of the classifier that distinguishes between *first examination* and *follow-up*. We can observe that the Random Forest approach provides in general the best results; moreover, Method 2 seems quite competitive except for the last level of classification, i.e. L.t. nonneopl., which is probably related to the low number of elements for this class in the training set.

Preliminary experimental results using a simple bag-of-words approach, which does not require the automatic annotation of the text, show that for the most specific levels (e.g. "Site", distinction between *stable* and *progressive relapse*) the automatic annotation of evidences in the text improves the classification whereas, for the upper levels, the bag-of-words approach obtains better results.

We performed an error analysis to understand better the behavior of the proposed methods. One cause of wrong classification is the low recall of the automatic annotation module (see Sect. 4.2), especially when using Method 2. With Method 1 there are many false positives for the "Lesion type" level. This is due to the fact that the information available in the text of a single report does not always enable the expert to identify the lesion type, whereas the system can find partial information and associates it to a type of lesion.

## 6    Conclusion

In this paper we have presented a system for the automatic classification of chest computed tomography reports in Italian. The approach is based on machine learning techniques and relies on a classification schema proposed by the radiologists involved in the project. We have compared the performance obtained by different machine learning techniques. The experiments performed on the reports annotated so far show encouraging results.

Recently, a sixth additional level has been added to the classification schema. It concerns only the follow-up examination and consists of the origin site of the follow-up, i.e. for which site a follow-up was recommended. In many cases, but not all, the origin site is the same as the site of the lesion. Currently, only part of the corpus is annotated according to such level, but soon we will extend the annotation of the origin site to the whole corpus and take it into consideration in the automatic classification system.

In the future we plan to extend the data set considering different radiology departments and different annotators. An Inter-Annotator Agreement phase between two annotators is ongoing. In this work we focused on reports of chest computed tomography. We plan to extend the classification to other parts of the body (e.g. encephalon) extending consequently the classification schema.

**Acknowledgments.** The research described in this paper has been partially supported by the Swiss National Science Foundation, grant number CR30I1_162758, and by the University of Brescia with the H&W SmartService project.

# References

1. Attardi, G., Cozza, V., Sartiano, D.: Annotation and extraction of relations from Italian medical records. In: Proceedings of the 6th Italian Information Retrieval Workshop (IIR 2015) (2015)
2. Cortes, C., Vapnik, V.: Support-vector networks. Mach. Learn. **20**(3), 273–297 (1995)
3. Esuli, A., Marcheggiani, D., Sebastiani, F.: An enhanced CRFs-based system for information extraction from radiology reports. J. Biomed. Inform. **46**(3), 425–435 (2013)
4. Friedman, N., Geiger, D., Goldszmidt, M.: Bayesian network classifiers. Mach. Learn. **29**(2–3), 131–163 (1997)
5. Garla, V., Taylor, C., Brandt, C.: Semi-supervised clinical text classification with laplacian SVMs: an application to cancer case management. J. Biomed. Inform. **46**(5), 869–875 (2013)
6. Girardi, C., Bentivogli, L., Farajian, M.A., Federico, M.: MT-EQuAl: a toolkit for human assessment of machine translation output. In: COLING 2014, 25th International Conference on Computational Linguistics, Proceedings of the Conference System Demonstrations, 23–29 August 2014, Dublin, Ireland, pp. 120–123 (2014)
7. Hall, M., Frank, E., Holmes, G., Pfahringer, B., Reutemann, P., Witten, I.H.: The WEKA data mining software: an update. SIGKDD Explor. Newsl. **11**(1), 10–18 (2009)
8. Haykin, S.: Neural Networks: A Comprehensive Foundation, 3rd edn. Prentice-Hall Inc, Upper Saddle River (2007)
9. Ho, T.K.: Random decision forests. In: Proceedings of the Third International Conference on Document Analysis and Recognition (Volume 1), ICDAR 1995, vol. 1, p. 278. IEEE Computer Society, Washington, DC (1995)
10. Kudo, T., Matsumoto, Y.: Fast methods for kernel-based text analysis. In: Proceedings of the 41st Annual Meeting of the Association for Computational Linguistics, ACL 2003, Stroudsburg, PA, USA, vol. 1, pp. 24–31 (2003)

11. Pianta, E., Girardi, C., Zanoli, R.: The TextPro tool suite. In: Proceedings of the Sixth International Conference on Language Resources and Evaluation (LREC 2008), Marrakech, Morocco, May 2008
12. Pons, E., Braun, L.M.M., Hunink, M.G.M., Kors, J.A.: Natural language processing in radiology: a systematic review. Radiology **279**(2), 329–343 (2016)
13. Rokach, L., Maimon, O.: Data Mining with Decision Trees: Theory and Applications. World Scientific Publishing Co. Inc., River Edge (2008)
14. Yetisgen-Yildiz, M., Gunn, M.L., Xia, F., Payne, T.H.: A text processing pipeline to extract recommendations from radiology reports. J. Biomed. Inform. **46**(2), 354–362 (2013)

# Learning Concept-Driven Document Embeddings for Medical Information Search

Gia-Hung Nguyen[1(✉)], Lynda Tamine[1], Laure Soulier[2], and Nathalie Souf[1]

[1] Université de Toulouse, UPS-IRIT, 118 Route de Narbonne,
31062 Toulouse, France
gia-hung.nguyen@irit.fr
[2] Sorbonne Universités-UPMC, Univ Paris 06, LIP6 UMR 7606,
75005 Paris, France

**Abstract.** Many medical tasks such as self-diagnosis, health-care assessment, and clinical trial patient recruitment involve the usage of information access tools. A key underlying step to achieve such tasks is the document-to-document matching which mostly fails to bridge the gap identified between raw level representations of information in documents and high-level human interpretation. In this paper, we study how to optimize the document representation by leveraging neural-based approaches to capture latent representations built upon both validated medical concepts specified in an external resource as well as the used words. We experimentally show the effectiveness of our proposed model used as a support of two different medical search tasks, namely health search and clinical search for cohorts.

**Keywords:** Medical information search · Representation learning · Knowledge resource · Medical concepts

## 1 Introduction

The importance of medical information access through a diversity of targeting tasks has attracted attention of many researchers from a variety of disciplines including health sciences, social psychology, and information retrieval (IR) [16]. More specifically, those tasks include evidence-based diagnosis, health-related search, and clinical trial patient recruitment; and beyond, diverse secondary tasks, such as population health management and translational research. Practically, from the IR area perspective, information access implies searching in large corpora of documents (e.g., electronic health records (EHRs), medical scientific reviews) for relevant information. The latter is retrieved by (1) matching user's queries, formulated through sets of keywords, with documents (e.g., search for diagnosis according to symptom description as a query input) and (2) matching documents with each other (e.g., identifying potential eligible patients for a clinical trial by matching their EHRs data to the textual description of the clinical trial requirements). However, numerous research studies have shown that such matching is complex, leading to system failure, mainly because of the gap

© Springer International Publishing AG 2017
A. ten Teije et al. (Eds.): AIME 2017, LNAI 10259, pp. 160–170, 2017.
DOI: 10.1007/978-3-319-59758-4_17

between low-level document features and high-level meaning. This is referred to as the semantic gap [6]. In the medical domain, the semantic gap is prevalent and could be implied by three core issues: [6,11]: (1) vocabulary mismatch: if the compared texts expressing similar word senses do not have overlapping keywords (e.g., *Melanome* vs. *skin cancer*); (2) granularity mismatch: if the compared texts contain instances of general entities (e.g., *Anti-inflammatory drug* vs. *Neodex*), and (3) logical implication: if the compared texts contain evidence allowing to infer implications that could not be automatically assessed (e.g., *anorexia* and *depression*). This problem has been faced so far by adopting two main approaches. The first one deals with the use of external domain knowledge resources, mainly to enhance text representations through document or query expansion [5,23]. However, previous research has shown that even if using concepts from controlled vocabularies (such as UMLS) leading to meaningful representations of texts, using only the latter is significantly less effective than keyword-based retrieval, or a combination of both [23]. This may be explained by errors in the concept extraction or the limitations of the hand-labelled concept vocabulary expressed in knowledge resources. The second approach relies on dimensionality reduction techniques that attempt to reduce the representation size of the document vocabulary using the hypothesis of the distributional semantic [9]. Recent research trends show that one effective and efficient way for dimensionality reduction is based on neural language models. The latter projects words in a latent semantic space, called embedding [17], by learning their semantic relationships from their context. However, it is well-known today that such representations do not allow to capture the different meanings of words [10].

In this paper, we address the issue of the medical document representation which is a critical step in the matching process. To cope with the vocabulary and granularity mismatch issues mentioned above, we advocate for the use of a neural-based approach to capture latent representations of documents built upon manually validated medical concepts specified in an external resource. To overcome the limitations of the concept vocabulary and to capture additional distributional semantic extracted from corpora that would face the issue of logical implication assessment, we attempt to achieve the optimal document representation through a refinement using both concept-based and keyword-based raw representations as inputs. The key contributions of this work are two-fold: (1) we develop a model for learning and refining neural based representations of documents using semantics from a medical knowledge resource; (2) we assess the effectiveness of the proposed model by conducting an extensive evaluation using two different medical tasks within a major evaluation benchmark (TREC[1]): (a) health-related search using a corpus of scientific reviews and (b) clinical search for cohorts using a corpus of patient discharge summaries.

The paper is structured as follows. Related work is discussed in Sect. 2. The model for learning the concept-driven document embeddings is detailed in Sect. 3. Section 4 presents the experimental evaluation based on two medical search tasks. Section 5 concludes the paper and suggests avenues for future work.

---

[1] Text Retrieval Conference (http://trec.nist.gov/).

## 2   On the Semantic Gap Problem in Medical Search

In the medical domain, the semantic gap problem is even more challenging for several reasons [11,23]: high variability of language and spelling, frequent use of acronyms and abbreviations, and inherent ambiguity for automated processes to interpret concepts according to contexts. The semantic gap is one of the critical factors that likely leads to dramatically decrease the IR effectiveness. We detail below the two lines of works that cope with this problem.

**Knowledge-Based Enhancement of Documents and Retrieval.** Early and intensive work has been undertaken to use resources, such as MeSH, UMLS, and SNOMED, to enhance the semantics of texts, either documents or queries, by performing smoothing techniques including query expansion [1,14,20] and/or document expansion techniques [8,13]. For instance, Lu et al. [14] investigate query expansion using MeSH to expand terms that are automatically mapped to the user query via the Pubmed's Automatic Term Mapping (ATM) service. In [8], authors combine both query expansion and document expansion using the MeSH thesaurus to retrieve medical records in the ImageCLEF 2008 collection. More concretely, for each MeSH concept, its synonyms and description are indexed as a single document in an index structure. The query is matched to the index to identify the best-ranked MeSH concepts. Finally, identified terms denoting MeSH concepts are used to expand both the document and the query. While knowledge-based document representations perform well in major benchmark evaluation campaigns [24], it is worth mentioning that they generally require a combination with keyword-based approaches [23]. The main underlying explanation is related to the limited expressiveness of concepts and/or the inaccuracy of the concept extraction method. More recently, an emerging line of work consists in enhancing the text-to-text matching model using evidence from external resources [11,15, 25]. All of these contributions share the same goal: making inferences about the associations between raw data and the concept layer in the resource by building a relevance model. For instance, in [11], the relevance model is built upon a graphical probabilistic approach in which nodes are information units from both raw data and the knowledge resource. Another graph-based approach [26] aims at representing concept using a spectral decomposition within a electric resistance network. The extended query is obtained according to a resistance distance-based metric. To the best of our knowledge, this work is the first one to integrate lower dimensional representations for query expansion.

**Representation Learning of Documents and Concepts from a Knowledge Resource.** Major approaches for unsupervised learning of the word representation from unlabelled data are the Skip-gram and CBOW models [17]. Basically, Skip-gram tries to predict the context of a given word, namely its collocated words, while jointly learning word representations. With the same objective of representation learning, CBOW rather relies on a prediction model of a word given its context. These models have been extended in several ways to represent documents [12], concepts of knowledge resources [19] as well as knowledge resources through concept-relation-concept triplets [2]. Beyond, several work focuses on the use of knowledge resources by leveraging concepts and their relations to updating (retrofitting) the latent representations of words [7,28]. For instance, Faruqui et al. [7] propose a "retrofitting" technique consisting in a leveraging lexicon-derived relational information, namely adjacent words of concepts, to refine their associated word embeddings. The underlying intuition is that adjacent concepts in the knowledge resource should have similar embeddings while maintaining most of the semantic information in their pre-learned distributed word representations. In the medical domain, an increasing number of work attempts to learn concept representations [3,4,18,27], with some of them [4,18,27] exploiting those representations within an information access task. Authors in [27] first extend the retrofitting of concepts proposed by [29] by weighting each word-to-word relation using its frequency in the corpus. Second, they (1) build the document representation by summing up the related word embeddings and then, (2) linearly combine relevance scores of documents obtained by matching the query with the bag-of-word representation and word embedding-based representation of documents. In [3], the authors propose the Med2vec model which leverages the sequence of patient visits to learn a latent representation of terminological concepts and visits using the Skip-gram model.

Unlike most of previous work [3,4,18,27,29] that learns contextless concept representations, our aim here is to jointly learn document representations by leveraging both the distributional semantic within text corpora and the concept word senses expressed in knowledge resources. In contrast to the closest previous work [3], we do not infer temporal dependencies between documents (namely visits) and concepts from a user point of view but rather address an information access task since our model leverages corpus-based semantics to cope with the problem of logical implication inference between words.

## 3   Model

### 3.1   Problem Formulation

The literature review highlights that: (1) using knowledge resources allow to enhance raw text representations while using them solely gives rise to both vocabulary limitation and inaccuracy; (2) neural language models explicitly model semantic relations between words but they are unable to highlight diverse word meanings. In this paper, we address these shortcomings by conjecturing that: (1) incorporating concepts in the learning process of high-quality document representations, rather than word representations, should build knowledge-based semantic document representations that cope with the limitations of both the resource vocabulary and the concept extraction process; (2) for a targeted document, the optimal representation expected to be achieved in the low dimensional space requires the closeness of the two distinct embeddings built upon the knowledge-based and corpus-based representations.

With this in mind and in order to achieve the goal of enhancing medical document representations with the perspective of performing effective matching, we propose a method to incorporate a semantic medical knowledge into the learning of document embeddings while leveraging from raw text representations. Given an optimal representation of documents highlighting activated concepts, a query expansion-based matching, for instance, could then be performed for retrieving relevant documents within an information search task.

Formally, document $d$ is a set of two elements $d = \{\mathcal{W}_d, \mathcal{C}_d\}$, where $\mathcal{W}_d$ and $\mathcal{C}_d$ express respectively sets of ordered words $w_i$ and ordered concepts $c_j$ in document $d$, namely $\{w_1, \ldots, w_i, \ldots, w_n\}$ and $\{c_1, \ldots, c_j, \ldots, c_m\}$. Using the *Distributed Version of the Paragraph Vector* model (PV-DM) [12], document $d$ is modeled through a word-based embedding vector $\hat{d}^{(PV-DM)}$. Our first objective is to build a concept-based embedding vector $\hat{d}_i^{(cd2vec)}$ that captures the explicit semantics expressed in knowledge resources. To do so, we propose the *cd2vec (conceptualDoc2vec)* model (Sect. 3.2). Assuming that the concept-based embeddings might suffer from limitations related to the vocabulary and the concept extraction process, our second objective is to find the optimal real-valued representation $\hat{d}$ of document $d$ such that the knowledge-based embedding $\hat{d}_i^{(cd2vec)}$ and the corpus-based embedding $\hat{d}^{(PV-DM)}$ are nearby in the latent space. This problem could be formulated as the minimization of this objective function:

$$\Psi(D) = \sum_{d \in D} \psi(d) = \sum_{d \in D} \left[ (1 - \beta) \times \|d - \hat{d}^{(cd2vec)}\|^2 + \beta \times \|d - \hat{d}^{(PV-DM)}\|^2 \right]$$

(1)

where $D$ is the document collection, $\|x - y\|$ the euclidean distance between $x$ and $y$ vector representations and $\beta$ is a weighting factor experimentally tuned.

## 3.2  Learning the Concept-Based Representation of Documents

Inspired by the Distributed Memory model (PV-DM)[12] which learns short text representations using the set of ordered words within each text, we propose the *conceptualDoc2vec* model that rather focuses on a set of ordered concepts. The PV-DM model originally stands for paragraphs and attempts to consider them as context meanings that are jointly learned with the word vectors. Considering the problem addressed in this paper, our intuition is that document vectors can play the role of context meaning and contribute to a prediction task about the next concept given many concept-based contexts sampled from documents in the corpus. The *conceptualDoc2vec* architecture is illustrated in Fig. 1. More particularly, document vectors $\hat{d}^{(cd2vec)}$ are learned so they allow predicting concepts in their context. More specifically, the goal of the *conceptualDoc2vec* is to maximize the following log-likelihood:

$$\varphi = \sum_{c_j \in \mathcal{C}_d} log P(c_j \mid c_{j-W} : c_{j+W}, d)$$

(2)

where $c_j$ is $j^{th}$ concept of ordered set $\mathcal{C}_d$. $W$ is the length of the context window. $c_{(j-W)} : c_{(j+W)}$ represents the set of concepts ranged between the $(j-W)^{th}$ and the $(j+W)^{th}$ positions in document $d$, without concept $c_j$.

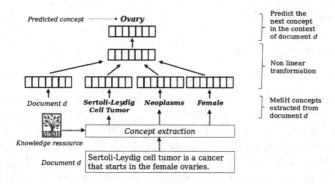

**Fig. 1.** Architecture of the *conceptualDoc2vec* model

The probability $P(c_j \mid c_{j-W} : c_{j+W}, d)$ is defined using a soft-max function:

$$P(c_j \mid c_{j-W} : c_{j+W}, d) = \frac{exp((\overline{v}_j^W)^\mathsf{T}.v_{c_j})}{\sum_{c_k \in \mathcal{C}_d} exp((\overline{v}_k^W)^\mathsf{T}.v_{c_k})} \tag{3}$$

where $v_{c_j}$ is the representation of concept $c_j$, and $\overline{v}_j^W$ is the averaged representation of concepts in window $[j - W; j + W]$, including document $d$.

### 3.3 Solving the Optimization Problem

Our objective is to solve the optimization problem (Eq. 1) that infers the optimal latent representation of document $d$ to be semantically close to the knowledge-based and keyword-based latent document representations in the low dimensional space. To do so, we learn the optimal document embeddings using a stochastic gradient descent. More particularly, this method updates, for each document $d$ its representation using the first derivative $\Delta = \frac{\partial \psi(d)}{\partial d}$ of function $\psi$ with respect to $d$ with a step size of $\alpha$, as illustrated in Algorithm 1.

---

**Algorithm 1.** Learning document representation using stochastic gradient descent

---

**Input:** $\hat{d}_i^{(PV-DM)}, \hat{d}_i^{(cd2v)}$
**Output:** $d$
  $d = randomVector()$
  $\psi(d) = (1 - \beta)\|d - \hat{d}^{(cd2v)}\|^2 + \beta\|d - \hat{d}^{(PV-DM)}\|^2$
  **while** $\psi(d) > \epsilon$ **do**
    $\Delta = 2 \times (1 - \beta) \times (d - \hat{d}^{(c2vec)}) + 2 \times \beta \times (d - \hat{d}^{(PV-DM)}))$
    $d = d - \alpha \times \Delta$
  **end while**
  **return** $d$

---

## 4  Experiments

### 4.1  Experimental Setup

**Tasks.** We evaluate our document representations on two medical search tasks:

• **Task1: health-related search.** This task refers to the situation where a physician seeks for relevant scientific articles providing with a fruitful assistance to achieve an accurate diagnosis/prognostic and/or to suggest a treatment considering the medical case. We use the standard OHSUMED collection consisting of a set of 348,566 references from MEDLINE and 63 queries. This dataset is known as a large-scale standard collection for ad-hoc medical IR [22]. An example of query is *"adult respiratory distress syndrome"*.

• **Task2: clinical search for cohorts.** The task consists in identifying cohorts in clinical studies for comparative effectiveness research. We use the standard TREC Med collection in which queries specify particular disease/condition sets and particular treatments or interventions, expressed by physicians in a natural language form; this document collection includes over $17,000$ de-identified medical visit reports and 35 queries. An example of query is *"find patients with gastroesophageal reflux disease who had an upper endoscopy"*.

**Query Expansion-Based Retrieval.** We inject the optimized document embeddings $d$ learned as detailed above in a text-to-text matching process according to a query expansion technique (noted $\mathbf{Exp_d}$). The latter enhances the input text with the top activated items (words or concepts) of the top-ranked documents within the closest embedding space (respectively, words or concepts depending on the $\beta$ value). To do so, we extract a relevance score for each pairwise item/top-ranked document expressing the probability of item to express the semantics of document within its embedding space. Then, we use a CombSum merging technique applied on those scores to identify the top activated items.

**Baselines.** We use the state-of the-art query expansion models:

• **Rocchio,** a query expansion model based on pseudo-relevance feedback [21].
• **LM-QE,** a language model applying a concept-based query expansion technique [20] in which candidate terms are ranked based on their similarity with descriptions in the knowledge resource. Default parameters are used.
  To show the value of our enhanced document representation over the conceptualDoc2vec $\hat{d}^{cd2vec}$ and over the text-based document embedding $\hat{d}^{PV-DM}$, we run the query expansion method on these two embeddings:
• $\mathbf{Exp_{\hat{d}^{cd2vec}}}$ the concept-based embedding estimated without the optimization of the conceptualDoc2vec embeddings (see Sect. 3.2).
• $\mathbf{Exp_{\hat{d}^{PV-DM}}}$ the text-based embedding obtained through PV-DM [12].

**Evaluation Metrics.** In order to measure the IR effectiveness, we use standard evaluation metrics, namely (1) *P@20* and *Recall@20* representing respectively the mean precision and recall values for the top 20 retrieved documents, and

(2) *MAP (Mean Average Precision)* calculated over all queries. The average precision of a query is computed by averaging the precision values computed for each relevant retrieved document at rank $k \in (1 \dots N)$, where $N = 1000$.

**Table 1.** Effectiveness of our $Exp_d$ model on two medical search tasks.

| Model | Health-related search | | | Clinical search for cohorts | | |
|---|---|---|---|---|---|---|
| | MAP | P_20 | Recall_20 | MAP | P_20 | Recall_20 |
| LM-QE | 0.0265 | 0.0686 | 0.0288 | 0.0793 | 0.1091 | 0.0519 |
| Rocchio | 0.0925 | 0.2262 | 0.0917 | 0.2096 | 0.2603 | 0.1701 |
| $Exp_{\hat{d}^{PV-DM}}$ | 0.1017 | 0.2556 | 0.1086 | 0.3254 | 0.3971 | 0.2278 |
| $Exp_{\hat{d}^{cd2vec}}$ | 0.0956 | 0.2365 | 0.0980 | 0.2255 | 0.2676 | 0.1319 |
| $Exp_d$ | 0.1020 | 0.2556 | 0.1086 | 0.2996 | 0.3426 | 0.1989 |

**Implementation Details.** We use the MeSH terminology which is mostly used in the biomedical domain [22] and the Cxtractor[2] tool to extract concepts. When learning the document representation, we set the minimum loss $\epsilon = 10^{-7}$ and the learning rate $\alpha = 0.01$. We set the size of word-based and concept-based embedding to 200. To tune the $\beta$ parameter (Eq. 1), we perform a cross-validation between both datasets and obtain optimal values in the training phase to 0.9 and 0.6 for respectively TREC Med and OHSUMED datasets. These values highlight that combining both words and concepts for representing documents is useful, with a higher prevalence of words for the TREC Med dataset. This could be explained by the fact that queries in this collection are more verbose.

## 4.2 Results

We present the performance of our model on two tasks: health-related search and clinical search for cohort. Table 1 shows the retrieval effectiveness in terms of *MAP, P@20* and *Recall@20* for our embedding model $Exp_d$ and the baselines as well. In general, we can observe that embedding-based expansion models ($Exp_{\hat{d}^{PV-DM}}$, $Exp_{\hat{d}^{cd2vec}}$, and $Exp_d$) allow to achieve better performance in both tasks than the two classic baselines, namely $LM-QE$ and *Rocchio* making use of raw-level representations of concepts and words respectively. For instance, the text-based embeddings-based expansion $\hat{d}^{PV-DM}$ achieves better results ($MAP = 0.2996$) than the Rocchio-based expansion which obtains a MAP value equal to 0.2096. Especially, our embedding expansion model $Exp_d$ reports significant better results in both tasks over all the three metrics compared with the $LM-QE$. These observations highlight the fact that the embedding models can improve the query expansion with help of learned latent semantics of words and/or concepts. Interestingly, by comparing the type of document embeddings

---

[2] https://sourceforge.net/projects/cxtractor/.

**Table 2.** Example of terms/concepts expanded for query 131 in Trec Med

| Query text | Patients underwent minimally invasive abdominal surgery |
|---|---|
| Extracted concepts | Patients; General surgery; |
| Added by $Exp_{\hat{d}^{PV-DM}}$ | myofascia; ultrasonix; overtube |
| Added by $Exp_{\hat{d}^{cd2vec}}$ | Mesna; Esophageal Sphincter, Upper; Ganglioglioma |
| Added by $Exp_d$ | *umbilical; ventral; biliary-dilatation* |

used in our query expansion framework, we could outline that our optimized vector allows to slightly increase the IR effectiveness with respect to text-based embeddings $\hat{d}^{PV-DM}$ or the concept-based ones $\hat{d}^{cd2v}$. This result shows that our document representation allows to overpass, on one hand, the raw level ambiguity challenge within texts, and on the other hand, the limitation underlying the vocabulary and/or the concept extraction. Table 2 shows an illustration (query 131 of the TREC Med) in which our model $Exp_d$ leveraging both evidences is able to find more relevant items for query expansion than other scenarios $Exp_{\hat{d}^{PV-DM}}$ and $Exp_{\hat{d}^{cd2vec}}$. More precisely, even if the high-level meaning of "abdominal surgery" is captured only by non-fined grained concept from MeSH ("General Surgery"), our model $Exp_d$ is able to identify relevant candidate words for query expansion ("ventral", "biliary dilatation"). Unlikely, the $Exp_{\hat{d}^{PV-DM}}$ model identifies less meaningful candidate words such as "myofascia". This observation reinforces our intuition about the usefulness of combining latent representations of texts and concepts for achieving IR tasks. This result is consistent with previous works [23].

## 5   Conclusion and Future Work

We propose to tackle the semantic gap issue underlying medical information access. Our contribution investigates how to leverage semantics from raw text and knowledge resources to achieve high-level representations of documents. We propose an optimization function that achieves the optimal document representation based on both text embedding and concept-based embedding which relies on an extension of the PV-DM model. We experimentally show the effectiveness of the learned document representations through query expansion within two medical search tasks. The overall obtained results reinforce the rationale of our proposed model. This work has some limitations. For instance, to identify the activated words and concepts reinjected in a query expansion method, we assume that the latent space built using our model is close to the initial embedding spaces. This could be addressed in the future by building a unified framework for learning the latent representations of documents. The latter would offer research opportunities to perform more accurate query expansion techniques fitting with other types of text mismatch faced in the medical domain.

# References

1. Abdou, S., Savoy, J.: Searching in MEDLINE: query expansion and manual indexing evaluation. Inf. Process. Manag. **44**(2), 781–789 (2008)
2. Bordes, A., Usunier, N., García-Durán, A., Weston, J., Yakhnenko, O.: Translating embeddings for modeling multi-relational data. In: NIPS (2013)
3. Choi, E., Bahadori, M.T., Searles, E., Coffey, C., Sun, J.: Multi-layer representation learning for medical concepts. In: KDD, pp. 1495–1504 (2016)
4. De Vine, L., Zuccon, G., Koopman, B., Sitbon, L., Bruza, P.: Medical semantic similarity with a neural language model. In: CIKM, pp. 1819–1822 (2014)
5. Dinh, D., Tamine, L.: Combining global and local semantic contexts for improving biomedical information retrieval. In: Clough, P., Foley, C., Gurrin, C., Jones, G.J.F., Kraaij, W., Lee, H., Mudoch, V. (eds.) ECIR 2011. LNCS, vol. 6611, pp. 375–386. Springer, Heidelberg (2011). doi:10.1007/978-3-642-20161-5_38
6. Edinger, N.T., Cohen, A.M., Bedrick, S., Ambert, K., Hersh, W.: Barriers to retrieving patient information from electronic health record data: failure analysis from the TREC medical records track. In: AMIA Annual Symposium, pp. 180–188 (2012)
7. Faruqui, M., Dodge, J., Jauhar, S.K., Dyer, C., Hovy, E., Smith, N.A.: Retrofitting word vectors to semantic lexicons. In: NAACL (2015)
8. Gobeill, J., Ruch, P., Zhou, X.: Query and document expansion with medical subject headings terms at medical Imageclef 2008. In: Peters, C., Deselaers, T., Ferro, N., Gonzalo, J., Jones, G.J.F., Kurimo, M., Mandl, T., Peñas, A., Petras, V. (eds.) CLEF 2008. LNCS, vol. 5706, pp. 736–743. Springer, Heidelberg (2009). doi:10.1007/978-3-642-04447-2_95
9. Hofmann, T.: Probabilistic latent semantic indexing. In: SIGIR, pp. 50–57 (1999)
10. Iacobacci, I., Pilehvar, M.T., Navigli, R.: Sensembed: learning sense embeddings for word and relational similarity. In: ACL, pp. 95–105 (2015)
11. Koopman, B., Zuccon, G., Bruza, P., Sitbon, L., Lawley, M.: Information retrieval as semantic inference: a graph inference model applied to medical search. Inf. Retrieval **19**(1–2), 6–37 (2016)
12. Le, Q.V., Mikolov, T.: Distributed representations of sentences and documents. In: ICML, pp. 1188–1196 (2014)
13. Le, T.-D., Chevallet, J.-P., Dong, T.B.T.: Thesaurus-based query and document expansion in conceptual indexing with UMLS. In: RIVF 2007, pp. 242–246 (2007)
14. Lu, Z., Kim, W., Wilbur, W.J.: Evaluation of query expansion using MeSH in PubMed. Inf. Retrieval **12**(1), 69–80 (2009)
15. Mao, J., Lu, K., Mu, X., Li, G.: Mining document, concept, and term associations for effective biomedical retrieval: introducing MeSH-enhanced retrieval models. Inf. Retrieval **18**(5), 413–444 (2015)
16. Marton, C., Choo, C.W.: A review of theroretical models on health information seeking on the web. J. Documentation **68**(3), 330–352 (2012)
17. Mikolov, T., Chen, K., Corrado, G., Dean, J.: Efficient estimation of word representations in vector space. arXiv preprint (2013). arXiv:1301.3781
18. Minarro-Gimenez, J., Marin-Alonso, O., Samwald, M.: Exploring the application of deep learning techniques on medical text corpora. Stud. Health Technol. Inf. **205**, 584–588 (2014)
19. Ni, Y., Xu, Q.K., Cao, F., Mass, Y., Sheinwald, D., Zhu, H.J., Cao, S.S.: Semantic documents relatedness using concept graph representation. In: WSDM (2016)

20. Pal, D., Mitra, M., Datta, K.: Improving query expansion using wordnet. JASIST **65**(12), 2469–2478 (2014)

21. Rocchio, J.J.: Relevance feedback in information retrieval. In: The SMART Retrieval System, pp. 313–323 (1971)

22. Stokes, N., Cavedon, Y., Zobel, J.: Exploring criteria for succesful query expansion in the genomic domain. Inf. Retrieval **12**, 17–50 (2009)

23. Trieschnigg, D.: Proof of concept: concept-based biomedical information retrieval. Ph.D. thesis. University of Twente (2010)

24. Voorhees, E., Hersh, W.: Overview of the TREC medical records track. In: TREC (2012)

25. Wang, C., Akella, R.: Concept-based relevance models for medical and semantic information retrieval. In: CIKM, pp. 173–182 (2015)

26. Wang, S., Hauskrecht, M.: Effective query expansion with the resistance distance based term similarity metric. In: SIGIR, pp. 715–716 (2010)

27. Liu, X., Nie, J.-Y., Sordoni, A.: Constraining word embeddings by prior knowledge – application to medical information retrieval. In: Ma, S., Wen, J.-R., Liu, Y., Dou, Z., Zhang, M., Chang, Y., Zhao, X. (eds.) AIRS 2016. LNCS, vol. 9994, pp. 155–167. Springer, Cham (2016). doi:10.1007/978-3-319-48051-0_12

28. Xu, C., Bai, Y., Bian, J., Gao, B., Wang, G., Liu, X., Liu, T.-Y.: Rc-net: a general framework for incorporating knowledge into word representations. In: CIKM (2014)

29. Yu, M., Dredze, M.: Improving lexical embeddings with semantic knowledge. In: ACL, pp. 545–550 (2014)

# Automatic Identification of Substance Abuse from Social History in Clinical Text

Meliha Yetisgen[1,2(✉)] and Lucy Vanderwende[1,3]

[1] Biomedical and Health Informatics, School of Medicine,
University of Washington, Seattle, WA, USA
melihay@uw.edu
[2] Department of Linguistics, University of Washington, Seattle, WA, USA
[3] Microsoft Research, Redmond, WA, USA
lucy.vanderwende@microsoft.com

**Abstract.** Substance abuse poses many negative health risks. Tobacco use increases the rates of many diseases such as coronary heart disease and lung cancer. Clinical notes contain rich information detailing the history of substance abuse from caregivers perspective. In this work, we present our work on automatic identification of substance abuse from clinical text. We created a publicly available dataset that has been annotated for three types of substance abuse including tobacco, alcohol, and drug, with 7 entity types per event, including status, type, method, amount, frequency, exposure-history and quit-history. Using a combination of machine learning and natural language processing approaches, our results on an unseen test set range from 0.51–0.58 F1 on stringent, full event, identification, and from 0.80–0.91 F1 for identification of the substance abuse event and status. These results indicate the feasibility of extracting detailed substance abuse information from clinical records.

**Keywords:** Clinical NLP · Machine learning · Information extraction

## 1 Introduction

Lifestyle and environmental factors play a significant role both in clinical research as well as clinical care. In clinical research, it has been established that 5–10% of cancers can be attributed to hereditary factors, while 90–95% have been found correlated with lifestyle and environmental factors such as smoking, diet and exercise [1]. For clinical care, it has long been practice to record social history as this history impacts not only diagnosis but also treatment options [2]. In this paper, we describe our current work on automatically identifying, for purposes of secondary use, three different types of substance abuse (tobacco, alcohol, and drug) documented in social history sections of clinical text. To accomplish this goal, we first identified a publicly available corpus of clinical notes, analyzed the language used to describe substance abuse, and formalized a substance abuse event definition. In our event definition, we extended Milton et al.'s analysis of social and behavior information [3] and Uzuner et al.'s information on smoking in discharge summaries [4]. We will make our annotated corpus for substance abuse events available. We next develop and experiment with several methodologies

© Springer International Publishing AG 2017
A. ten Teije et al. (Eds.): AIME 2017, LNAI 10259, pp. 171–181, 2017.
DOI: 10.1007/978-3-319-59758-4_18

for identifying substance abuse events and the related information. The results illustrate the feasibility of extracting substance abuse events and will serve as a benchmark for future research.

## 2   Related Work

Informatics for Integrating Biology & the Bedside (i2b2) aims to support automated identification of clinically relevant information to yield insights that can directly impact healthcare improvement. The first i2b2 challenge, alongside de-identification of clinical records, was the smoking challenge dataset, comprised of annotations on discharge records as "past smoker", "current smoker", "smoker", "non-smoker" and "unknown" [4]. Since its release, many systems have been developed to automatically annotate clinical text with these smoking annotations [5–7]. Methods range from lexicon-based, to rule-based and to supervised machine-learning models.

Melton et al. reviewed three widely used public health surveys, i.e. practitioners' questions to measure behaviors that may be relevant to clinical care; they propose an information model for survey items related to alcohol, drug and tobacco use, as well as occupation [3]. Included are dimensions of temporality, degree of exposure, and frequency. We follow Melton's model in our guidelines for annotating substance use as this model reflects the practioners' needs and insights into how these substances impact patient health. Most of the dimensions discussed in the Carter et al. study of drug use are also annotated in our dataset, except "start age" and "quit age", which is computable from the exposure history and quit history, together with a patient's birth information [8]. Chen et al. survey the use of free-text describing alcohol use, coding for "type", "status", "temporal" and "amount", also captured in our annotation guideline [9]. Dimensions similar to our annotation guidelines are explored in Wang et al., who use a rule-based method to identify fine grained substance information from the social history in clinical text; a comparison with their study, however, is impossible as they do not make a dataset available [10].

## 3   Dataset

We created a corpus from the MTSamples website (http://www.mtsamples.com/), a large collection of publicly available transcribed medical records. From this resource, we identified 516 reports of history and physical notes which were expected to contain very rich social history information. We then further applied our in-house statistical section chunker and identified 364 sections tagged as social history from these 516 reports [11]. We used OpenNLP sentence chunker on the 364 social history sections, which identified 1234 sentences.

### 3.1   Annotation Process

We created a detailed annotation guideline to annotate substance abuse events. Our substance abuse event definition captured tobacco, alcohol, and drug use of patients with the following 7 different dimensions: (1) *status* (possible discrete values: past, current, none), (2) *type* (e.g., cigarettes, wine, cocaine), (3) *method* (e.g., inhale, chew), (4) *amount* (e.g., 2 packs, 3–4 glasses), (5) *frequency* (e.g., a day), (6) *exposure history* (e.g., since 1980), and (7) *quit history* (e.g., 3 years ago). Using the BRAT rapid annotation tool, two annotators each annotated 20 social history sections. In the first round, inter-rater agreement was 0.59 F1 for the 3 substance abuse types and their 7 dimensions. The annotators then met and resolved all the conflicts, after which the annotation guideline was updated. A single annotator annotated the remainder of the corpus based on the revised annotation guideline. A summary of the revised annotation guideline is presented in Table 1 and an example social history section annotated based on this guideline is presented in Fig. 1. The complete annotation guideline and the dataset will be released at our group website (http://depts.washington.edu/bionlp/index.html?corpora).

**Table 1.** A summary of revised annotation guideline

| Type | Entity | Example |
|---|---|---|
| Tobacco | Status | Possible discrete values: *current, past, none* |
| | Type | Default value is *tobacco*. We decided to annotate type if mention is more specific than *tobacco*. e.g., *cigarette* |
| | Method | Default value is *smoking*. We decided to annotate method only if mention is different than *smoking*. e.g., *chew* |
| | Amount | *e.g., minimal, significant, <#> packs* |
| | Frequency | *e.g., daily, occasionally, heavy* |
| | Exposure history | *e.g., since 1990* |
| | Quit history | *e.g., 3 years ago* |
| Alcohol | Status | Possible discrete values: *current, past, none* |
| | Type | Default value is *alcohol*. We decided to annotate method only if mention is more specific than *alcohol*. e.g., *beer, hard liquor* |
| | Method | Default value is *drinking*. We decided not to annotate method for alcohol since there is no other alternative method. |
| | Amount | *e.g., Significant, minimal, <#> [of glasses/drinks/bottles]* |
| | Frequency | *e.g., a week, on social occasions, heavy* |
| | Exposure history | *e.g., for many years, for the last 50 years* |
| | Quit history | *e.g., several years ago, in 1984* |
| Drug | Status | Possible discrete values: *current, past, none* |
| | Type | *e.g., illicit, cocaine, recreational, illegal, caffeine* |
| | Method | *e.g., iv, smoking* |
| | Amount | *e.g., abuse, significant* |
| | Frequency | *e.g., chronic, per day* |
| | Exposure history | *e.g., approximately 1 year ago* |
| | Quit history | *e.g., when he was 25, in 2005* |

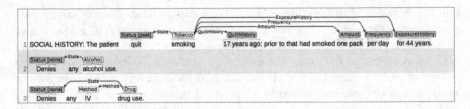

**Fig. 1.** Annotation examples for tobacco, alcohol, and drug abuse events.

## 3.2 Annotation Statistics

Of the 364 social history sections, 282 sections include substance abuse event annotations. Many sections included annotations of multiple types of substance abuse events, yielding in total 278 tobacco, 254 alcohol, and 154 drug event mentions. The annotations for Type and Method are only for non-default values. Annotation statistics are presented in Table 2 at the entity levels.

**Table 2.** Substance abuse entity annotation statistics.

| Entity | Frequency | | |
|---|---|---|---|
| | Tobacco | Alcohol | Drug |
| Status | 278 | 254 | 154 |
| Type | 50 | 26 | 112 |
| Method | 4 | 0 | 10 |
| Amount | 78 | 69 | 25 |
| Frequency | 59 | 65 | 6 |
| Exposure history | 37 | 7 | 10 |
| Quit history | 37 | 6 | 2 |

Annotators assigned a discrete status value (current, past, none) to each substance abuse event annotated in text. If there is no mention of a substance abuse event, we will assume status is unknown for the patient when we use this information for secondary use applications. For this dataset, the tobacco status is known for approximately 76% of patients, and alcohol status for 70% of patients, which suggests that the social history provides reasonably good coverage for secondary use.

## 4 Methods

Our method for automatic identification of substance abuse events involves three main steps. The first step identifies sentences that contain substance abuse events. The second step extracts the 7 entities used to describe events, and the final step creates event templates from the extracted entities.

## 4.1    Step 1 - Identification of Sentences with Substance Abuse Events

The annotated corpus contained 454 sentences with substance abuse events, of which 174 sentences included information about multiple events of the same or different types. We trained a two-layer classifier to identify sentences with substance abuse events. The first layer classifier identifies sentences with any of the substance abuse events. The second layer of binary classifiers identifies the types of events described in those sentences identified by the first layer classifier. We represented the content of sentences with n-grams (uni, bi, tri) and defined a binary feature to identify sentences that include terms related to smoking, alcohol consumption, or drug abuse. We automatically compiled a list of 640 abuse terms by recursively traversing the WordNet hierarchy [12] for all the hyponyms of the root phrases "tobacco" for tobacco, "alcoholic drink" for alcohol, and "sedative", "narcotic", and "controlled substance" for drug abuse. There were 46 terms for tobacco (e.g., cigarette, cigar), 324 terms for alcohol (e.g., wine, porter, scotch), and 270 terms for drug (e.g., heroin, cocaine).

We manually added the verbs that describe the relevant actions (e.g., smoke) and their different forms (e.g., smokes, smoking, smoked). Our final list included 668 phrases. As new substances come into use, such as e-cigarette use or a new street drug, it would be possible to add such terms manually. We defined two types of features from this list. The first type is a binary feature (WordNet) to indicate a given sentence includes any term from the list of 668 phrases regardless of event type. We used this feature in the first layer classifier. We also defined 3 more binary features refined with event types to indicate the presence of such terms in a given sentence: tobacco (WordNet_tobacco), alcohol (WordNet_alcohol), and drug (WordNet_drug). We used these features in the second layer classifiers.

## 4.2    Step 2 - Extraction of Entities for Substance Abuse Events

After identifying sentences that describe substance abuse events, the next step is to extract the seven entities we used in our event definitions. We explored the training set and grouped the seven entities into three categories. The first group included *status*. We defined the *status* extraction problem as a multi-class classification into three possible values for status: *past, current*, and *none*. The second group included *amount, frequency, exposure history*, and *quit history*. For this group of entities, the language used was quite rich so we defined the extraction problem as a statistical named-entity recognition (NER) task to identify the spans of text that represent these entities. The final group included *type* and *method*. Because the vocabulary used to describe these entities in clinical notes is very limited, we used a rule-based approach for extraction.

**Group 1 – *Status* Extraction with Multi-class Classification:** For each event type, we trained a classifier that determines the status of the event (past, current, none). We used n-gram features and a binary assertion feature to represent the sentences in our experiments. We applied our in-house statistical assertion classifier to the substance abuse terms to establish their presence or absence [13]. While this assertion analyzer returns 6 values (absent, present, conditional, possible, hypothetical, and associated with someone else), the binary assertion feature for our experiments is positive when present, conditional, possible or hypothetical, and negative when absent or associated with someone else.

**Group 2 – *Amount, Frequency, Exposure History,* and *Quit History* Extraction with Sequential Classification:** We used NER with sequential classification to extract text spans. Due to the small size of the dataset and the similarities in the language used to describe amount and frequency, we merged annotations from all three substance abuse types in order to train these models. We used conditional random fields (CRF) with unigram and gazetteer features [14]. We manually curated gazetteers for each entity type based on annotations of the training set (Table 3). Each gazetteer includes very small number of frequent words. We experimented with various window sizes to identify the optimum configuration for each entity type.

**Table 3.** Gazetteers for entity types.

| Entity Type | Terms included in gazetteer |
| --- | --- |
| *Amount* | Abuse, social, mild, moderate, heavy, light, significant, pack, glass, bottle, cup |
| *Frequency* | Occasional, rare, chronic, daily, monthly, yearly, everyday |
| *Exposure and quit history* | Day, month, year |

**Group 3 – *Type* and *Method* Extraction with Rules:** For *type* and *method* we defined a rule-based approach. For *type*, we used WordNet lemmas. All appearances of WordNet lemmas were identified as type of the corresponding substance abuse.

For *method*, a list of tokens used to describe the method of abuse was obtained by processing the training set. Because the annotation guidelines did not annotate default values *smoke* for tobacco and *drink* for alcohol, annotations for method are quite sparse. Terms included are *chew* and *dip* for tobacco and *IV* for drug.

### 4.3    Step 3 – Event Template Creation

We used a basic approach to create the event templates. For each sentence identified as positive for a substance abuse event, we created an event template with 7 entities. Because our annotation guideline required a status assignment for each annotated event, we filled the status with the automatically extracted status value. We filled the remaining 6 entities with any corresponding text spans that were identified by either the NER or the rule-based methods.

## 5    Results

We divided our corpus of social history sections into a training set (80%, viz. 291 sections) and a test set (20%, viz. 73 sections). There were 993 sentences in the training set and 241 sentences in the test set. 358 sentences in the training set (36%)

and 96 sentences (40%) in the test set included substance abuse event. We used cross-validation on the training set to select the best performing features and algorithms to develop our models. After model development was complete, we calculated the overall extraction performance on the test set only, using precision, recall, and F1-score. In the following sections, we detail the experimental results.

## 5.1 Performance of Sentence and Entity Extraction Steps on the Training Set

Table 4 presents the results of 5-fold cross-validation on the training set for identifying sentences with substance abuse events. We first experimented with a number of classifiers and selected the MaxEnt implementation in Mallet (http://mallet.cs.umass.edu/), for its good performance on the task. The simplest model represents the text using only unigram features. As can be seen from Table 4, adding bigrams and trigrams did not improve the performance over the simple model. Using only the WordNet feature resulted in higher recall but lower precision compared to the simple model. A model which combines the unigram model with the WordNet features resulted in the best classification performance. Thus, the MaxEnt model with unigram and WordNet features was selected as the first layer classification, for identifying sentences with substance abuse events.

**Table 4.** Sentence classification results.

| Features | TP | TN | FP | FN | Precision | Recall | F1-Score |
|---|---|---|---|---|---|---|---|
| Unigram (baseline) | 335 | 623 | 12 | 23 | 0.9654 | 0.9358 | 0.9504 |
| Uni+bigram | 328 | 625 | 10 | 30 | 0.9704 | 0.9162 | 0.9425 |
| Uni+bi_trigram | 327 | 626 | 9 | 31 | 0.9734 | 0.9134 | 0.9424 |
| WordNet | 358 | 605 | 30 | 0 | 0.9227 | 1.0 | 0.9598 |
| Unigram+WordNet | 355 | 620 | 15 | 3 | 0.9595 | 0.9916 | **0.9753** |

For each sentence identified as including a substance abuse event, the second layer of binary classifiers identified the type of substance. Table 5 includes 5-fold cross validation results for MaxEnt with different feature combinations based on human-annotations of substance-event sentences (i.e. gold annotations). As can be seen from the table for all three types of events, adding WordNet features improved the performance over the model using unigrams alone. Thus, models using both unigram and WordNet features were selected for this second step in identifying substance abuse events.

**Table 5.** Substance abuse event classification results on gold annotations.

| EventType | Features | TP | TN | FP | FN | Precision | Recall | F1-Score |
|---|---|---|---|---|---|---|---|---|
| Tobbaco | Unigram | 209 | 135 | 7 | 7 | 0.9676 | 0.9676 | 0.9676 |
| Tobacco | Unigram+WordNet_Tobacco | 216 | 140 | 2 | 0 | 0.9908 | 1.0 | **0.9954** |
| Alcohol | Unigram | 184 | 148 | 6 | 20 | 0.9684 | 0.9020 | 0.9340 |
| Alcohol | Unigram+WordNet_Alcohol | 203 | 150 | 4 | 1 | 0.9807 | 0.9951 | **0.9878** |
| Drug | Unigram | 106 | 241 | 0 | 11 | 1.0 | 0.9060 | 0.9507 |
| Drug | Unigram+WordNet_Drug | 114 | 238 | 3 | 3 | 0.9744 | 0.9744 | **0.9744** |

**Step 2. Performance of Entity Extraction on the Training Set:** Table 6 includes multiclass classification results for *status*, based on human-annotations of substance abuse events (i.e. gold annotations). We show here only results for the unigrams models and those using both unigrams and assertion features. As can be seen in Table 6, models that included the assertion feature display improved classification performance with the exception of Tobacco – *status: past*. For alcohol – *status: past* (0.35 F1) and drug – *status: past*, the number of positive cases was insufficient to train a classifier.

**Table 6.** Status classification results on gold annotations.

| EventType | Status | Features | TP | TN | FP | FN | Precision | Recall | F1-Score |
|---|---|---|---|---|---|---|---|---|---|
| Tobacco | None | Unigram | 122 | 78 | 14 | 5 | 0.8971 | 0.9606 | 0.9278 |
| | | Unigram+assertion | 122 | 84 | 8 | 5 | 0.9385 | 0.9606 | **0.9494** |
| Tobacco | Current | Unigram | 35 | 161 | 11 | 12 | 0.7609 | 0.7447 | 0.7527 |
| | | Unigram+assertion | 37 | 164 | 8 | 10 | 0.8222 | 0.7872 | **0.8043** |
| Tobbaco | Past | Unigram | 31 | 173 | 1 | 14 | 0.9688 | 0.6889 | **0.8052** |
| | | Unigram+assertion | 31 | 171 | 3 | 14 | 0.9118 | 0.6889 | 0.7848 |
| Alcohol | None | Unigram | 123 | 67 | 9 | 5 | 0.9318 | 0.9609 | 0.9462 |
| | | Unigram+assertion | 123 | 70 | 6 | 5 | 0.9535 | 0.9609 | **0.9572** |
| Alcohol | Current | Unigram | 47 | 137 | 5 | 15 | 0.9038 | 0.7581 | 0.8246 |
| | | Unigram+assertion | 52 | 131 | 11 | 10 | 0.8254 | 0.8387 | **0.8320** |
| Alcohol | Past | Unigram | 3 | 190 | 0 | 11 | 1.0 | 0.2143 | **0.3529** |
| | | Unigram+assertion | 3 | 190 | 0 | 11 | 1.0 | 0.2143 | **0.3529** |
| Drug | None | Unigram | 101 | 17 | 0 | 1 | 0.9099 | 0.9902 | 0.9484 |
| | | Unigram+assertion | 102 | 22 | 5 | 0 | 0.9533 | 1.0 | **0.9761** |
| Drug | Current | Unigram | 13 | 103 | 4 | 9 | 0.7647 | 0.5909 | 0.6667 |
| | | Unigram+assertion | 14 | 103 | 4 | 8 | 0.7778 | 0.6364 | **0.7000** |
| Drug | Past | Unigram | 0 | 124 | 0 | 5 | – | 0 | – |
| | | Unigram+assertion | 0 | 124 | 0 | 5 | – | 0 | – |

To identify text spans for *amount, frequency, exposure history*, and *quit history*, we used Conditional Random Fields (CRF) implementation in CRFSuite (http://www.chokkan.org/software/crfsuite/). We ran an extensive number of experiments to determine the proper window size k (k = 1 − 5). For *amount* and *frequency*, we achieved the best performance when k = 1, for *exposure history* and *quit history*, we achieved the best performance when k = 2. As can be seen in Table 7, adding gazetteer features improved performance for *amount, frequency* and *quit history*.

Table 8 includes performance values for *type* and *method*. There were only two annotations of *method* for tobacco in the training set and recall was 0.5 due to an annotation error.

**Table 7.** Statistical named entity extraction token-level results on gold annotations.

| Event | Entity | Features | Precision | Recall | F1-Score |
|---|---|---|---|---|---|
| Merged | Amount | Unigram | 0.8944 | 0.4086 | 0.5610 |
| | | Unigram+gazetteer | 0.8858 | 0.5711 | **0.6944** |
| Merged | Frequency | Unigram | 0.9151 | 0.448 | 0.6016 |
| | | Unigram+gazetteer | 0.9002 | 0.5757 | **0.7023** |
| Merged | Exposure History | Unigram | 0.8361 | 0.3542 | **0.4976** |
| | | Unigram+gazetteer | 0.8361 | 0.3542 | **0.4976** |
| Merged | Quit History | Unigram | 0.8837 | 0.4578 | 0.6031 |
| | | Unigram+gazetteer | 0.8750 | 0.5341 | **0.6633** |

**Table 8.** Rule-based named entity extraction results on gold annotations.

| Event | Entity | Method | Precision | Recall | F1-Score |
|---|---|---|---|---|---|
| Tobacco | Type | Rule-based – WordNet | 0.9706 | 0.9166 | 0.9428 |
| Alcohol | Type | Rule-based – WordNet | 0.7586 | 0.88 | 0.8148 |
| Drug | Type | Rule-based – WordNet | 0.9783 | 0.9474 | 0.9626 |
| Tobacco | Method | Rule-based – manually curated list | 1 | 0.5 | 0.6667 |
| Drug | Method | Rule-based – manually curated list | 1 | 0.9 | 0.9474 |

## 5.2 Performance of the Event Extraction Pipeline on the Test Set

We applied the event extraction pipeline with its statistical models trained on the whole training set to the test set for end-to-end performance measurement. Table 9 includes the performance values for template extraction. When we filled all the seven entities for events, the performance values were 0.51 F1 for *tobacco*, 0.58 F1 for *alcohol*, and 0.53 for *drug*. This relatively low performance was expected because exact match is a very stringent measure between the system generated templates and human annotations. When we relaxed the evaluation and evaluated the templates over only *status*, the performance increased to 0.80 F1 for *tobacco*, 0.91 F1 for *alcohol*, and 0.85 F1 for *drug*.

**Table 9.** Template extraction results on test set.

| EventType | Template | TP | FP | FN | Precision | Recall | F1-Score |
|---|---|---|---|---|---|---|---|
| Tobbaco | All entities | 30 | 28 | 29 | 0.5172 | 0.5085 | 0.5128 |
| | Only status | 47 | 11 | 12 | 0.8103 | 0.7966 | 0.8034 |
| Alcohol | All entities | 30 | 23 | 20 | 0.566 | 0.6 | 0.5825 |
| | Only status | 47 | 6 | 3 | 0.8868 | 0.94 | 0.9126 |
| Drug | All entities | 13 | 11 | 12 | 0.5417 | 0.52 | 0.5306 |
| | Only status | 21 | 3 | 4 | 0.875 | 0.84 | 0.8571 |

### 5.3   Error Analysis

When we performed error analysis, we analyzed the performance of each extraction step to identify the main cause of false positive and negative templates. In the first step of extraction, sentences that contained substance abuse events were identified with high performance values (tobacco: 0.97 F1, alcohol: 0.97 F1, drug: 0.98 F1). Extraction of the specific entities related to substance abuse event caused the majority of the errors. Our statistical named entity extractors identified *amount* with 0.76 F1, *frequency* with 0.75 F1, *quit history* with 0.75 F1, and *exposure history* with 0.63 F1. Our error analysis indicated that the training set size was not enough to capture the characteristics of the entities that were extracted with statistical approaches. We plan to increase the size of our annotated corpus from other publicly available clinical corpora and clinical reports from our institution. Finally, in the third step of extraction, we created the templates by using a very simple heuristic that sometimes let to incorrect entity assignments to events. As part of future work, we plan to use statistical approaches to template filling after we extend the size of our annotated corpus.

## 6   Conclusion

In this paper, we present an annotated corpus for tobacco, alcohol, and drug abuse and an approach to extract substance abuse events automatically from clinical notes. Our annotation includes not only the status of the substance abuse (past, current or none), but also a rich annotation of other pertinent information, such as type, frequency, method, amount, quit history and exposure history, wherever that information has been included in the clinical record. To our knowledge, our corpus is the first publicly available clinical corpus for all three types of substance abuse. Our results on an unseen test set range from 0.51–0.58 F1 on the identification of the full event template, which is a stringent measure. Our results range 0.80–0.91 F1 for identification of the substance abuse event and its status, which represents the majority of the available data. Given that up to 76% of social history sections include an indication of smoking status, and up to 70% an indication of alcohol status, we hope to have demonstrated that automatic identification methods have potential to enable research in substance abuse events using the information already contained in the clinical records for secondary use research.

## References

1. Anand, P., Kunnumakara, A.B., Sundaram, C., et al.: Cancer is a preventable disease that requires major lifestyle changes. Pharm. Res. **25**(9), 2097–2116 (2008)
2. Srivastava, R.: Complicated lives – taking the social history. NEJM **265**(7), 587–589 (2011)
3. Melton, G.B., Manaktala, S., Sarkar, I.N., Chen, E.S.: Social and behavioral history information in public health datasets. In: AMIA Annual Symposium Proceedings 2012, pp. 625–634 (2012)

4. Uzuner, Ö., Goldstein, I., Luo, Y., Kohane, I.: Identifying patient smoking status from medical discharge records. J. Am. Med. Inform. Assoc. **15**(1), 15–24 (2008)
5. Cohen, A.M.: Five-way smoking status classification using text hot-spot identification and error-correcting output codes. J. Am. Med. Inform. Assoc. **15**(1), 32–35 (2008)
6. Clark, C., Good, K., Jezierny, L., Macpherson, M., Wilson, B., Chajewska, U.: Identifying smokers with a medical extraction system. J. Am. Med. Inform. Assoc. **15**(1), 36–39 (2008)
7. Jonnagaddala, J., Dai, H.J., Ray, P., Liaw, S.T.: A preliminary study on automatic identification of patient smoking status in unstructured electronic health records. In: ACL-IJCNLP 2015, pp. 147–151, 30 July 2015
8. Carter, E.W., Sarkar, I.N., Melton, G.B., Chen, E.S.: Representation of drug use in biomedical standards, clinical text, and research measures. In: AMIA Annual Symposium Proceeding 2015, pp. 376–385 (2015)
9. Chen, E., Garcia-Webb, M.: An analysis of free-text alcohol use documentation in the electronic health record: early findings and implications. Appl. Clin. Inform. **5**(2), 402–415 (2014)
10. Wang, Y., Chen, E.S., Pakhomov, S., Arsoniadis, E., Carter, E.W., Lindemann, E., Sarkar, I.N., Melton, G.B.: Automated extraction of substance use information from clinical texts. In: AMIA Annual Symposium Proceeding 2015, pp. 2121–2130, 5 November 2015
11. Tepper, M., Capurro, D., Xia, F., Vanderwende, L., Yetisgen-Yildiz, M.: Statistical section segmentation in free-text clinical records. In: Proceedings of LREC, Istanbul, May 2012
12. Millet, G.A.: WordNet: a lexical database for English. Commun. ACM **38**(11), 39–41 (1995)
13. Bejan, C.A., Vanderwende, L., Xia, F., Yetisgen-Yildiz, M.: Assertion modeling and its role in clinical phenotype identification. J. Biomed. Inform. **46**(1), 68–74 (2013)
14. McCallum, A., Li, W.: Early results for named entity recognition with conditional random fields, feature induction and web-enhanced lexicons. In: Proceedings of CONLL at HLT-NAACL, pp. 188–191 (2003)

# Analyzing Perceived Intentions of Public Health-Related Communication on Twitter

Elena Viorica Epure(✉), Rébecca Deneckere, and Camille Salinesi

Université Paris 1 Panthéon-Sorbonne, CRI, 90 Rue de Tolbiac,
75013 Paris, France
Elena.Epure@malix.univ-paris1.fr,
{Rebecca.Deneckere,Camille.Salinesi}@univ-paris1.fr

**Abstract.** The increasing population with chronic diseases and highly engaged in online communication has triggered an urge in healthcare to understand this phenomenon. We propose an automatic approach to analyze the perceived intentions behind public tweets. Our long-term goal is to create high-level, behavioral models of the health information consumers and disseminators, relevant to studies in narrative medicine and health information dissemination. The contributions of this paper are: (1) a validated intention taxonomy, derived from pragmatics and empirically adjusted to Twitter public communication; (2) a tagged health-related corpus of 1100 tweets; (3) an effective approach to automatically discover intentions from text, using supervised machine learning with discourse features only, independent of domain vocabulary. Reasoning on the results, we claim the transferability of our solution to other healthcare corpora, enabling thus more extensive studies in the concerned domains.

**Keywords:** Intention mining · Text mining · Natural language processing · Classification · Machine learning · Twitter · Speech acts · Linguistics

## 1 Introduction

The Internet has nurtured a highly available and accessible environment for disseminating health information. While in 2001, 70 000 websites contained health-related information [4], the order of magnitude for *healthcare*-verbatim websites only, has increased by three by 2013 in the United States [7]. Correlated to the massive production of online health-related content, the number of online health information seekers has doubled in this period, reaching 100 million [4,7]. Nowadays, the dissemination and consumption of health information have been also impacted by the tremendous adoption of social media [18]. This has led to the creation of communities, fostered by the interpersonal interactions with acknowledged advantages such as anonymity and 24-hour availability [4].

Social media allows to disseminate health information, express beliefs, feelings about health matters and react to existing content. One's beliefs are built on or altered by the information to which he or she is exposed [4]. This could be

© Springer International Publishing AG 2017
A. ten Teije et al. (Eds.): AIME 2017, LNAI 10259, pp. 182–192, 2017.
DOI: 10.1007/978-3-319-59758-4_19

further reflected in new behavioral intentions and eventually new behaviors [1]. Social media has even a stronger influence on consumers because of the social norms: the attitudes and behaviors of the community towards health matters are transparent in this online environment. This aspect could be exploited to promote healthy behavior such as quitting smoking. However, inaccurate information, available to a very large and generally vulnerable target–people impacted directly or indirectly by chronic diseases, could become harmful and have mass consequences [4]. Therefore, there is an urge in healthcare *to understand the effects* of the exposure to health information disseminated through social media, on consumers. The reason is *to predict these effects* as potential immediate or long-term behaviors [3,4]. However, suitable methods are necessary for this.

In the current paper, we propose a means *to discover automatically the perceived intentions of the Twitter public posts*. We call them *perceived* because they are interpreted from the stance of the information consumer. The link between perceived intentions and revealing or predicting behavior is as follows. First, the perceived intentions are a behavioral component of health disseminators. Second, the perceived intentions allow to create automatic techniques to measure the impact of various message formulation on health information consumers, similar to message framing [16]. For instance, is a consumer more likely to read an online, health-related article if its link is tweeted in a rhetorical question or in an informative fact? Third, we want to support the automatic discovery of collective narratives concerning healthcare, as they can be strong drivers for behaviors [11]. Automatic techniques already exist for identifying components of the crowd narratives such as topics or events [19]. However, disseminators' disposition towards presented events is necessary for a thorough narrative's representation [14].

## 2   Related Work

Related works to analyze the exposure to health information disseminated on social media rely on traditional research methods [3,4,16]. First, data is gathered through interviews and questionnaires. In order to reach people consuming online health information, clinics, hospitals or online communities are targeted. Then, the data is analyzed using statistics or qualitative methods. Although, these methods empowered the healthcare community to gain valuable insights, there are several limitations. The research results are mainly descriptive. However, for acting on the existing knowledge, *predictive models* are required too. Further, an extensive study is rarely possible, the reported samples being at best of several thousand subjects. These studies also come with localization and time-span constraints. In reality an online community could include worldwide members. Moreover, the questions of how certain results change over time, or how the discovered knowledge depends on current temporal trends, or how predictions could be made in real-time are challenging to answer.

Computer science methods to complement existing studies, by exploiting automatically social media in creating predictive models, exist [5,10,23]. However, predicting behavior in social media requires deep consideration of underlying processes, of how the cause–the disseminated health information–leads to the effects–the behavior changes. Hence, a solution should go beyond the black-box approach often employed in data mining and incorporate also theoretical knowledge from humanities. In conflict resolution, a model for predicting behavior changes from Twitter, based on narrative theory, is proposed [11]. Focused on perception, Myslin *et al.* [15] discover narrative-related elements from tweets about smoking: genres (e.g. first-hand experience, opinion) and themes (e.g. cessation, pleasure). Priesto *et al.* [19] identify health topics from tweets, similar to narrative's themes (e.g. depression, flue). Though, a solid start for our long-term goal, these works can be augmented with more high-level behavioral cues.

## 3   Research Design

Our objective is to analyze the perceived intentions of the publicly disseminated tweets. Specifically, several research questions are identified:

1. Could a *valid taxonomy of perceived intentions, representative* for Twitter public communication be defined?
2. How do various *supervised machine learning algorithms* with various configurations of features compare for discovering perceived intentions?
3. What are *the most predictive features* for each type of intention?

**Data Collection.** We collected two sets of tweets via Twitter Streaming API. The first consists of 2714 tweets and the second of 43153 tweets. The first set was fetched by the keyword *autoimmune*. The second set was collected with commonly used medical terms and jargon, proposed by Ridpath et al. [20]. Even though the communication on autoimmune diseases was initially targetted, we included the second set for data diversity in evaluation, thus aiming at the input generalizability. Further, we sampled randomly 600 tweets from the first set and 500 tweets from the second set, with no message format duplicates. The selected 1100 tweets were used in the intention taxonomy's validation and in creating the ground-truth corpus for the machine learning experiment.

**Research Method.** For the first research question, we considered as *valid* an intention taxonomy that is consistently applied by raters, showing thus an alignment in the tweets' perception and ensuring experimental reproducibility. For this, the selected corpus was tagged in parallel by two raters: one expert involved in defining the taxonomy, the other seeing the taxonomy for the first time (researcher in computer science). The taxonomy was briefly presented as the inexperienced rater was expected to rather rely on its intuition when tagging. At least one intention had to be chosen per tweet. More intentions were allowed when the tweet had multiple sentences or one intention could not be clearly conveyed. After the tagging, the Fleiss' Kappa statistical test [6] was used for evaluating the validity. When multiple tags were used, an alignment of the sets was necessary. For instance, $[t_1, t_2]$ and $[t_2, t_1, t_3]$ would align as: $t_1$-$t_1$,

$t_2$-$t_2$, $t_3$-*unknown*. The test considers the degree of agreement for each such pair and the probability of agreeing by chance. A score of the Fleiss' Kappa test over 0.6 is considered a good result, specifically between 0.61 and 0.8 *substantial* and between 0.81 and 0.99 *almost perfect* [6]. Further, for assessing if the taxonomy was *representative*, a tag *other* was created to be used when none of the proposed intentions was a suitable choice and its frequency was computed.

For the next research questions, several steps were required. First, the truth set was created based on the tagged corpus. For each non-agreement, a discussion took place between the raters and a collective final decision was made. In the final corpus[1], 89% of the tweets were single-tag and the rest had associated 2–3 intentions. Second, text processing and Tweet NLP [17] were applied for feature extraction. Two types of features were defined: *Content* and *Discourse* features. *Content* features consisted of standard text mining features: *BagOfWords* and *OpinionKeywords*. These were computed after the corpus' lower case conversion and lemmatisation. *BagOfWords* was a dictionary of tokens and their frequencies. The tokens were extracted from the pre-processed Twitter corpus before the classification. *OpinionKeywords* included the frequencies and ratios of negative and positive opinion words from a predefined lexicon [13]: *freqPositiveWords*, *ratioPositiveWords*, *freqNegativeWords*, *ratioNegativeWords*. *Discourse* features are novel and defined by considering linguistic means to express intentions. They are described in Sect. 5, after introducing the proposed taxonomy.

*Logistic Regression, Linear SVM, Random Forest* and *Multinomial Naive Bayes* were the selected classifiers. The *scikit-sklearn*[2] implementations were used with default parameters. *Multinomial Naive Bayes* was selected as its library's implementation allowed continuous features too. *Naive Bayes* and *Random Forest* handled inherently multiple classes while *Logistic Regression* and *Linear SVM* in an on-versus-all strategy. Various configurations of features were evaluated: discourse features only (*DiscF*), content features only (*ContF*), all features (*AllF*). Also, the features were scaled beforehand. The metrics for performance evaluation of the *single-tag* corpus were *precision*, *recall* and *f-score*. These were computed as *macro scores*, weighted by the support of each intention and averaged over *10 folds*. Same evaluation decisions were applied within each fold in cross-validation. The *hamming loss* was used to evaluate the accurate prediction of *multi-tag* corpus. For this, the single-tag corpus was the train set and the multi-tag corpus the test set. The *hamming loss* is the ratio of intentions in average that are incorrectly predicted. Finally, with no interaction in models, the most predictive features per intention were found by analyzing the weights of the best classifier, trained on standardized features' values [8].

## 4   An Intention Taxonomy for Public Tweets

Human behavior is intrinsically intentional as thoroughly discussed in philosophy [2] and psychology [1]. Though, behavior is not necessarily linked to only physical

---

[1] http://tinyurl.com/hk9t83y.

[2] http://scikit-learn.org/stable/supervised_learning.html.

human acts but also to language. Generally, people communicate with various intentions. Utterances are considered thus as acting through words while their leading intentions are called *speech acts* [21]. For example, *This hospital has a nonstop emergency service* asserts the speaker' belief about the world while *Could you please give me a painkiller?* requires the listener to act. Searle [21] proposed five *classes* of speech acts. *An assertive* is used for stating information being true or false about the state of affairs in the speaker's world. *A commissive* denotes the engagement of the speaker to a future course of action. *A directive* implies the listener carrying out an action as a result of the speaker's utterance. *An expressive* is used to express the speaker's feelings towards the state of affairs in the world. Finally, *a declarative* is the type of utterance, changing the world' state such as firing someone. *The speech act theory* emerged over time, as a highly-adopted framework to extract or predict behavior from text.

However, we aimed at more granular intentions than Searle's classes [21]. This was enabled by the work of Vanderveken [22] who proposed a lexicalization of the intentions through 300 English verbs. These verbs are organized in five hierarchies where the roots are the speech act classes and each level is a specialization of the parents. Our taxonomy emerged from the Vanderveken's theoretical work [22] but its refinement was based on manual corpus analysis. Thus, we empirically discovered that *assertive* and *directive* classes cover most of Twitter public communication (see Table 1). These findings are not surprising considering that public tweets rarely contain personal feelings (*expressive*) or personal goals (*commissive*). Such communication takes place rather privately. Moreover, declarative speech acts are very rare even in live settings.

**Table 1.** Identified intentions from Twitter public communication.

| Class | Intention | Tweet example |
|---|---|---|
| Assertive | assert | New study reveals autoimmune/inflammatory syndrome triggered by HPV vaccine URL |
| Assertive | hypothesize | Vitamin B1 may help relieve fatigue in Hashimoto's thyroid patients URL |
| Directive | propose | The gut microbiota and inflammatory bowel disease #microbiology #autoimmune URL #gutmicrobiota |
| Directive | direct | Are you a Cure Champion? Sign up for the Walk to Cure Psoriasis in a city near you... URL |
| Directive | advise | How To Avoid Holiday Autoimmune Flares URL |
| Directive | warn | Why you shouldn't be going from competition to competition URL #thyroid #metabolism #autoimmune |

The rational behind the association of intentions to tweets is presented further. An *assert* is a tweet that clearly conveys the message such as news or personal public declarations. Though often it has a url, the linked resource appears with the role to sustain or detail the message. A *hypothesize* is a tweet containing a weak assertion such as probable statements or hypothetical questions. Compared to the assertives, a *propose* is a tweet that always references an external resource, which must be accessed in order to consume the message. The *propose* tweets usually provide key or opinion words about the resource's content and could be considered a weak attempt to make the reader access the url. In contrast, a *direct* is a strong attempt to make the reader act to demands, requests, invitations, encouragements or questions. Finally, *advise* and *warn*, which are very similar, are directives to a future action or resource consumption that is supposed to be good or bad for the reader.

## 5   Discourse Features for Intention Discovery

A person could use Twitter for addressing utterances to the community or to specific users. In a live communication, the utterance's intention is implicitly understood from the utterance's content, speaker's gestures and voice. Though not as rich as this case, the written tweets have also characteristics that convey their intentions. We relate them to *Discourse* features, which are further presented.

*PronominalKeywords* are frequencies of various pronominal forms. The first person singular (*freq1stPersonSg*) is chosen because it could be a sign of personal declarations specific to assert tweets (e.g. *I'm for vaccination*). Similarly, the third person (*freq3rdPerson*) could be linked to assert tweets when reporting. By contrary, the first person plural form ( *freq1stPersonPl*) or second person (*freq2ndPerson*) could be cues of direct tweets (e.g. *We must vaccinate our kids! You should too*). Further, *PunctuationMarks* are indicators of the discourse' functions: the presence of exclamation (*hasExclamation*), interrogation (*hasQuestion*), ellipsis (*hasEllipsis*; for omissions, hesitations); colon (*hasColon*; for titles, explanations) or quotes (*hasQuotes*). *QuestionKeywords* complements the punctuations for revealing discourse functions (e.g. questions). We separate the frequency of *what, when, where, why, who* (*freq5W*), from that of *how* (*freq1H*) because *how* is also used in advice or proposals (e.g. *How I Gave Up Smoking*).

The frequency of *EmoticonCues* (*freqEmoticons*) could be inversely correlated with the impersonal reporting; hence possibly linked to assert and propose tweets. Both ASCII and Unicode emoticons were checked. *TitleCues* (*hasTitle*) seems to be an often marker for news, being thus a potential discriminator for the assert and propose tweets (e.g.: *New Releases in Science*). *VerbPhrases* (*hasVerb*, present and past participles not considered) could be an indicator of weak directives when false. These tweets often lack the subject-predicate form. *VerbMoods* contains *hasImperative*, which is frequent in requests, demands; thus being an indicator of direct tweets. For identifying imperatives, we created a rule-based algorithm, having as input the part-of-speech (POS) tags. *VerbKeywords* encompasses features regarding the modals (*hasCan, hasCould, hasMust, hasMay, hasMight, hasShould*). Modals could show various intentions: hypothesize, advise, direct etc. For verb features, negative forms were identified too.

*SyntaticConstructs* features are created with the goal of incorporating syntactic characteristics of the discourse. The assumption was that tweets with same intention might share similar discourse form. We used the POS tags in order to dynamically discover representative POS-related features. The way we proceeded was: POS-tag the corpus using a dedicated tweet NLP parser [17]; compute the normalized frequencies of each two consecutive POS tags from the output; select those with a score of at least *0.5* per intention. The final syntactic features are: *hasNV, hasNN, hasAN, hasNComma, hasPN, hasDN, hasVN, hasCommaU, hasNP*. The encoding of these features is: *N* nouns; *V* verbs, *A* adjectives, *Comma* punctuation, *P* pre-, post-position or subordinating conjunctions, *D* determiners, and *U* urls. Compared to the original output of the parser [16], two changes were made. *N* incorporates also the proper nouns (the symbol ^) and pronouns (the symbol *O*). *Comma* replaces *,*.

## 6    Results and Discussion

The first research question sought to answer if the proposed intention taxonomy was *representative for Twitter public communication* and *valid*. The condition of being *representative* could be considered fulfilled through artifact design, by being both theory- and corpus-driven. Relying on theory, we ensured that the taxonomy was linguistically representative for written utterances. Relying on corpus analysis, we ensured that the taxonomy was mapped on the actual Twitter public communication. Further, we agreed that a high frequency of the tag *other* would denote a lack of representativeness. However, *other* was used only in 0.005 of the cases by both raters, supporting thus our taxonomy's design. The tag *other* replaced *expressives* such as greetings or *commissive* such as public promises. The Fleiss' Kappa test was performed to assess the taxonomy's *validity* (see Table 2). The overall *intention-wise* Fleiss' Kappa score was 71.5% ($z = 40.9$, $p = 0.001$) being considered a *substantial* result. All intentions apart from *advise* and *other* achieved substantial scores. The result for *other* is not surprising given its very low frequency. The overall *class-wise* Fleiss' Kappa score was 78.3% ($z = 28.1$, $p = 0.001$), showing thus that mismatches occurred sometimes between intentions of the same class. Related works of speech act tagging for tweets reported similar results (Kappa scores between 0.6 and 0.85) [5,10,23].

**Table 2.** Fleiss' Kappa reported as *result, z-score* for intentions and classes.

| Assertive: 0.8,27.2 | | Directive: 0.8,27.1 | | | | Other: 0.41,13.9 |
|---|---|---|---|---|---|---|
| assert | hypothesize | propose | advise | direct | warn | other |
| **0.77**, 26.4 | **0.78**, 26.7 | **0.70**, 24.1 | 0.49, 16.7 | **0.68**, 23.3 | **0.76**, 25.9 | 0.35, 12.1 |

Several decisions were made for the final corpus creation. 82% of mismatches between raters concerned the *advise* tweets, specifically with propose (55%), direct (25%), assert (15%), warn (5%). By further manual analysis, we concluded a tweet was an advice either because it redirected the reader to an external source that actually contained an advice or it directly contained the advice.

The first case corresponded to mismatches involving *propose* (e.g. *4 Steps to Heal Leaky Gut and Autoimmune Disease URL*), while the second case to mismatches regarding *direct* (e.g. *#TipTuesday: Vitamin D deficiency is linked to autoimmune diseases. Add mushrooms to Thanksgiving!*) and *assert* (*The quicker you receive treatment, the better your chances for a good recovery from #Stroke are*). It might be surprising an advice is stated as an assert but this is an example of *indirect* speech acts. Nevertheless, what emerges is that an *advise* tweet could be ultimately a *propose*, *direct* or *assert* too. Considering its low Fless' Kappa score, we decided for now to transform the *advise* tweets in their secondary speech acts. Moreover, we transformed the *warn* tweets too because of insufficient instances (0.02% of the corpus) and for maintaining consistency with *advise*.

The next research question looked into the comparison of classifiers with different features' configuration. The results are summarized in Table 3. Statistically, Logistic Regression and Linear SVM are comparable ($p > 0.5$, two-tailed t-test) and both outperformed Random Forest. Multinomial Naive Bayes is not reported as yielded results similar to Random Forest for *DiscF* and *ContF*, and worse for *AllF*. We can notice that *DiscF* systemically leads to similar results as *AllF* (apart from *hypothesize* with Logistic Regression) and improves results over *ContF* up to 5 times with Random Forest. For multi-label classification, the minimum *hamming loss* scores are obtained when using *DiscF* with Logistic Regression (0.287) and Linear SVM (0.264).

The last research question aimed at the evaluation of the features' predictive power in relation to each intention. This was assessed based on the features' weights estimated for Linear SVM. As expected, for *assert*, the most predictive discourse features are related to reporting information or making public declarations: hasVerb, hasNV, hasQuotes, freq3rdPerson, freq1stPersonPl, freq1stPersonSg. The most representative content features seem to be related to news and particularly to scientific ones ("implications", "future", "bacteria", "influence"). For *hypothesize*, both the discourse and content features reveal the importance of modals ("might", "may" and "could"). Apart from these, the most important discourse feature is hasQuestion. For *direct*, hasImperative is the most discriminatory feature. This also emerges from top predictive content features that contain multiple verbs ("know", "learn", "read", "check", "enjoy"). Then, features linked to requests, encouragements and questions follow in importance in the classification of *direct*: for discourse features–hasQuestion, hasExclamation, freq2ndPerson, freq1H, freq5W; for content features –"what", "please", "let". For *propose*, the top most important discourse features are related to news summaries (hasTitle, freq5W, hasColon) and advice or warnings (freq1H, hasMust, hasShould, freq2ndPerson). These features are correlated to the content ones, which incorporate interrogation cues ("why", "how"), impersonal scientific words ("epidemic", "lupus") and advice/warning words ("step", "recipes", "good"). Finally, the POS-related features appear highly predictive for all intentions, in particular for *assert*, *direct* and *hypothesize*.

In conclusion, the proposed *discourse features* improved significantly the discovery of intentions and we often observed that they were also correlated to the

**Table 3.** Results of the classification experiment using various feature sets.

| Intention | Metric | LogisticRegression | | | RandomForest | | | LinearSVM | | |
|---|---|---|---|---|---|---|---|---|---|---|
| | | DiscF | ContF | AllF | DiscF | ContF | AllF | DiscF | ContF | AllF |
| assert | *precision* | 0.75 | 0.67 | 0.77 | 0.70 | 0.57 | 0.66 | 0.77 | 0.69 | 0.79 |
| sup. = 416 | *recall* | 0.83 | 0.71 | 0.84 | 0.79 | 0.64 | 0.86 | 0.82 | 0.66 | 0.83 |
| | *f-score* | 0.79 | 0.69 | **0.80** | **0.74** | 0.60 | **0.74** | **0.80** | 0.68 | **0.81** |
| hypothesize | *precision* | 0.62 | 0.62 | 0.77 | 0.65 | 0.50 | 0.60 | 0.71 | 0.55 | 0.75 |
| sup. = 49 | *recall* | 0.41 | 0.10 | 0.47 | 0.49 | 0.06 | 0.12 | 0.76 | 0.33 | 0.55 |
| | *f-score* | 0.49 | 0.18 | **0.58** | **0.56** | 0.11 | 0.20 | **0.73** | 0.41 | 0.64 |
| direct | *precision* | 0.71 | 0.66 | 0.78 | 0.70 | 0.59 | 0.67 | 0.75 | 0.50 | 0.77 |
| sup. = 119 | *recall* | 0.79 | 0.18 | 0.75 | 0.70 | 0.11 | 0.35 | 0.82 | 0.27 | 0.79 |
| | *f-score* | 0.75 | 0.28 | **0.76** | **0.70** | 0.18 | 0.46 | **0.78** | 0.35 | **0.78** |
| propose | *precision* | 0.80 | 0.63 | 0.79 | 0.74 | 0.57 | 0.75 | 0.80 | 0.62 | 0.78 |
| sup. = 391 | *recall* | 0.72 | 0.79 | 0.75 | 0.66 | 0.71 | 0.70 | 0.72 | 0.76 | 0.76 |
| | *f-score* | 0.76 | 0.70 | **0.77** | 0.70 | 0.64 | **0.72** | 0.76 | 0.68 | **0.77** |

top most important content features. However, the discourse features benefit of being much fewer (30 vs. 4608) and corpus-independent, allowing thus reproducibility on other medical English corpora. In the analyzed corpus, the most popular intentions are *assert* (46%) and *propose* (41%) revealing thus that Twitter is publicly used for information dissemination. However, it is quite interesting that the strategies are different, half of the tweets' messages being self-standing (*assert*) while the other half requiring external redirection (*propose*). The ratio of *direct* tweets (18%) shows also a significant expected reaction from consumers, by replying or following advice, warnings, requests or invitations. Similar to us, Godea *et al.* [9] identify tweets' purposes in healthcare (advertising, informational, positive or negative opinions). However, we focus on intentions as established by pragmatics, ensuring thus a domain-independent, general approach.

## 7   Conclusion and Future Work

An approach for analyzing the perceived intentions in the Twitter public communication was proposed. An *intention taxonomy* for public tweets was defined and validated. Its automatic discovery proved effective, with f-scores between 0.73 and 0.8 using *Linear SVM* and *discourse features* only. The most predictive content features were often linked to the discourse ones and intentions, acting thus as a positive feedback loop to the proposed taxonomy and features' decisions.

Future work must address several limits. As *advise* and *warn* had low scores, the taxonomy must be revised and experiments deployed with more raters. Twitter private communication should be researched too, including intentions from *expressive* and *commissive*. The impact of parameters' values on classifiers' performance must be assessed. Finally, as the proposed approach is domain-independent, we envision a large-scale analysis of public, health-related tweets.

Information dissemination in different communities (e.g. diseases) can be compared. The public's reactions to the same, but differently formulated messages can be analyzed (e.g. which formulation is most re-tweeted?). Then, within a community, conflicting narratives could emerge. The work of Houghton *et al.* [11] can be extended to identify them while also considering disseminators' disposition as perceived intentions. The current work is our first attempt in joining narrative medicine, for *bringing the patient as a subject back into medicine* [12].

# References

1. Ajzen, I.: The theory of planned behavior. Organ. Behav. Hum. Decis. Process. J. **50**(2), 179–211 (1991)
2. Bratman, M.: Intention, Plans, and Practical Reason. Harvard University Press, Massachusetts (1987)
3. Calvert, J.K.: An ecological view of internet health information seeking behavior predictors: findings from the chain study. Open AIDS J. **7**(1), 42–46 (2013)
4. Cline, R.J.W.: Consumer health information seeking on the internet: the state of the art. Health Educ. Res. **16**(6), 671–692 (2001)
5. Ding, X., Liu, T., Duan, J., Nie, J.Y.: Mining user consumption intention from social media using domain adaptive convolutional neural network. In: The 29th AAAI Conference, pp. 2389–2395. AAAI Press (2015)
6. Fleiss, J.L., Cohen, J., Everitt, B.S.: Large sample standard errors of kappa and weighted kappa. Psychol. Bull. **72**(5), 323–327 (1969)
7. Fox, S., Duggan, M.: Health online 2013. Technical report, PEW (2017). http://www.pewinternet.org/2013/01/15/health-online-2013/
8. Gelman, A.: Scaling regression inputs by dividing by two standard deviations. Stat. Med. **27**(15), 2865–2873 (2008)
9. Godea, A.K., Caragea, C., Bulgarov, F.A., Ramisetty-Mikler, S.: An analysis of Twitter data on E-cigarette sentiments and promotion. In: Holmes, J.H., Bellazzi, R., Sacchi, L., Peek, N. (eds.) AIME 2015. LNCS (LNAI), vol. 9105, pp. 205–215. Springer, Cham (2015). doi:10.1007/978-3-319-19551-3_27
10. Hemphill, L., Roback, A.J.: Tweet acts: how constituents lobby congress via Twitter. In: CSCW 2014, pp. 1200–1210. ACM, New York (2014)
11. Houghton, J., Siegel, M., Goldsmith, D.: Modeling the influence of narratives on collective behavior case study. In: International System Dynamics Conference (2013)
12. Kalitzkus, V.: Narrative-based medicine: potential, pitfalls, and practice. Perm. J. **13**(1), 80 (2009)
13. Liu, B., Hu, M., Cheng, J.: Opinion observer: analyzing and comparing opinions on the web. In: The 14th International Conference on WWW, pp. 342–351. ACM, New York (2005)
14. Murphy, J.: The role of clinical records in narrative medicine: a discourse of message. Perm. J. **20**, 103 (2016)
15. Myslin, M., Zhu, S.H., Chapman, W., Conway, M.: Using Twitter to examine smoking behavior and perceptions of emerging tobacco products. Med. Internet Res. **15**(8), e174 (2013)
16. O'Keefe, D., Jensen, J., Jakob, D.: The relative persuasiveness of gain-framed loss-framed messages for encouraging disease prevention behaviors. Health Commun. J. **12**(7), 623–644 (2007)

17. Owoputi, O., Dyer, C., Gimpel, K., Schneider, N., Smith, N.: Improved part-of-speech tagging for online conversational text with word clusters. In: NAACL (2013)
18. Perrin, A., Duggan, M., Greenwood, S.: Social media update 2016. Technical report, PEW (2017). http://www.pewinternet.org/2016/11/11/social-media-upda te-2016/
19. Prieto, V.M., Matos, S., Alvarez, M., Cacheda, F., Oliveira, J.L.: Twitter: a good place to detect health conditions. PLOS ONE 9(1), 1–11 (2014)
20. Ridpath, J.R., Wiese, C.J., Greene, S.M.: Looking at research consent forms through a participant-centered lens. J. Health Promot. 23(6), 371–375 (2009)
21. Searle, J.R.: Speech Acts, vol. 1. Cambridge University Press, Cambridge (1969)
22. Vanderveken, D.: Meaning and Speech Acts. Cambridge University Press, Cambridge (1990)
23. Wang, J., Cong, G., Zhao, W.X., Li, X.: Mining user intents in Twitter. In: The 29th AAAI Conference, pp. 318–324. AAAI Press (2015)

# Exploring IBM Watson to Extract Meaningful Information from the List of References of a Clinical Practice Guideline

Elisa Salvi[1(✉)], Enea Parimbelli[1], Alessia Basadonne[1], Natalia Viani[1],
Anna Cavallini[2], Giuseppe Micieli[2], Silvana Quaglini[1],
and Lucia Sacchi[1(✉)]

[1] Department of Electrical Computer and Biomedical Engineering,
University of Pavia, Pavia, Italy
lucia.sacchi@unipv.it
[2] IRCCS Istituto Neurologico C. Mondino, Pavia, Italy

**Abstract.** In clinical practice, physicians often need to take decisions based both on previous experience and medical evidence. Such evidence is usually available in the form of clinical practice guidelines, which elaborate and summarize the knowledge contained in multiple documents. During clinical practice the synthetic format of medical guidelines is an advantage. However, when guidelines are used for educational purposes or when a clinician wants to gain deeper insight into a recommendation, it could be useful to examine all the guideline references relevant to a specific question. In this work we explored IBM Watson services available on the Bluemix cloud to automatically retrieve information from the wide corpus of documents referenced in a recent Italian compendium on emergency neurology. We integrated this functionality in a web application that combines multiple Watson services to index and query the referenced corpus. To evaluate the proposed approach we use the original guideline to check whether the retrieved text matches the actions mentioned in the recommendations.

**Keywords:** Information retrieval · Clinical decision support · Natural language processing

## 1 Introduction

During decision-making processes, physicians often rely on clinical guidelines, which are documents summarizing the bio-medical evidence available in the literature, to get an advice that can be immediately used for a specific clinical condition. However, in some cases it may be useful for physicians to gain deeper insight into a recommendation provided by the guideline, and examine in depth the relevant content of the corpus of documents that were used to generate the examined suggestion. For example, given their limited practical experience, junior physicians use online information resources for expanding their medical knowledge beyond clinical guidelines and improving their performances in daily practice [1]. In such cases, question-answering systems (QASs) are interesting tools for extracting the information of interest from a wide corpus of relevant documents [2].

© Springer International Publishing AG 2017
A. ten Teije et al. (Eds.): AIME 2017, LNAI 10259, pp. 193–197, 2017.
DOI: 10.1007/978-3-319-59758-4_20

One of the most known domain-independent QASs is IBM Watson$^{TM}$ [3], a system that in 2011 proved to be able to successfully compete and win against human players in the popular quiz show "Jeopardy!®". To achieve this goal, IBM Watson$^{TM}$ exploited multiple components implementing solutions in the fields of natural language processing (NLP), machine learning, and information retrieval to process natural language questions, elaborate evidence from multiple sources of textual information (e.g., encyclopedias, books, and articles), and return the most plausible answer. After the result achieved in the "Jeopardy!®" challenge, a lot of attention has been focused on IBM Watson$^{TM}$, since the potential of such a system clearly extends beyond winning a quiz game.

Several efforts have been dedicated to the development of QASs for the clinical domain [4–6]. Among these, some systems integrate IBM Watson$^{TM}$ components [6].

To support developers in the implementation of QASs, IBM has published on the IBM Bluemix cloud a set of services named "Watson services", which allow the integration of some basic Watson$^{TM}$ functionalities into custom applications (https://www.ibm.com/watson/developercloud/services-catalog.html).

Watson services are freely available for researchers at several universities thanks to the IBM Academic Initiative (https://developer.ibm.com/academic/). In this work, we explore how Watson services can be used for the automatic retrieval of relevant information from a corpus of documents on the basis of user-defined natural language questions. As a case study, we focus on a recent Italian compendium that collects recommendations on emergency neurology [7], and we use the content of the clinical documents referenced in the volume as a corpus for questions answering.

## 2  Methods

To assess how IBM Watson could be exploited for the considered task, we started by exploring the available services. Each service on the Bluemix cloud provides a Representational State Transfer (REST) Application Programming Interface (API) that allows developers to exploit the functionalities of the service in custom applications. In addition, IBM provides an online user interface to test such functionalities without coding. However, such console can only be used to test each service separately and with IBM default configurations.

We identified two services suitable to deal with our use-case: the "Document Conversion", and the "Retrieve and Rank" services. Using IBM APIs, we integrated these services into a Java-based web-application that processes a corpus of documents to retrieve meaningful content related to user-defined requests. Differently from the IBM console, our application combines multiple services into an integrated environment. The application workflow is shown in Fig. 1. For each step of the workflow, we list the service used for implementation.

The input of the workflow is a corpus of documents of interest. The first step is dedicated to document pre-processing, where each document is divided into portions known as answer units (AUs), which represent the minimum unit of information that can be returned to the user.

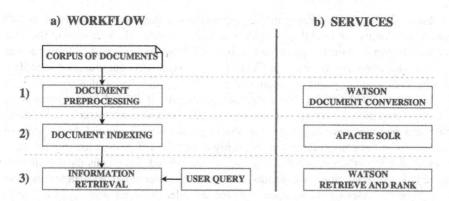

**Fig. 1. Application workflow.** Steps of the workflow (a) and related Watson services (b).

To perform this task, we exploited the methods of the Document Conversion service, that splits HTML and PDF documents into AUs on the basis of rules defined on heading tags or font size, respectively.

To carry out the Document Indexing step, we used Apache SolR (http://lucene. apache.org/solr/), a search platform that supports information retrieval relying on different features. For example, it tags and stores documents into a structured collection that can be easily queried. In our case, we applied SolR features on the collection of the obtained AUs to transform them in a format suitable to perform the Information Retrieval step.

In the Information Retrieval step, the system asks for a user query (i.e. a free-text question), and uses the Retrieve and Rank (RR) service to search for the most relevant AUs related to that question. This search is performed using NLP techniques such as tokenization, stop word removal, stemming. The RR service methods exploit SolR functionalities, which are domain independent. To the best of our knowledge, the Retrieve service does not provide methods to customize the search in order to make it domain-specific. For any customization (e.g. declaration of synonyms), expertise on Apache SolR principles is required. Unlike retrieval, the ranking of returned contents can be tuned by the developer by training the provided ranking algorithm on a set of sample questions. Two approaches are available for training. On the one hand, the IBM online console provides a 5-stars rating system for manually ranking the AUs returned for each training question. As an alternative, methods of the RR service can be applied to automatically train the algorithm on a set of questions-answers pairs collected in a text file. To explore IBM Watson services as they are, we decided not to perform any training for the ranking step.

## 3   Results

To test the proposed system, we used the 49 documents referenced in the first chapter of the considered compendium [7]. This chapter assists physicians in optimizing the management of patients who present to the neurology emergency department complaining headache. Pre-processing resulted in a total of 743 AUs on the corpus.

Since the recommendations in the compendium represent a synthesis of the 49 source documents, we used them as gold standard to evaluate the relevance of the AUs returned by the system. We prepared 6 sample questions to be given as an input to our system. Questions are reported in Table 1. To design each question, we identified specific compendium paragraphs containing interesting information that could be used as the subject of the query. We formulated questions of several types, going from questions requiring a very specific answer, to questions whose answer is more complex and includes several points. For creating questions 2, 3, and 5 we selected paragraphs in the compendium that explicitly cited the documents used to write their content. For the other questions (1, 4, and 6), the related paragraphs did not include any citation. Among these, we decided to include also a complex paragraph (question 1), where the relevant information (i.e. the intended answer) was organized as a bullet list.

To evaluate the system, for each question we compared the content of the returned AUs with the paragraph content. With such approach, we identified the AUs that contained the same information (*matching AUs*), and the AUs that did not match the paragraph content exactly, but were consistent with the expected answer (*consistent AUs*). Such AUs may interest physicians, since they supply additional material beyond the synthetic compendium content. For example, for question 2 the AU containing the arteritis prevalence value mentioned in the compendium ("133 per 100000 population aged 50 and older") was considered as matching AU. Two AUs reporting prevalence values that are different from the mentioned one were considered as consistent AUs. Any other AU was labelled as irrelevant. Table 1 shows the obtained results. Relevant AUs are the sum of matching and consistent AUs.

The best result was obtained for question 5: 87.5% of the returned AUs were indeed relevant. For two questions (3 and 6) we obtained 75%. For the remaining 3 questions, performances were lower, reaching 55% in the worst case.

**Table 1.** Qualitative evaluation of the AUs returned for 6 sample queries.

| Query string | Returned AUs | Matching AUs | Consistent AUs | Relevant AUs |
|---|---|---|---|---|
| 1. Signs or symptoms of secondary headache | 8 | 4 (50%) | 1(12.5%) | 5(62.5%) |
| 2. Temporal arteritis prevalence | 5 | 1 (20%) | 2(40%) | 3 (60%) |
| 3. Differential diagnosis for acute headache with fever and neck stiffness | 8 | 2 (25%) | 4(50%) | 6 (75%) |
| 4. Decreased cerebrospinal fluid glucose differential diagnosis | 9 | 5 (55.6%) | – | 5(55.6%) |
| 5. Drugs for treating episodes of moderate or severe migraine | 8 | 6 (75%) | 1(12.5%) | 7(87.5%) |
| 6. Procedure and tests for patients with severe headache with fever or neck stiffness | 8 | 1 (12.5%) | 5(62.5%) | 6 (75%) |

# 4 Discussion and Conclusion

The aim of the proposed work was to explore IBM Watson services as they are, and test their applicability to information retrieval processes in the clinical context. The preliminary results are encouraging, although highlighting some limitations. Some of the returned AUs were long and some effort was required to get through them. Besides their complexity, extensive AUs were more likely to be inconsistent with the provided question. The system retrieved such AUs probably because their text showed high incidence of recurring medical terms (e.g. "patient", "diagnosis", "drug"), that were also mentioned in the question. To mitigate these effects, we are integrating Watson services with custom solutions to improve the quality of the obtained AUs, and to target information retrieval to the considered domain. In particular, we aim to apply additional criteria to split the AUs during document pre-processing, and to optimize the search for the medical context. For example, we plan to filter out from the user query words that are less informative, such as words with high term frequency in the corpus (e.g. patient). We will also focus on designing domain-specific criteria for sorting the retrieved AUs, and then assess to what extent such service customization impacts on the obtained results.

# References

1. Chong, H.T., Weightman, M.J., Sirichai, P., Jones, A.: How do junior medical officers use online information resources? A survey. BMC Med. Educ. **16**, 120 (2016)
2. Strzalkowski, T., Harabagiu, S.: Advances in Open Domain Question Answering. Springer Science & Business Media, Heidelberg (2006)
3. Ferrucci, D.A.: Introduction to "This is Watson". IBM J. Res. Dev. **56**, 1:1–1:15 (2012)
4. CLINIcal Question Answering System. http://www.isical.ac.in/∼zahid_t/Cliniqa/
5. Cairns, B.L., Nielsen, R.D., Masanz, J.J., Martin, J.H., Palmer, M.S., Ward, W.H., Savova, G. K.: The MiPACQ clinical question answering system. In: AMIA Annual Symposium Proceeding, vol. 2011, pp. 171–180 (2011)
6. Lally, A., Bachi, S., Barborak, M.A., Buchanan, D.W., Chu-Carroll, J., Ferrucci, D.A., Glass, M.R., Kalyanpur, A., Mueller, E.T., Murdock, J.W., et al.: WatsonPaths: scenario-based question answering and inference over unstructured information. IBM Research, Yorktown Heights (2014)
7. Consoli, D., Sterzi, R., Micieli, G.: La neurologia dell'emergenza-urgenza. Algoritmi clinici per il neurologo che opera in Pronto Soccorso e nei dipartimenti di emergenza. Il Pensiero Scientifico (2014)

# Recurrent Neural Network Architectures for Event Extraction from Italian Medical Reports

Natalia Viani[1]([✉]), Timothy A. Miller[2], Dmitriy Dligach[3],
Steven Bethard[4], Carlo Napolitano[5], Silvia G. Priori[5,6],
Riccardo Bellazzi[1,5], Lucia Sacchi[1] [iD], and Guergana K. Savova[2]

[1] Department of Electrical, Computer and Biomedical Engineering,
University of Pavia, Pavia, Italy
natalia.viani01@universitadipavia.it
[2] Boston Children's Hospital and Harvard Medical School, Boston, MA, USA
[3] Department of Computer Science, Loyola University Chicago,
Chicago, IL, USA
[4] School of Information, University of Arizona, Tucson, AZ, USA
[5] IRCCS Istituti Clinici Scientifici Maugeri, Pavia, Italy
[6] Department of Molecular Medicine, University of Pavia, Pavia, Italy

**Abstract.** Medical reports include many occurrences of relevant events in the form of free-text. To make data easily accessible and improve medical decisions, clinical information extraction is crucial. Traditional extraction methods usually rely on the availability of external resources, or require complex annotated corpora and elaborate designed features. Especially for languages other than English, progress has been limited by scarce availability of tools and resources. In this work, we explore recurrent neural network (RNN) architectures for clinical event extraction from Italian medical reports. The proposed model includes an embedding layer and an RNN layer. To find the best configuration for event extraction, we explored different RNN architectures, including Long Short Term Memory (LSTM) and Gated Recurrent Unit (GRU). We also tried feeding morpho-syntactic information into the network. The best result was obtained by using the GRU network with additional morpho-syntactic inputs.

**Keywords:** Information extraction · Natural language processing · Neural network models

## 1  Introduction

In medical reports it is frequent to find occurrences of clinical events, such as diseases and diagnostic tests, in the form of free-text. To make data easily accessible and improve medical decisions, clinical information extraction (IE) is crucial.

Over the past years, interest in the application of natural language processing (NLP) methods to the analysis of clinical texts has increased [1]. The creation and release of corpora annotated with complex information has greatly supported the

© Springer International Publishing AG 2017
A. ten Teije et al. (Eds.): AIME 2017, LNAI 10259, pp. 198–202, 2017.
DOI: 10.1007/978-3-319-59758-4_21

development of new approaches, mostly based on elaborately designed features. However, especially for languages other than English, progress has been limited by the scarce availability of shared tools and resources. In this work, we propose the use of neural network (NN) models for clinical IE from reports written in Italian, our language of interest. Such models are generalizable, and are in principle able to automatically extract features of interest without requiring language-specific NLP tools.

NNs have been applied to natural language text with promising results [2]. Among other NLP tasks, they have been used to identify relevant entities in different domains and languages [3–7]. In the clinical domain, Li and Huang have applied convolutional neural networks to clinical event extraction [3]. As regards the Italian language, Bonadiman et al. explored NNs to perform named entity recognition (NER) [7].

For standard NER, recurrent neural networks (RNNs) have been widely explored [4–7]. As shown by Graves, they are in general a good choice for sequence labeling problems, as they make a flexible use of context information [8]. Among existing RNN models, Long Short-Term Memory (LSTM) architectures [9] have been often used [4, 5], as they are able to store and access information over long periods of time.

In this work, we explore LSTM architectures for clinical event extraction from Italian medical reports. This problem can be addressed as a BIO (Beginning-Inside-Outside) tagging task, where sequences of tokens have to be transcribed with labels denoting inclusion (B, I) or exclusion (O) in event spans. To the best of our knowledge, this is the first time LSTM models are used to extract information from clinical text in Italian. We performed experiments on a set of documents provided by the Molecular Cardiology Laboratories of the ICS Maugeri hospital in Pavia, Italy.

## 2 Materials and Methods

In this work we explore different RNN-based models for clinical event extraction. The proposed architecture includes two main components: (i) a word-embedding layer, which transforms each token into a real-valued vector, and (ii) an RNN layer, which outputs one predicted class for every token. The developed code leverages the work of Mesnil et al. [6]: the text is split into sentences, each with a variable number of tokens, and each sentence represents a separate input. To run experiments, we used the Keras library with Theano backend (https://keras.io/).

### 2.1 Dataset

Our dataset includes about 11,000 reports belonging to patients with inherited or acquired arrhythmias. A subset of 50 documents has been annotated with BIO notation for clinical events. For each B or I token, one of four semantic types is specified (problem, test, treatment or occurrence), resulting in a total of 9 possible classes. As an example of the annotation process, the sentence "*Patient with Brugada Syndrome*" is transcribed to the sequence: "O, O, B-problem, I-problem".

## 2.2    The Model

**Embedding Layer.** To represent words at the syntactic and semantic level, we transform tokens into real-valued vectors called embeddings, which are able to capture lexical-level features of words [10]. Since using pre-trained embeddings often yields good results [11], we initialized embeddings with 200-dimensional vectors pre-trained on a large corpus, obtained by merging our arrhythmias dataset (32300 words) with 3000 general domain documents gathered from the web (103800 words) [12].

Considering that output BIO sequences do not necessarily represent syntactically meaningful text spans, we also tried to provide additional morpho-syntactic information as an input. As shown in Fig. 1, for each token we transformed the part of speech (POS) tag into a vector, and concatenated it to the standard word-embedding. POS tagging is the only language-dependent component of the proposed approach.

**RNN Layer.** Given an input sequence $x = (x_1, x_2, ..., x_n)$, at each step $i = 1...n$ RNNs are able to predict an output $y_i$ based on the history of previous inputs. In this work, we consider two commonly used RNN models: the standard LSTM architecture, and one of its most popular variations, the Gated Recurrent Unit (GRU), which relies on a simpler model [13]. In our experiments, we explored both the standard LSTM and GRU models, and their bidirectional variants. Experiments were run with 100 neural network hidden units.

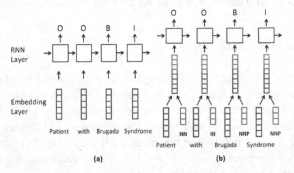

**Fig. 1.** RNN models with word embeddings only (a), and word plus POS embeddings (b). NN: noun, IN: preposition, NNP: proper noun.

## 2.3    Evaluation Metrics

In this work, we focused on the correct identification of BI sequences denoting clinical events. Performances of BIO tagging were evaluated by using the F1 score. We defined true positives as BI sequences that are present both in the gold standard and in the predicted output, regardless of the semantic type. We defined false negatives and false positives in a similar way. Semantic type accuracy is computed as the number of correct semantic types over the total number of true positives.

# 3  Results

To find the best architecture for clinical event extraction, we explored several network models. The annotated dataset was split into training (80%) and test (20%) sets, and we ran a 5-fold cross-validation on the training set to select the best model. For each model and fold, we considered the average F1 score obtained with three runs. In Table 1, we report the results of the performed experiments. For comparison purposes, we also include the results obtained using a support vector machine (SVM) (row 5). For each token, the SVM uses features that do not rely on language-dependent tools: its word embedding, and the word embeddings and tags of the two previous tokens.

**Table 1.** Comparison among different models for event extraction.

| Row | Model | F1 score mean value | F1 score standard deviation | Sem. Type accuracy |
|---|---|---|---|---|
| 1 | LSTM | 83.9% | 4.7% | 98.8% |
| 2 | GRU | 84.2% | 4.2% | 99.3% |
| 3 | Bidirectional GRU | 83.6% | 4.5% | 99.1% |
| 4 | GRU + POS | 85.9% | 4.1% | 98.9% |
| 5 | SVM | 78.4% | 5.8% | 99.2% |

**RNN Layer.** We started tuning the model by exploring different architectures for the RNN layer, keeping standard word embeddings (without morpho-syntactic information) at the embedding layer. First, we compared the performances of LSTM and GRU networks. Then, we explored the bidirectional variants of both networks. For the forward networks, using GRU instead of LSTM obtained slightly better results, with an average F1 score of 84.2%. Using the bidirectional version of GRU, adding a backward pass to the standard forward pass, led to a decrease in the F1 score (83.6%).

**Embedding Layer.** In an attempt to improve performance, we tried feeding POS information into the network at the embedding level. We ran experiments with both the GRU forward and bidirectional networks, which performed better than the LSTM. Results for the forward version are displayed in Table 1 (row 4). Feeding POS information into the network allowed the F1 score to increase to 85.9%. Also in this case, using the bidirectional network led to lower results (not shown in Table 1).

# 4  Discussion and Conclusion

In this work, we explored different RNN models to extract events from Italian texts. Unlike previous work on the Italian language [7], we focused on the clinical domain.

Italian is more morphologically complex than English, and allows more flexibility in word order. Developing algorithms that correctly identify event spans is a challenging task. Nevertheless, the models we explored showed good performance in event detection. In particular, all RNN models outperformed the considered SVM baseline.

Using GRU networks in the recurrent layer resulted in slightly better performance than standard LSTM networks. In this case, the use of simpler models has proven to be suitable for the task under consideration. On the other hand, adding a backward pass to the standard forward pass was not useful for the considered task. As a final consideration, feeding POS information into the network resulted in the best F1 score across all experiments. This indicates that the integration of easily obtainable morphosyntactic features into RNNs models can lead to promising results.

One major limitation of the proposed approach concerns the small size of the considered dataset. In future work, more documents will be annotated. As another weakness of the proposed approach, we did not perform an exhaustive search for parameters. To overcome this limitation, further experiments will be run in the future.

In conclusion, we proposed an event extraction system that mostly leverages NN models, without requiring elaborately designed features or accurate pre-processing tools. This is particularly important for languages such as Italian, for which shared NLP tools and resources might not be easily accessible.

# References

1. Meystre, S.M., Savova, G.K., Kipper-Schuler, K.C., Hurdle, J.F.: Extracting information from textual documents in the electronic health record: a review of recent research. Yearb Med. Inf. **35**, 128–144 (2008)
2. Goldberg, Y.: A primer on neural network models for natural language processing. arXiv: 1510.00726 (2015)
3. Li, P., Huang, H.: Clinical information extraction via convolutional neural network. arXiv: 1603.09381 (2016)
4. Hammerton, J.: Named entity recognition with long short-term memory. In: Proceedings of the Seventh Conference on Natural Language Learning at HLT-NAACL 2003-Volume 4, pp. 172–175 (2003)
5. Lample, G., Ballesteros, M., Subramanian, S., Kawakami, K., Dyer, C.: Neural architectures for named entity recognition. arXiv:1603.01360 (2016)
6. Mesnil, G., He, X., Deng, L., Bengio, Y.: Investigation of recurrent-neural-network architectures and learning methods for spoken language understanding. In: Proceedings of Interspeech 2013 (2013)
7. Bonadiman, D., Severyn, A., Moschitti, A.: Deep neural networks for named entity recognition in Italian. In: Proceedings of CLiC-IT 2015 (2015)
8. Graves, A.: Supervised Sequence Labelling with Recurrent Neural Networks. Studies in Computational Intelligence. Springer, Heidelberg (2012)
9. Hochreiter, S., Schmidhuber, J.: Long short-term memory. Neural Comput. **9**, 1735–1780 (1997)
10. Mikolov, T., Yih, W., Zweig, G.: Linguistic regularities in continuous space word representations. In: Proceedings of NAACL-HLT 2013, pp. 746–751 (2013)
11. Erhan, D., Bengio, Y., Courville, A., Manzagol, P.-A., Vincent, P., Bengio, S.: Why does unsupervised pre-training help deep learning? J. Mach. Learn. Res. **11**, 625–660 (2010)
12. Lyding, V., Stemle, E., Borghetti, C., Brunello, M., Castagnoli, S., Dell'Orletta, F., Dittmann, H., Lenci, A., Pirrelli, V.: The PAISA corpus of italian web texts. In: Proceedings of the WaC-9 Workshop, pp. 36–43 (2014)
13. Cho, K., van Merrienboer, B., Bahdanau, D., Bengio, Y.: On the properties of neural machine translation: encoder-decoder approaches. arXiv:1409.1259 (2014)

# Numerical Eligibility Criteria in Clinical Protocols: Annotation, Automatic Detection and Interpretation

Vincent Claveau[1]([⊠]), Lucas Emanuel Silva Oliveira[2], Guillaume Bouzillé[3], Marc Cuggia[3], Claudia Maria Cabral Moro[2], and Natalia Grabar[4]

[1] IRISA - CNRS, Rennes, France
vincent.claveau@irisa.fr
[2] PUCPR - Pontifícia Universidade Católica do Paraná, Curitiba, Brazil
{lucas.oliveira,c.moro}@pucpr.br
[3] INSERM/LTSI, HBD; CHU de Rennes; Université Rennes 2, Rennes, France
{guillaume.bouzille,marc.cuggia}@chu-rennes.fr
[4] CNRS, Univ. Lille, UMR 8163 - STL - Savoirs Textes Langage, 59000 Lille, France
natalia.grabar@univ-lille3.fr

**Abstract.** Clinical trials are fundamental for evaluating therapies and diagnosis techniques. Yet, recruitment of patients remains a real challenge. Eligibility criteria are related to terms but also to patient laboratory results usually expressed with numerical values. Both types of information are important for patient selection. We propose to address the processing of numerical values. A set of sentences extracted from clinical trials are manually annotated by four annotators. Four categories are distinguished: $C$ (concept), $V$ (numerical value), $U$ (unit), $O$ (out position). According to the pairs of annotators, the inter-annotator agreement on the whole annotation sequence $CVU$ goes up to 0.78 and 0.83. Then, an automatic method using CFRs is exploited for creating a supervised model for the recognition of these categories. The obtained F-measure is 0.60 for $C$, 0.82 for $V$, and 0.76 for $U$.

**Keywords:** Natural language processing · Supervised learning · Clinical trials · Patient eligibility · Numerical criteria

## 1 Introduction

In the clinical research process, recruitment of patients for clinical trials (CTs) remain an unprecedented challenge, while they are fundamental for evaluating therapies or diagnosis techniques. They are the most common research design to test the safety and efficiency of interventions on humans. CTs are based on statistical inference and require appropriate sample sizes from well identified population. The challenge is to enroll a sufficient number of participants with suitable characteristics to ensure that the results demonstrate the desired effect with a limited error rate. Hence, CTs must define a precise set of inclusion and

© Springer International Publishing AG 2017
A. ten Teije et al. (Eds.): AIME 2017, LNAI 10259, pp. 203–208, 2017.
DOI: 10.1007/978-3-319-59758-4_22

exclusion criteria (e.g., age, gender, medical history, treatment, biomarkers). With paper files and EHRs as the main sources of information, only human operators are capable to efficiently detect the eligible patients [3]. This is a laborious and costly task, and it is common that CTs fail because of the difficulty to meet the necessary recruitment target in an acceptable time [6]: almost half of all trial delays are caused by participant recruitment problems. Only 18% in Europe, 17% in Asia-Pacific, 15% in Latin America, and 7% in the USA complte enrollment on time [4]. The existing enrollment systems are facing the gap between the free text representation of clinical information and eligibility criteria [11,17]. Most of them propose to fill in this gap manually, while automatic NLP methods may help to overcome this issue.

The traditional NLP work is dedicated to the recognition and extraction of terms. Yet, there is an emerging work on detection of temporality, certainty, and numerical values. Such information has the purpose to complete, enrich and more generally make more precise the terminological information. In the general language, framework for automated extraction and approximation of numerical values, such as height and weight, has been proposed [5]. It uses relation patterns and WordNet and shows the average precision up to 0.84 with exact matching and 0.72 with inexact matching. Another work proposes two extraction systems: rule based extractor and probabilistic graphical model [9] for extraction of life expectancy, inflation, electricity production, etc. It reaches 0.56 and 0.64 average F-measure for the rule-based and probabilistic systems, respectively. On the basis of a small set of clinical annotated data in French, a CRF model is trained for the recognition of concepts, values, and units. Then, a rule-based system is designed for computing the semantic relations between these entities [2]. The results obtained show average F-measure 0.86 (0.79 for concepts, 0.90 for values and 0.76 for units). On English data, extraction of numerical attributes and values from clinical texts is proposed [13]: after the extraction of numerical attributes and values with CRFs, relations for associating these attributes to values are computed with SVMs. The system shows 0.95 accuracy with entities and 0.87 with relations. Yet another work is done on cardiology radiological reports in English [10] and achieves 93% F1-measure. In contrast with these studies, here we focus on clinical trial protocols written in English.

## 2    Material

**Clinical Trials.** In December 2016, we downloaded protocols of the whole set of CTs from www.clinicaltrials.com. The corpus counts 211,438 CTs. We focus on inclusion and exclusion criteria (more than 2M sentences).

**Reference Annotations.** 1,500 randomly selected sentences are annotated by 3 annotators with different backgrounds (medical doctor and computer scientists). Each sentence is annotated by at least two of them. On such typical sentences:

– *Absolute neutrophil count $\geq$ 1,000 cells/$\mu l$.*

– *Exclude if T3 uptake is less than 19%; T4 less than 2.9 (g/dL); free T4 index is less than 0.8.*

the annotators have to mark up three categories of entities: $C$ (concepts *Absolute neutrophil count, T3 uptake, T4, free T4 index*), $V$ (numerical values $\geq$ *1,000, less than 19, less than 2.9, less than 0.8*), $U$ (units *cells/μl, %, g/dL*).

## 3   Methods

The main objective of the methods is to create an automatic model for the detection of numerical values (concept, value and unit).

**Inter-annotator Agreement.** In order to assess the inter-annotator agreement, we compute Cohen's $\kappa$ [1] between each pair of annotators. The final version is obtained after a consensus is reached among the annotators.

**Automatic Annotation.** Conditional Random Fields (CRFs) [7] are undirected graphical models that represent the probability distribution of annotation $y$ on observations $x$. They are widely used in NLP thanks to their ability to take into account the sequential aspect and rich descriptions of text sequences. CRFs have been successfully used in many tasks casted as annotation problems: information extraction, named entity recognition, tagging, etc. [12,14,18]. From training data, CRF models learn to assign a label to each word of a sentence such that the tag sequence is the most probable given the sentence given as input. We want the CRFs to learn to label words denoting a concept with the tag $C$, values with $V$, units with $U$, while every other words will receive a void label noted $O$. In order to handle multi-word concepts, values and units, we adopt the so-called BIO scheme: the first word of a multi-word concept is labeled as $B\_C$ ($B$ stands for *beginning*), next words are labeled as $I\_C$ ($I$ stands for *inside*), the same for values and units. To find the most probable label of a word at position $i$ in a sentence, CRFs exploit features describing the word (for example, Part-of-Speech tags, lemmas [15], graphemic clues) and its context (words and features at positions $i-1$, $i+1$, $i-2$) up to 4 words. The CRF implementation used for our experiments is Wapiti [8], which is known for its efficiency.

**Evaluation of Automatic Annotation.** The evaluation is performed against the reference data and is measured with token errors (percentage of words wrongly labeled with respect to the human annotation) and F-measure [16].

## 4   Results and Discussion

**Inter-annotator Agreement.** In Table 1, we indicate the inter-annotator agreement for each pair of annotators and taking into account two tagsets: whole set of tags and the tagset without concepts. The figures indicate that A1 and A2 show the highest agreement: both have important experience in medical area. When concepts are not taken into account the agreement is even better: manual

**Table 1.** Inter-annotator agreement (Cohen's $\kappa$) on the whole and reduced tag sets

|  | A1 vs. A2 | A1 vs. A3 | A2 vs. A3 |
|---|---|---|---|
| $\kappa$ whole tagset | 0.78 | 0.51 | 0.47 |
| $\kappa$ without 'concept' | 0.83 | 0.60 | 0.64 |

annotation of concepts is more complicated than annotation of the two other categories. With annotations from A1 and A2, the consensual annotation is built. This version of data is used for training and evaluation of the supervised model.

**Automatic Annotation.** In Fig. 1, we present the evaluation of automatic annotation in terms of token errors and F-score for each category. In order to estimate the ideal amount of the required training data, we also display the evolution of the performance according to the size of the training data used. First, the global error rate tends to decrease. Since its decrease continues, more training data would help reaching better results. Among the categories aimed, the best performance is obtained with units, while the concept category is the most difficult to detect. For these two categories, the performance continues to grow up: a larger set of annotated data would be helpful. As for the value category, its evolution is less linear and finally it seems to find a "plateau" with no more apparent evolution. Otherwise, the detection efficiency of this category is in between the two other categories. The obtained F-measure is 0.60 for $C$, 0.82 for $V$, and 0.76 for $U$.

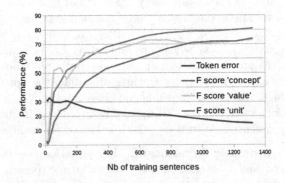

**Fig. 1.** CRF annotation performance (globally in terms of token errors, and by category in terms of F-score) according to the amount of training data (number of sentences)

## 5    Conclusion and Future Work

Recruitment of patients for CTs is a difficult task. We proposed a contribution to this task. We generate an automatic model for the detection of numerical values, composed of thee items (concept $C$, value $V$ and unit $U$), in narrative text in

English. These results are evaluated against reference data and show F-measure 0.60 for $C$, 0.82 for $V$, and 0.76 for $U$. We have several directions for future work: to normalize the units; to build resources and rules for their standardization (*cell/mm3* instead of *cell/cm3*); to prepare a larger set of reference annotations; to complete these annotations with temporal information; to apply the models for enrollment of patients in French and Brazilian hospitals.

**Acknowledgements.** This work was partly funded by CNRS-CONFAP project FIGTEM for Franco-Brazilian collaborations and a French government support granted to the CominLabs LabEx managed by the ANR in Investing for the Future program under reference ANR-10-LABX-07-01.

# References

1. Artstein, R., Poesio, M.: Inter-coder agreement for computational linguistics. Comput. Linguist. **34**(4), 555–596 (2008)
2. Bigeard, E., Jouhet, V., Mougin, F., Thiessard, F., Grabar, N.: Automatic extraction of numerical values from unstructured data in EHRs. In: MIE (Medical Informatics in Europe) 2015, Madrid, Spain (2015)
3. Campillo-Gimenez, B., Buscail, C., Zekri, O., Laguerre, B., Le Prisé, E., De Crevoisier, R., Cuggia, M.: Improving the pre-screening of eligible patients in order to increase enrollment in cancer clinical trials. Trials **16**(1), 1–15 (2015)
4. Center Watch: State of the clinical trials industry: a sourcebook of charts and statistics. Technical report, Center Watch (2013)
5. Davidov, D., Rappaport, A.: Extraction and approximation of numerical attributes from the web. In: 48th Annual Meeting of the Association for Computational Linguistics, pp. 1308–1317 (2010)
6. Fletcher, B., Gheorghe, A., Moore, D., Wilson, S., Damery, S.: Improving the recruitment activity of clinicians in randomised controlled trials: a systematic review. BMJ Open **2**(1), 1–14 (2012)
7. Lafferty, J., McCallum, A., Pereira, F.: Conditional random fields: probabilistic models for segmenting and labeling sequence data. In: International Conference on Machine Learning (ICML) (2001)
8. Lavergne, T., Cappé, O., Yvon, F.: Practical very large scale CRFs. In: Proceedings the 48th Annual Meeting of the Association for Computational Linguistics (ACL), pp. 504–513. Association for Computational Linguistics, July 2010. http://www.aclweb.org/anthology/P10-1052
9. Madaan, A., Mitta, A., Mausam, Ramakrishnan, G., Sarawagi, S.: Numerical relation extraction with minimal supervision. In: Thirtieth AAAI Conference on Artificial Intelligence (2016)
10. Nath, C., Albaghdadi, M., Jonnalagadda, S.: A natural language processing tool for large-scale data extraction from echocardiography reports. PLoS One **11**(4), 153749–153764 (2016)
11. Olasov, B., Sim, I.: Ruleed, a web-based semantic network interface for constructing and revising computable eligibility rules. In: AMIA Symposium, p. 1051 (2006)
12. Pranjal, A., Delip, R., Balaraman, R.: Part of speech tagging and chunking with HMM and CRF. In: Proceedings of NLP Association of India (NLPAI) Machine Learning Contest (2006)

13. Sarath, P.R., Mandhan, S., Niwa, Y.: Numerical atrribute extraction from Clinical Texts. CoRR 1602.00269 (2016). http://arxiv.org/abs/1602.00269
14. Raymond, C., Fayolle, J.: Reconnaissance robuste d'entités nommées sur de la parole transcrite automatiquement. In: Actes de la conférence Traitement Automatique des Langues Naturelles. Montréal, Canada (2010)
15. Schmid, H.: Probabilistic part-of-speech tagging using decision trees. In: Proceedings of International Conference on New Methods in Language Processing, pp. 44–49 (1994)
16. Sebastiani, F.: Machine learning in automated text categorization. ACM Comput. Surv. **34**(1), 1–47 (2002)
17. Shivade, C., Raghavan, P., Fosler-Lussier, E., Embi, P.J., Elhadad, N., Johnson, S.B., Lai, A.M.: A review of approaches to identifying patient phenotype cohorts using electronic health records. J. Am. Med. Inform. Assoc. **21**(2), 221–230 (2014)
18. Wang, T., Li, J., Diao, Q., Hu, W., Zhang, Y., Dulong, C.: Semantic event detection using conditional random fields. In: IEEE Conference on Computer Vision and Pattern Recognition Workshop (CVPRW 2006), p. 109 (2006)

# Enhancing Speech-Based Depression Detection Through Gender Dependent Vowel-Level Formant Features

Nicholas Cummins[1]([✉]), Bogdan Vlasenko[1], Hesam Sagha[1],
and Björn Schuller[1,2]

[1] Chair of Complex and Intelligent Systems, University of Passau,
Passau, Germany
nicholas.cummins@uni-passau.de

[2] Department of Computing, Imperial College London, London, UK

**Abstract.** Depression has been consistently linked with alterations in speech motor control characterised by changes in formant dynamics. However, potential differences in the manifestation of depression between male and female speech have not been fully realised or explored. This paper considers speech-based depression classification using gender dependant features and classifiers. Presented key observations reveal gender differences in the effect of depression on vowel-level formant features. Considering this observation, we also show that a small set of hand-crafted gender dependent formant features can outperform acoustic-only based features (on two state-of-the-art acoustic features sets) when performing two-class (depressed and non-depressed) classification.

**Keywords:** Depression · Gender · Vowel-level formants · Speech motor control · Classification

## 1 Introduction

With the aim of enhancing current diagnostic techniques, investigations into new approaches for objectively detecting and monitoring depression based on measurable biological, physiological, or behavioural signals is a highly active and growing area of research [1]. In this regard, possible key markers of depression are changes in paralinguisitic cues [1]. Formant features, representing the dominant components in the speech spectrum and capturing information on the resonance properties of the vocal tract, are one such feature [2–4]. They are strongly linked to changes in speech motor control associated with depression [2].

Whilst there is evidence for differentiation in depression symptoms between men and women (e.g., appetite and weight [5]), possible potential acoustic differences have received very little attention. Investigations into the similarity and differences between speech affected by depression or fatigue suggest that the effect of depression on formant features may differ between the genders [3].

© Springer International Publishing AG 2017
A. ten Teije et al. (Eds.): AIME 2017, LNAI 10259, pp. 209–214, 2017.
DOI: 10.1007/978-3-319-59758-4_23

This result is supported, in part, by studies which show the usefulness of performing gender dependent classification when using formant and spectral features [4]. Such results are not unexpected; formant distributions should differ between genders due to physiological differences and variations in emotionality [6,7].

Herein, we investigate the effects of depression on formant dynamics analysed on a per gender basis. Performing vowel-level formant analysis, we extract a set of gender dependent formant features, and then test their suitability for detecting depression state. The presented results are generated from a subset of *The Distress Analysis Interview Corpus – Wizard of Oz* (DAIC-WOZ) database [8], containing speech from 12 males and 16 females clinically diagnosed with depression, as assessed by the widely used *Patient Health Questionnaire* (PHQ-8) self-assessed depression questionnaire. As a control group, speech from 67 males and 47 females without clinical depression is also provided.

## 2    Vowel-Level Formant Analysis

In the first stage of our evaluation, we automatically estimated the phoneme boundaries. These were determined, using *forced alignment* provided by $HTK^1$. Mono-phone *Hidden Markov Models* (HMMs) were trained on acoustic material presented in the DAIC-WOZ corpus. To execute a vowel-level analysis, a phoneme level transcription is needed; which requires a corresponding lexicon containing a phonetic transcription of words presented in the corpus. As the DAIC-WOZ corpus does not provide such a lexicon, phonetic transcriptions were taken from the *CMU Pronouncing Dictionary*.

Upon automatic extraction of phoneme borders, we estimate the average of the *first formant* ($F1$) and the *second formant* ($F2$) values for each vowel instance. Formant contours were extracted via the Burg algorithm using *PRAAT* [9]. The following setup was used: the maximum number of formants tracked was five, the maximum frequency of the highest formant $= 6\,kHz$, the effective duration of the analysis window $= 25\,ms$, the time step between two consecutive analysis frames $= 10\,ms$, and the amount of pre-emphasis $= 50\,Hz$.

As can be seen in Fig. 1, the vowel-level mean values for the $F1$ and $F2$ are different for depressed and non-depressed speech. As expected, the results differ for each gender; for male speakers we see displacement of mean values to the left (i.e., lower $F1$) for depressed speech, as in the case with low-arousal emotional speech described in [7]. For female speakers, on the other hand, we see an opposite tendency; displacement to the right side (i.e., higher $F1$). This observation forms the basis for our decision to perform gender-dependent analysis for more reliable depression detection analysis.

To characterise the changes of the vowels' quality under the influence of the speaker's depressive state, we estimated the mean of the first and the second formants for each vowel (15 vowels in the ARPAbet non-stressed phoneme set) individually. This resulted in $2 \times 15 = 30$ pairs of mean and standard deviations

---

[1] http://htk.eng.cam.ac.uk/.

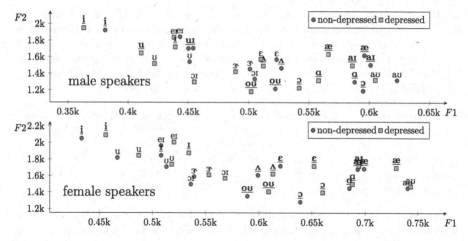

**Fig. 1.** Positions of average $F1$ and $F2$ values of English vowels in $F1/F2$ [Hz/Hz] space in the training and development partitions of the DAIC-WOZ depression corpus. Abbreviations: $F1$ – first formant, $F2$ – second formant. The formants values for indicative vowels selected by our analysis are underlined.

for average $F1$ and $F2$ values extracted. The random variables which represent average $F1$ and $F2$ features are approximately normally distributed. Finally, two sets (one per gender) of 10 gender-dependent vowel-level formant features, which are highly indicative of the effects of depression in speech, were selected using the z-test; these vowels are underlined in Fig. 1.

## 3 Classification Experiments

### 3.1 Set-Up

We compared the efficacy of our extracted *vowel-level formant features* (VL-Formants) for classifying speech affected by depression with two commonly used audio feature sets: the *extended Geneva Minimalistic Acoustic Parameter Set* (eGeMAPS) [10], and the COVAREP feature set [11]. Both feature representations are extracted at the *turn-level*. The VL-Formants are extracted on a *participant-level* as per Sect. 2. With the turn-level eGeMAPS and COVAREP features, we simply concatenate the per participant features with the matching turn-level features and treat this new feature representation as an *enhanced* turn-level feature.

All classifications are performed using the *Liblinear* package [12], with the cost parameter tuned separately for each experiment via a grid search. As the number of depressed participants is very low, we perform a series of speaker independent 4-fold participant-based cross-validation tests (make-up of folds available upon request), noting we use 4 as it is a divisor of the total number of depressed participants available.

**Table 1.** Results for depression classification using either eGeMAPS, COVAREP, our gender dependent VL-Formants, and early fusion combination thereof. Performance is given in terms of $F_1$-score for *depressed* (*not-depressed*) classes. Scores are the average $F_1$-score of four-fold cross validation on the training and development partitions of the DAIC-WOZ Corpus. Gender dependent: GD. Note, gender independent (GI) testing is not performed on the VL-Formants as extracted on a per gender basis.

| $F_1$ | eGeMaps | | COVAREP | | VL-Formants | VL-Formants and COVAREP | VL-Formants and eGeMaps |
|---|---|---|---|---|---|---|---|
| | GI | GD | GI | GD | GD | GD | GD |
| Male | .07 (.67) | .13 (.46) | .17 (.51) | .08 (.91) | .37 (.81) | .42 (.87) | **.49** (.85) |
| Female | .19 (.45) | .23 (.44) | .33 (.21) | .28 (.20) | .55 (.86) | .52 (.62) | **.55** (.80) |
| Overall | .15 (.68) | .36 (.71) | .26 (.41) | .28 (.69) | .38 (.73) | .45 (.75) | **.63** (.89) |

All results are reported in terms of $F_1$-*score* for both the *depressed* and *non-depressed* classes. Results for the eGeMAPs and COVAREP features performed in both a *Gender Independent* (classifier trained with instances from both genders) or *Gender Dependent* (classifier trained with only instances from the target gender) scenario. As the VL-Formants representations differ for each gender (cf. Sect. 2), the results for this feature set and for feature fusion are reported for the *Gender Independent* scenario only.

## 3.2   Results

Results from our classification analysis indicate that performance gains can be found by performing gender dependent depression classification while using eGeMAPS features (cf. Table 1). These results provide support for our decision to extract the VL-formant features on a per-gender basis. The advantages of gender dependent classifiers when using COVAREP are not as obvious; the male gender dependent results are in particular poor (cf. Table 1). The weaker performance of the COVAREP in the gender dependent setting is not unexpected; the voice quality features in COVAREP have been shown to be gender independent in relation to detecting depression [13].

VL-Formants perform the strongest out of the feature sets tested (cf. Table 1), highlighting their suitability for capturing depression information. The results provide a strong evidence in support of our decision to perform gender dependent feature extraction and classification. As indicated in Table 1, the early fusion of VL-Formants with the other two feature sets improves the overall $F_1$ scores for depression when compared to the individual feature set alone. The biggest gain was obtained when combining VL-Formants and eGeMAPS. The early fusion of all feature sets did not result in any further improvements in system accuracy (result not given). We also tested late fusion of the different feature sets; however, the improvements gained did not outperform the early fusion set-up.

## 4   Conclusions

This paper investigated the effects of depression on formant dynamics analysed on a per gender basis. Our analysis indicates that, indeed, the effects of depression may manifest differently in formant measures for male and females. Based on this finding, we extracted two sets of gender dependant vowel-level formant features which showed promising performance improvement for classifying depression from speech. This result matches with two key results presented in the literature: firstly, depression manifests at the phoneme level of speech [14]; and secondly, the effects of depression in speech can be captured by features which characterise speech motor control [2,15]. In future work, we aim to verify these findings on other depression-speech databases.

 **Acknowledgements.** The research leading to these results has received funding from the European Community's Seventh Framework Programme through the ERC Starting Grant No. 338164 (iHEARu), and IMI RADAR-CNS under grant agreement No. 115902.

# References

1. Cummins, N., Scherer, S., Krajewski, J., Schnieder, S., Epps, J., Quatieri, T.: A review of depression and suicide risk assessment using speech analysis. Speech Commun. **71**, 1–49 (2015)
2. Scherer, S., Lucas, G.M., Gratch, J., Rizzo, A.S., Morency, L.-P.: Self-reported symptoms of depression and PTSD are associated with reduced vowel space in screening interviews. IEEE Trans. Affect. Comput. **7**, 59–73 (2016)
3. Hönig, F., Batliner, A., Nöth, E., Schnieder, S., Krajewski, J.: Automatic modelling of depressed speech: relevant features and relevance of gender. In: Proceedings of INTERSPEECH, pp. 1248–1252. ISCA, Singapore (2014)
4. Alghowinem, S., Goecke, R., Wagner, M., Epps, J., Breakspear, M., Parker, G.: From joyous to clinically depressed: mood detection using spontaneous speech. In: Proceedings of FLAIRS, pp. 141–146. AAAI, Marco Island (2012)
5. Young, M.A., Scheftner, W.A., Fawcett, J., Klerman, G.L.: Gender differences in the clinical features of unipolar major depressive disorder. J. Nerv. Ment. Dis. **178**(3), 200–203 (1990)
6. Kring, A.M., Gordon, A.H.: Sex differences in emotion: expression, experience, and physiology. J. Pers. Soc. Psychol. **74**(3), 686–703 (1998)
7. Vlasenko, B., Prylipko, D., Philippou-Hübner, D., Wendemuth, A.: Vowels formants analysis allows straightforward detection of high arousal acted and spontaneous emotions. In: Proceedings of INTERSPEECH, pp. 1577–1580. ISCA, Florence (2011)
8. Valstar, M., Gratch, J., Schuller, B., Ringeval, F., Lalanne, D., Torres, M.T., Scherer, S., Stratou, G., Cowie, R., Pantic, M.: AVEC 2016 - depression, mood, and emotion recognition workshop and challenge. In: Proceedings 6th ACM International Workshop on Audio/Visual Emotion Challenge, pp. 3–10. ACM, Amsterdam (2016)
9. Boersma, P., Weenink, D.S.: Praat, a system for doing phonetics by computer. Glot Int. **5**(9/10), 341–345 (2002)
10. Eyben, F., Scherer, K.R., Schuller, B., Sundberg, J., Andre, E., Busso, C., Devillers, L.Y., Epps, J., Laukka, P., Narayanan, S.S., Truong, K.P.: The Geneva minimalistic acoustic parameter set (GeMAPS) for voice research and affective computing. IEEE Trans. Affect. Comput. **7**, 190–202 (2016)

11. Degottex, G., Kane, J., Drugman, T., Raitio, T., Scherer, S.: COVAREP - a collaborative voice analysis repository for speech technologies. In: Proceedings of ICASSP, pp. 960–964. IEEE, Florence (2014)

12. Rong-En, F., Chang, K.-W., Hsieh, C.-J., Wang, X.-R., Lin, C.-J.: LIBLINEAR: a library for large linear classification. J. Mach. Learn. Res. **9**, 1871–1874 (2008)

13. Scherer, S., Stratou, G., Gratch, J., Morency, L.-P.: Investigating voice quality as a speaker-independent indicator of depression and PTSD. In: Proceedings of INTERSPEECH, pp. 847–851. ISCA, Lyon (2013)

14. Trevino, A., Quatieri, T., Malyska, N.: Phonologically-based biomarkers for major depressive disorder. EURASIP J. Adv. Sig. Proc. **2011**, 1–18 (2011)

15. Cummins, N., Sethu, V., Epps, J., Schnieder, S., Krajewski, J.: Analysis of acoustic space variability in speech affected by depression. Speech Commun. **75**, 27–49 (2015)

# A Co-occurrence Based **MedDRA** Terminology Generation: Some Preliminary Results

Margherita Zorzi[1(✉)], Carlo Combi[1], Gabriele Pozzani[2], Elena Arzenton[2], and Ugo Moretti[2]

[1] Department of Computer Science, University of Verona, Verona, Italy
`margherita.zorzi@univr.it`
[2] Department of Diagnostics and Public Health, University of Verona, Verona, Italy

**Abstract.** The generation of medical terminologies is an important activity. A flexible and structured terminology both helps professionals in everyday manual classification of clinical texts and is crucial to build knowledge bases for encoding tools implementing software to support medical tasks. For these reasons, it would be nice to "enforce" medical dictionaries such as MedDRA with sets of locutions semantically related to official terms. Unfortunately, the manual generation of medical terminologies is time consuming. Even if the human validation is an irreplaceable step, a significant set of "high-quality" *candidate terminologies* can be automatically generated from clinical documents by statistical methods for linguistic. In this paper we adapt and use a *co-occurrence* based technique to generate new MedDRA locutions, starting from some large sets of narrative documents about adverse drug reactions. We describe here the methodology we designed and results of some first experiments.

## 1 Introduction

The processing and classification of narrative text into medical dictionaries is a challenging task that plays a crucial role in everyday clinical activities. In general, written language describing medical conditions is informal and its manual processing is expensive. An expert reviewer has to map inaccurate information into a strict technical terminology. In the era of web, spontaneous reporting grew exceptionally and the experts' manual annotation work has become unbearable. Thus, a number of software tools have been defined to support this delicate clinical task [1]. The automatic processing of technical language is not trivial and pure syntactical matching can easily fail [2]. Synonymical dictionaries are the most basic tool for improving software performances. Therefore, the problem of *terminology acquisition* induces a problem of *terminology generation*.

According to the described scenario, it seems to be interesting to equip medical dictionaries with reasoned sets of locutions, effectively used in everyday clinical language, semantically related to official terms. In this paper, we will focus on the MedDRA terminology [3], which aims to contain the majority of locutions used by professionals in describing clinical pathologies and adverse reactions

© Springer International Publishing AG 2017
A. ten Teije et al. (Eds.): AIME 2017, LNAI 10259, pp. 215–220, 2017.
DOI: 10.1007/978-3-319-59758-4_24

related to drug consumption (ADRs). Notwithstanding, the use of medical jargon in ADR reports represents a gap between what effectively people say and the official terminology, making encoding harder. Moreover, in the majority of MedDRA translations, a number of terms are lost in translation. A "poor" dictionary does not allow a completely accurate encoding of ADR descriptions. Experts of the domain can generate synonyms and variations of MedDRA terms in everyday activities, during the manual revision of spontaneous reports. This is a natural and good *modus operandi*, but it is not trivial and it requires a huge amount of work by an expert of the domain. Moreover, for human experts it is more difficult to analyse and filter locutions by frequency (a meaningful medical synonym is useless if, *de facto*, it is never used in free-text reporting). In this paper we adapt and use a *co-occurrence* based technique to generate new Italian MedDRA terms, starting from two large sets of narrative documents about adverse drug reactions, the set of Summary of Products Characteristics (SPCs) of Italian drugs and a set of ADRs spontaneous reports. Our technique, based on the Mutual Information principle, performs effectively in medical synonyms detection. Statistical methods are able to retrieve and filter large and refined sets of *candidate terms*. The construction of the new terminology is semi-automatic. Candidate terms are automatically generated from some corpora of narrative documents. Then, new MedDRA terms are verified and eventually approved by a domain expert, who links the new terminology to the official one. We describe here the methodology we designed and results from first experiments. We show that co-occurrence based techniques are able to generate an effective extension of MedDRA terminology. The paper is organized as follows. In Sect. 2 we recall related work and some background notions. In Sect. 3 we introduce the techniques and experiments we adopted. In Sect. 4 conclusions and future work are reported.

## 2    Background and Related Work

In [4,5] some techniques based on Pointwise Mutual Information (PMI) and Information Retrieval (IR) are proposed for recognizing synonyms in specific linguistic domains. In [6], authors use co-occurrence based techniques to build a knowledge base for a patient-centered question-answering system. In [7] authors propose to use association mining and Proportional Reporting Ratio to mine associations between drugs and adverse reactions from the user contributed content in social media. The Consumer Health Vocabulary is exploited to generate new *ADR terminology*. Another ontological based approach can be found in [8], where the authors build a semantic resource based on formal description logic to improve MedDRA term retrieval, mapping MedDRA terms to concepts from terminologies or ontologies from the Unified Medical Language System. The Medical Dictionary for Regulatory Activities (MedDRA) [3] is a multilingual medical terminology used to encode adverse events associated with the use of biopharmaceuticals and other medical products. MedDRA terms are organised into a hierarchy. The SOC (System Organ Classes) level includes the most

general terms. The LLT (Low Level Terms) level includes more specific terminologies. Between SOC and LLT there are three intermediate levels (HLGT, HLT,
and PT).

# 3   Methods and Experiments

Our goal is to generate a set of new locutions, called *pseudo*-LLT*s*, in order to
*extend the official* MedDRA *terminology*. The construction of the set of new terms
is *semi-automatic*. Given the MedDRA dictionary and a corpus of documents,
*a set of candidate locutions is generated by statistical methods*; then, the *new
terms are submitted to pharmacovigilance experts, who can refuse or approve
them, linking accepted* pseudo-LLTs *to already existing official* LLTs.

To automatically generate new MedDRA locutions, we adapt to the healthcare domain some techniques used in linguistics for quantitative analysis and in
IR methods [2,4,5,9]. The main idea is to collect "frequent pairs" of words, called
*co-occurrence*, appearing in the same contexts (documents) starting from some
*pivot* words. To control the generation of MedDRA terminology, we "guide" the
generation of co-occurrences in the following way. We start from a controlled set
of words, i.e., MedDRA words having an autonomous semantical meaning. This
set of "prefixes" are successively extended by frequently co-occurring words in
significative sets of documents. This way, we try to enlarge and specialise medical concepts following what documents (and people) effectively say in formal
and colloquial narrative texts. One of the most useful features of co-occurrence
based techniques is the strong independence from the language and from the linguistic domain one considers. This has been proved in [4,5] for English language
and different technical terminologies. We show here how co-occurrence methods
perform very well in a different language (Italian) and for the ADR terminology
generation. In the paper, we use a general definition of **co-occurrence** [9]. Two
different words $w_i$ and $w_j$ are *co-occurring*, or, equivalently, form a *co-occurrence*
(pair), when they appear in the same document $d \in D$ (where $D$ is a collection
of documents). In the definition of co-occurrence, *the order of words does not
matter*. The co-occurrence of two words $w_i$ and $w_j$ in the same document within
a span of maximally $k$ words is represented by $w_i$ near$^k$ $w_j$. If $k = 0$, $w_i$ and $w_j$
are contiguous. The co-occurrence of two words $w_i$ and $w_j$ in the same document
without regard to their distance is represented by $w_i$ near$^\infty$ $w_j$.

The first step for the generation of co-occurrences is the choice of the *guide
words* or pivots. To deal with clinical terminology we "tune" the set of pivots
to be a significative set. In particular, pivot words come from MedDRA. We use
some auxiliary definitions. A **unary-LLT** is a single word LLT; the set of unary-
LLTs will be dubbed as U. A **pseudo-LLT** is a locution semantically equivalent
(or semantically related) to an LLT. In this paper, we consider only set U as
pivot words. Examples of terms in U are *febbre* (in English, *fever*) and *cefalea*
(in English, *headache*). To discover frequent co-occurrences we adopt a version
of the technique used in [4,5]. It is based on the (pointwise) Mutual Information
principle (MI) introduced in [4,10]. We compute co-occurring pairs with two
different scores closely related to the ones proposed in [4]:

$$s_1 : \frac{\text{hits}(\mathsf{p} \; \text{near}^\infty \mathsf{w}_i)}{\text{hits}(\mathsf{w}_i)} \qquad s_2 : \frac{\text{hits}(\mathsf{p} \; \text{near}^k \mathsf{w}_i)}{\text{hits}(\mathsf{w}_i)}$$

where $\text{hits}(E)$, for any expression $E$, is the number of documents in which $E$ holds. Score values range from 0 to 1: pairs $(\mathsf{p}, \mathsf{w})$ evaluated as maximally co-occurring are scored with 1. We denote pivot words with $\mathsf{p}$ (possibly indexed) and co-occurring words with $\mathsf{w}$ (possibly indexed). Informally, $s_1$ computes the score of the "simple" co-occurrence of the pair $(\mathsf{p}, \mathsf{w}_i)$ and $s_2$ computes the score of the co-occurrence of the pair $(\mathsf{p}, \mathsf{w}_i)$ up to a given proximity threshold (or window) $k$. We notice that, even if the order of co-occurring words in the document does not matter by definition, the roles played by $\mathsf{p}$ and $\mathsf{w}_i$ are not interchangeable.

We consider two sets of documents. The first one, $\mathcal{D}_{SPC}$, is built on 18k SPCs of Italian drugs[1]. The second one, $\mathcal{D}_{Vigi}$, is built on about 244k spontaneous reports from VigiSegn, a data warehouse system for the Italian pharmacovigilance activities. Pivot words $\mathsf{p}$ range over $\mathsf{U}$ while co-occurring words $\mathsf{w}_i$ range over $\mathcal{D}_{SPC}$ or $\mathcal{D}_{Vigi}$. We set $k = 8$[2]. We compute MI scores $s_1$ and $s_2$ on both $\mathcal{D}_{SPC}$ and $\mathcal{D}_{Vigi}$. For each run we obtain a set of pairs labeled with a value according to the score we computed. As in most IR problems, a crucial point is to discriminate, among the whole set of outputs, a "small" subset of potentially good solutions. The outputs of the $s_1$ and $s_2$ runs are filtered taking into account only pairs that obtained a score value $t$ greater than 0.4 and that occur with $f > 1$. Threshold $f$ rejects well-scored pairs having a negligible frequency in the whole set of documents. Score $s_1$ retrieved 2495 pairs from $\mathcal{D}_{SPC}$ and 2232 pairs from $\mathcal{D}_{Vigi}$. Score $s_2$ retrieved 279 pairs from $\mathcal{D}_{SPC}$ and 217 pairs from $\mathcal{D}_{Vigi}$. Score $s_1$ retrieved a large set of results. Therefore, to create a gold standard, we focus on $s_2$. Output pairs of $s_2$ have been manually validated by a pharmacovigilance expert. For each co-occurrence, expert refused or accepted it as a pseudo-LLT, a concept representing a synonym of a MedDRA LLT. Each pseudo-LLT has been also linked to the corresponding official MedDRA term. The sets of manually approved terms from $\mathcal{D}_{SPC}$ and $\mathcal{D}_{Vigi}$ are denoted as $\mathsf{S}_{SPC}$ and $\mathsf{S}_{Vigi}$, respectively. An example of term in $\mathsf{S}_{SPC}$, scored with value 1, is *allergia iatrogena* (in English, *iatrogenic allergy*), built by $s_2$ score from the pivot *allergia* (id $= 1001738$). The pseudo-LLT *allergia iatrogena* has been mapped to the LLT *allergia a farmaco* (in English, *drug allergy*, id $= 10013661$). The following table summarizes our results.

| Source (no. of docs) | Score | $t$ | $f$ | Retrieved | Approved | % of approved pairs |
|---|---|---|---|---|---|---|
| SPCs (18K) | $s_2$ | $0.4 < t \leq 1$ | $f > 1$ | 279 | 42 | 15.05% |
| Reports (244K) | $s_2$ | $0.4 < t \leq 1$ | $f > 1$ | 217 | 116 | 53.45% |

---

[1] SPCs are available at https://farmaci.agenziafarmaco.gov.it/bancadatifarmaci/.
[2] The choice of $k$ depends on the average length of the documents in the datasets.

Both $\mathcal{D}_{SPC}$ and $\mathcal{D}_{Vigi}$ generate new MedDRA terms. Only 15.05% of pairs computed on $\mathcal{D}_{SPC}$ are recognised as pseudo-LLTs. This percentage is due to the narrative style exploited in SPCs. SPCs are formal and well written and the used language is closed to official MedDRA terminology. Dataset $\mathcal{D}_{Vigi}$ performs differently. About half of retrieved terms have been recognized to be useful synonyms. The amount of documents plays a role in this performance. Notwithstanding, the generation of the new terminology is again strictly dependent on the narrative text in reports. Spontaneous reports are informally written (w.r.t. SPCs) and medical concepts are expressed, by people (heterogeneous users and professionals), in terms of *real variations of official locutions*.

We use now sets $S_{SPC}$ and $S_{Vigi}$ as control sets to evaluate the performances of $s_1$. We test the percentage of terms in $S_{SPC}$ and $S_{Vigi}$ also retrieved by $s_1$. Score $s_1$ is able to retrieve 30 co-occurrences in $S_{SPC}$ (71.43%) and 108 co-occurrences in $S_{Vigi}$ (93.1%). $s_1$ performs satisfactorily in synonyms detection task but the set of retrieved solutions is too large to be easily manually validated. It is preferable to use a more restrictive score, as $s_2$, able to filter efficiently good solutions.

## 4   Conclusions

In this paper, we adapted methods from statistical linguistics to extend MedDRA terminology. This is an ongoing project, open to several directions. First, we are computing MI-based co-occurrences on different sets of pivot words. MI scores can be refined taking into account negations in the narrative descriptions and contexts (other words we require to appear in the document together with the co-occurring pair) [4]. Moreover, we are extending MI techniques to the generation of $n$-uple of co-occurring words. Finally, we are also developing and adapting to healthcare terminologies other statistical techniques, such as the *cosine similarity* [2]. The technique we proposed in this paper has some interesting potential applications, such as the semi-automatic generation of knowledge bases for NLP tools [1], and the proposal of additions to the official MedDRA terminology.

## References

1. Zorzi, M., Combi, C., Lora, R., Pagliarini, M., Moretti, U.: Automagically encoding adverse drug reactions in MedDRA. In: 2015 IEEE International Conference on Healthcare Informatics, ICHI 2015, pp. 90–99 (2015)
2. Schütze, H., Pedersen, J.O.: A cooccurrence-based thesaurus and two applications to information retrieval. Inform. Process. Manag. **33**(3), 307–318 (1997)
3. ICH: MedDRA data retrieval and presentation: points to consider (2016)
4. Turney, P.D.: Mining the web for synonyms: PMI-IR versus LSA on TOEFL. In: Raedt, L., Flach, P. (eds.) ECML 2001. LNCS, vol. 2167, pp. 491–502. Springer, Heidelberg (2001). doi:10.1007/3-540-44795-4_42
5. Baroni, M., Bisi, S.: Using cooccurrence statistics and the web to discover synonyms in a technical language. In: Proceedings of LREC (2004)
6. Schulz, S., Costa, C.M., Kreuzthaler, M., et al.: Semantic relation discovery by using co-occurrence information. In: Proceedings of BioTxtM (2014)

7. Yang, C.C., Yang, H., Jiang, L., Zhang, M.: Social media mining for drug safety signal detection. In: Proceedings of SHB, pp. 33–40. ACM (2012)
8. Souvignet, J., Declerck, G., Asfari, H., Jaulent, M.C., Bousquet, C.: OntoADR a semantic resource describing adverse drug reactions to support searching, coding, and information retrieval. J. Biomed. Inform. **63**, 100–107 (2016)
9. Manning, C.D., Schütze, H.: Foundations of Statistical Natural Language Processing. MIT Press, Cambridge (1999)
10. Church, K.W., Hanks, P.: Word association norms, mutual information, and lexicography. In: Proceedings of ACL 1989, Stroudsburg, PA, USA, pp. 76–83 (1989)

# Health Care Processes

# Towards Dynamic Duration Constraints for Therapy and Monitoring Tasks

Carlo Combi, Barbara Oliboni, and Francesca Zerbato[(✉)]

Department of Computer Science, University of Verona, Verona, Italy
{carlo.combi,barbara.oliboni,francesca.zerbato}@univr.it

**Abstract.** Duration constraints are among the most subtle and important aspects of clinical practice. When designing healthcare processes, it is important to incorporate such constraints into process diagrams and to provide suitable mechanisms for managing their violations during process run-time. Nonetheless, the Business Process Model and Notation (BPMN 2.0) fails in providing proper support for duration constraint modeling. In this paper, we propose a set of BPMN patterns to foster the modeling and management of *shifted duration*, that constitutes an unexplored kind of duration constraints, recurrent in medical practice. Specifically, this constraint is suitable for dealing with the real-time adjustments required by pharmacological therapies and monitoring tasks.

## 1 Introduction

Duration is one of the most frequently used temporal constraints in Process-Aware Information Systems (PAIS) [14]. In the medical domain, clinical and healthcare processes are likely to observe strict temporal guidelines and constraints, with the aim of improving treatment success and service quality [2]. Among these, duration constraints are intrinsic of medical tasks and, in many circumstances, they are strongly related with the quality of care activities. For instance, treatment duration often determines drug therapy effectiveness, hospital length of stay is an important factor for evaluating chances of patient readmission, and duration of symptoms can be employed to establish a diagnosis.

Capturing such fine but crucial aspects is desirable for improving design, management, and execution of healthcare processes. In addition, explicitly modeling duration constraints eases the detection and handling of temporal violations, thus reducing the costs associated with their management and, in general, with process monitoring and execution [11].

Activity duration is supported by most process execution languages, workflow management systems and simulation tools, which often allow specifying initiation and completion times of tasks. Besides, information on activity duration can be easily extracted from process execution logs.

Nonetheless, the Business Process Model and Notation (BPMN 2.0) [17], the leading standard for process modeling, does not properly support the representation of activity duration at modeling time, within process diagrams [4,8,14].

© Springer International Publishing AG 2017
A. ten Teije et al. (Eds.): AIME 2017, LNAI 10259, pp. 223–233, 2017.
DOI: 10.1007/978-3-319-59758-4_25

In real clinical (and non-clinical) scenarios, it is common to deal with violations of duration constraints, that is, activities last either less or longer than expected [2]. For example, pharmacological therapy may be interrupted whenever severe drug side effects occur, or treatments may be prolonged if patients present reduced response to medications. In general, violations of duration constraints have a proper semantics, that must be considered for their management.

Especially in the context of pharmacological therapy or monitoring tasks, it is often interesting to constrain the duration of a process activity by considering related conditions. In other words, duration is measured only after a certain condition is met during activity execution and, in many cases, it holds until the activity is completed. We refer to this kind of duration constraints as *shifted duration* to describe the situation in which the evaluation of activity duration starts at a particular (shifted) moment. For example, when treating a patient experiencing puerperal endometritis, a relatively common complication that may follow Cesarean section, physicians know that antimicrobial drugs should be administered for 24–48 h after fever has resolved [15]. In the presented scenario, even though activity "administration of antimicrobial drugs" begins as soon as the infection is detected, it is interesting to measure its duration starting from a successive temporal moment, highlighted by the occurrence of some expected, relevant *milestone event* happening during process execution, represented in this case by "fever resolution". This milestone event can occur once or many times in the process environment, or it can be the result of other process activities.

In this paper, we propose a set of BPMN patterns for modeling shifted duration constraints in healthcare processes, aimed at fostering the run-time management of duration violations. In particular, we consider the possibility of violating either the minimum or maximum duration of a medical task. This proposal complements the patterns presented in [8], and contains significant added value in terms of modeling these yet undiscovered kinds of duration constraints.

The remainder is structured as follows. Section 2 discusses related work. Section 3 motivates the need of shifted duration in real clinical scenarios. Section 4 presents the BPMN patterns for specifying the studied constraint and for managing its violations. Finally, Sect. 5 outlines conclusions and future work.

## 2    Background and Related Work

Duration constraints are a central aspect of healthcare process modeling and management, being often used as an indicator of patient safety and care quality [6,9]. In this section, we introduce the Business Process Model and Notation and present a few research efforts that address the modeling and management of duration constraints, within the workflow and business process community.

BPMN is the leading standard for process modeling, aimed at representing processes as graphical diagrams [17]. A process diagram is mostly composed of *flow objects*, connected by a *sequence flow* that denotes their ordering relations. *Activities* represent work that is performed within the process and can either be atomic (*tasks*) or compound (*sub-processes*). Graphically, all activities are depicted as rounded-corner rectangles, and *collapsed sub-processes* are distinguished by a small "+" maker, located at the bottom-center of the shape. *Events* denote facts that occur instantaneously during process execution and have an effect on its flowing. Graphically, events are depicted as circles, with a single or double border, depending on their kind. *Start* events initiate a process instance, while *end* events are used to conclude it. Both start and end events are depicted with a single border. Conversely, *intermediate* events occur at any point in time between the process start/end and are characterized by a double border. Intermediate events can also be attached to the boundary of an activity. Boundary interrupting events are used to interrupt an on-going activity, whereas non-interrupting events initiate a new process path, which is executed concurrently with respect to the activity itself. Depending on their kind and semantics, all events can either catch a trigger or throw a result. Among intermediate catching events, *timer* events are triggered by a specific time or date, whereas *conditional* events are triggered whenever a specific condition becomes true. Throwing and catching *signal* events are used for broadcast communication within the process. *Gateways* are used to control the branching and merging of the sequence flow based on data or events occurrences. Graphically, they are depicted as diamonds with an internal marker that differentiates their routing behavior. A "+" sign is used for *parallel* branching, whereas a "x" sign identifies a data-based *exclusive* gateway, i.e., a point in the process where a condition must be evaluated in order to choose one path out of more. Finally, in *event-based* gateways, which are distinguished by an encircled pentagon marker, the alternative paths that follow the gateway are chosen based on event occurrence.

Despite its diffusion, BPMN still lacks of support towards the representation of temporal information and time constraints, as temporal properties, when present, remain hidden within process diagrams [14]. Accordingly, most BPMN-oriented proposals found in literature focus on extending the notation to support the specification of temporal and duration constraints [4,12]. In [12], BPMN 1.2 is extended to support temporal constraint modeling, but activity duration is not addressed. In [4], the duration of tasks and sub-processes is defined through ad-hoc attributes and model checking approaches are used for formal verification. Finally, in [10], the authors present an approach to dynamically check temporal constraints of multiple concurrent workflows with resource constraints, which succeeds in detecting where temporal violations occurred within the process.

In the context of clinical guidelines design, BPMN is suitable to represent knowledge-intensive clinical pathways [3,7]. An approach for capturing relevant temporal aspects during the conceptual modeling of workflows, has been introduced in [5]. Durations are expressed as nodes attributes, having the form of intervals with a defined temporal granularity. In [2], the authors propose a controlled approach for managing violations of time constraints in temporal workflows belonging to the medical domain. Temporal constraints are expressed by means of formal expressions with the aim of supporting consistency checking, and trade-offs between constraint relaxation and violation penalties are discussed. Various deadline escalation strategies for PAIS are defined in [1], to detect and manage predictable temporal exceptions. Finally, a set of time patterns is proposed in [14], to foster comparison of PAIS and to evaluate the suitability of different approaches with respect to the introduced patterns.

## 3    Motivating Clinical Scenarios

In this section, we consider a few examples, taken from real clinical domains and guidelines, to motivate why considering shifted duration constraints in healthcare processes is helpful for everyday medical and organizational practice.

Firstly, let us consider Hospitalization as a general medical task. In order to safely dismiss a patient, physicians must ensure that he/she has had no fever (NF) within the 24 h prior to the planned end of hospitalization. Otherwise, the patient is kept in hospital until fever resolves and the patient has remained afebrile for 24 h. Figure 1(a) shows a case of Hospitalization, where a shifted duration of 24 h is measured after the occurrence of event NF, which denotes the beginning of apyrexia. However, fever solution may be followed by future episodes of fever (F), occurring during hospitalization. In this scenario, physicians must wait until fever resolves again, before starting measuring the 24-hour shifted duration, required for properly concluding hospitalization. Figure 1(b) shows the explained setting: as fever F occurs within the 24 h following NF, shifted duration is reset, and physicians must wait until fever resolves (NF) in order to restart measuring the length of hospitalization.

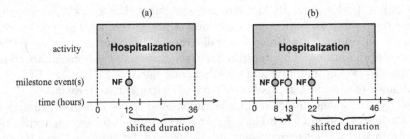

**Fig. 1.** (a) Example of shifted duration, started by event NF, which denotes the beginning of apyrexia. (b) Example of simple shifted duration with reset, caused by F.

Secondly, shifted durations are suitable to capture fine temporal aspects related to the administration of effective antibiotic therapies [16]. For instance, let us consider the treatment of infective endocarditis, a deadly disease affecting cardiac valves [13]. Successful treatment relies on pathogens eradication by antimicrobial drugs. Treatment administration begins as soon as a diagnosis of endocarditis is made and lasts 2–6 weeks. However, the (shifted) duration of the treatment is based on the first day of effective antibiotic therapy, i.e., it is measured starting from the first day of negative blood cultures, not from the day of diagnosis [16].

In some cases, antibiotics alone are insufficient to solve endocarditis. If the infection persists, that is, blood cultures remain positive after 7 days of antibiotic treatment, the patient must undergo surgery [13]. In this case, as the milestone event (e.g., blood cultures turn negative) that initiates effective therapy does not occur within a certain amount of time, alternative care is enacted for patient safety. In other words, the occurrence of the milestone event is constrained to prevent undesired process deadlocks. Figure 2 represents the described scenario. Specifically, if blood cultures do not turn negative (NC?) within 7 days (MAX START), Surgery is performed, and shifted duration is evaluated starting from the moment this is concluded. MIN and MAX duration limits are also highlighted on the time line, to show that Antibiotic Treatment must last between 2–6 weeks.

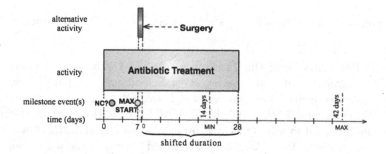

**Fig. 2.** Example of shifted duration with constrained start.

In real practice, it is important to ensure that minimum and maximum duration constraints are also observed, that is, shifted durations for medical activities must be bound according to the specific (therapeutic) goal. Besides, violations of such constraints must be properly detected and managed, as small changes in the duration of treatments and monitoring may strongly impact patient lives.

## 4   Modeling and Managing Shifted Durations in BPMN

In this section, we introduce a set of BPMN-based patterns to specify shifted durations constraints, considering also resetting, and constraints on the occurrence of milestone events. Then, we propose a few mechanisms to foster the management of duration violations.

### 4.1   Shifted Duration Specification and Detection of Violations

Activity duration is the amount of time needed for its execution. Activities are often constrained in terms of minimum and maximum duration, and such constraints also hold for shifted durations. Figure 3 shows the process diagram proposed for modeling shifted duration and for detecting violations of the minimum and maximum duration constraints.

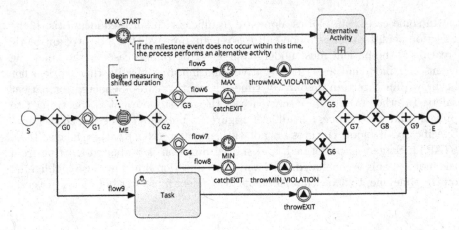

**Fig. 3.** BPMN process for the specification of activity shifted duration.

The shifted duration of the Task represented at the bottom is modeled by a BPMN diagram, composed by start event S, end event E, parallel gateways (G0, G2, G7, G9), event-based gateways (G1, G3, G4), exclusive gateways (G5, G6, G8) timer events (MIN, MAX, MAX_START), conditional event (ME) and the remaining signal events. Once the process is started at S, the flow is split into two parallel branches at G0: flow9 is directed towards Task, which can begin executing, whilst the other process branch reaches event-based gateway G1. The flowing of the tokens is assumed not to consume time.

At this point, while Task is already executing, the process waits until either a milestone event occurs, shown in Fig. 3 by conditional event ME, or the maximum time (MAX_START) set for its occurrence expires. This timer event prevents the process from waiting indefinitely for a milestone event. When ME occurs, events MAX on flow5, catchEXIT on flow6, MIN on flow7 and catchEXIT on flow8 are activated, whilst the process branch containing MAX_START is withdrawn. Timer events MIN and MAX are used to define the minimum and, respectively, maximum values for shifted duration.

The detection of shifted duration violations is realized by means of a simple signal-based communication pattern. When Task completes its execution, signal event throwEXIT is broadcast to be caught by any of the corresponding catchEXIT events that are active at the moment of broadcasting. Based on when Task ends, any of the following mutually exclusive scenarios can occur.

Minimum shifted duration is violated if Task completes earlier the minimum set duration limit (MIN). In this case, both signals catchEXIT on flow6 and flow8 are triggered, whereas the other process branches outgoing of event-based gateways G3 and G4 are withdrawn. Then, signal event throwMIN_VIOLATION, located within flow8, is used to capture the violation. Finally, once synchronization has occurred at G9, the process can conclude.

If Task completes anytime between MIN and MAX, signal event throwEXIT is caught only by the corresponding catchEXIT located on flow6, as timer event

MIN has already been triggered, thus causing flow8 to be withdrawn. In this case, no violation occurs.

Maximum shifted duration is violated whenever the execution of Task lasts longer than MAX. In this case, right after MAX, signal throwMAX_VIOLATION is broadcast to detect that maximum shifted duration has been violated. Trivially, as timer event MIN had also been triggered before MAX, the process can complete soundly, once synchronization has occurred at G9.

As a practical example, let us consider effective therapy for endocarditis, introduced in Sect. 3 and depicted in Fig. 2. In this case, the BPMN Task of Fig. 3 corresponds to antibiotic treatment, MIN corresponds to 2 weeks (14 days), and MAX corresponds to 6 weeks (42 days). Milestone event ME is negative blood culture (NC), whilst event MAX_START can be set to 7 days, that is the time limit after which Surgery is considered as an Alternative Activity.

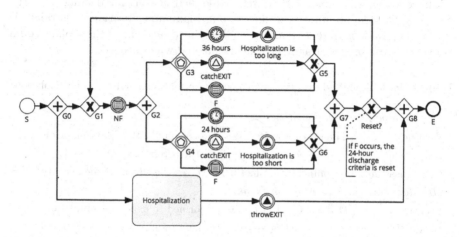

**Fig. 4.** BPMN process for the specification of activity shifted duration with reset.

In real scenarios, it is quite common that shifted durations are reset by the occurrence of other milestone events. To deal with these situations, we introduce a BPMN pattern suitable for modeling and detecting shifted durations with reset. Without loss of generality, in Fig. 4, we show the proposed process pattern adapted to the example of the 24-hour criterion for hospitalization, discussed in Sect. 3 and represented in Fig. 1(b). With respect to the pattern shown in Fig. 3, this process handles the eventuality that another conditional event F, which represents fever, occurs during shifted duration. In this case, if F occurs before 24 h, the patient cannot be dismissed and, consequently, the shifted duration for Hospitalization is reset. Similarly, if F occurs between 24 h and 36 h, the physician can decide either to keep the patient in hospital, that is, to reset shifted duration, or to end hospitalization.

## 4.2  Management of Violated Shifted Duration Constraints

Violations of shifted durations can be easily detected with the approach introduced in Sect. 4.1. However, proper management mechanisms must be enacted during process execution to ensure that violations are handled properly.

In this direction, signal events throwMIN(MAX)_VIOLATION of Figs. 3 and 4 are used to trigger process fragments that are specifically modeled to handle duration violations. According to [8], the management of maximum duration constraints can be either weak or strong, depending on how constraints are defined with respect to activity interruption. The management of weak maximum duration constraints does not require activities to be interrupted, but it rather focuses on performing side activities, entitled to solve violations and to expedite activity completion. Conversely, the management of strong maximum duration constraints is meant to promptly interrupt activities whenever their execution is taking longer than expected. A set of planned actions for dealing with the management of minimum, weak and strong maximum duration constraints is proposed in Table 1. In real organizational settings, each one of these plans could correspond to a process fragment, that is enacted when needed.

**Table 1.** Possible shifted duration management actions, enacted in case of violation.

| (A) Minimum duration violation | |
| --- | --- |
| Waiting | The process waits before proceeding to the following element |
| Repetition | The whole activity or part of it is repeatedly executed |
| Compensation | A compensation handler reverses the effects of the activity |
| (B) Maximum duration violation (weak) | |
| Escalation | Dedicated activities are performed to expedite completion |
| Extra Workforce | The activity is assigned to more resources that execute it in parallel |
| (C) Maximum duration violation (strong) | |
| Skipping | The activity is interrupted and the remaining is skipped |
| Undoing | The activity is interrupted and its effects are reversed |

Figure 5 shows the sub-process Strong shifted duration manager, proposed to manage violations of minimum and strong maximum durations, which require prompt activity interruption, and adapted to suit the previously discussed example of antibiotic treatment for endocarditis.

If minimum duration is violated, that is, signal catchEXIT is triggered before MIN, compensation event throwMIN_VIOLATION detects the violation and activates sub-process Handle MIN constraint violation. The latter is attached to Antibiotic Treatment (AT) though event catchMIN_VIOLATION and it is represented in Fig. 5 as a "black-box". Ideally, any of the actions defined in Table 1(A) can be executed by the mentioned compensation sub-process. In the considered case, AT must be repeated, as shorter antibiotic treatment is not sufficient no achieve cure [16].

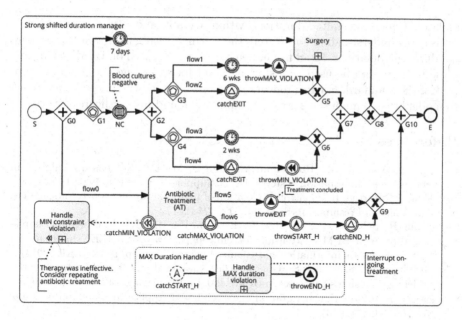

**Fig. 5.** Process pattern for managing strong maximum shifted duration violations.

Conversely, if maximum duration is violated, signal throwMAX_VIOLATION detects the violation and triggers boundary event catchMAX_VIOLATION that promptly interrupts AT. Following interruption, flow6 is activated and escalation event throwSTART causes event sub-process MAX Duration Handler to start. This is designed to take care of maximum duration violation and any of the actions described in Table 1(C) can be enacted. In the considered scenario, AT is interrupted. Start event catchSTART_H is non-interrupting in order to prevent the termination of the parent process. Synchronization between the latter one and MAX Duration Handler is achieved through signal-based communication. Signal event catchEND_H on flow6 waits for end event throwEND_H before proceeding. This solution prevents process Strong shifted duration manager to complete while the handler is still executing in order to prevent undesired handler termination.

The process for managing weak maximum duration violation differs from the one of Fig. 5 in event catchMAX_VIOLATION, which is non-interrupting in this latter case. For this reason, sub-process MAX Duration Handler and AT run concurrently, and their execution is already synchronized.

## 5    Conclusions

In this paper, we presented an approach for modeling and managing shifted duration constraints in healthcare processes, by using known business process modeling techniques. Physicians constantly deal with such temporal constraints and, consequently, they must be able to enact corrective measures if these are violated. Capturing shifted durations within readable and re-usable process patterns eases understanding and fosters the finding of valuable means for managing duration violations. In addition, being exclusively based on BPMN elements, these patterns can be used as a starting point for extending the temporal support of the standard. For future work, we plan to extend the proposed approach to model other temporal constraints commonly found in healthcare settings.

# References

1. Van der Aalst, W.M., Rosemann, M., Dumas, M.: Deadline-based escalation in process-aware information systems. Decis. Support Syst. **43**(2), 492–511 (2007)
2. Akhil, K., Barton, R.: Controlled violation of temporal process constraints models, algorithms and results. Inf. Syst. **64**, 410–424 (2017)
3. Braun, R., Schlieter, H., Burwitz, M., Esswein, W.: BPMN4CP: design and implementation of a BPMN extension for clinical pathways. In: IEEE International Conference on Bioinformatics and Biomedicine (BIBM), pp. 9–16. IEEE (2014)
4. Cheikhrouhou, S., Kallel, S., Guermouche, N., Jmaiel, M.: Toward a time-centric modeling of business processes in BPMN 2.0. In: Proceedings of International Conference on Information Integration and Web-based Applications and Services, pp. 154–166. ACM (2013)
5. Combi, C., Gozzi, M., Juarez, J., Oliboni, B., Pozzi, G.: Conceptual modeling of temporal clinical workflows. In: 14th International Symposium on Temporal Representation and Reasoning, pp. 70–81. IEEE (2007)
6. Combi, C., Keravnou-Papailiou, E., Shahar, Y.: Temporal Information Systems in Medicine. Springer Science & Business Media, Heidelberg (2010)
7. Combi, C., Oliboni, B., Zardini, A., Zerbato, F.: Seamless design of decision-intensive care pathways. In: International Conference on Healthcare Informatics (ICHI 2016), pp. 35–45. IEEE (2016)
8. Combi, C., Oliboni, B., Zerbato, F.: Modeling and handling duration constraints in BPMN 2.0. In: Proceedings of the 32nd Annual ACM Symposium on Applied Computing, SAC 2017, pp. 727–734. ACM (2017)
9. Dadam, P., Reichert, M., Kuhn, K.: Clinical workflows the killer application for process-oriented information systems? BIS 2000. Springer, London (2000)
10. Du, Y., Xiong, P., Fan, Y., Li, X.: Dynamic checking and solution to temporal violations in concurrent workflow processes. IEEE Trans. Syst. Man Cybern. - Part A: Syst. Hum. **41**(6), 1166–1181 (2011)
11. Eder, J., Panagos, E., Rabinovich, M.: Workow time management revisited. In: Bubenko, J., Krogstie, J., Pastor, O., Pernici, B., Rolland, C., Sølvberg, A. (eds.) Seminal Contributions to Information Systems Engineering, pp. 207–213. Springer, Berlin (2013). doi:10.1007/978-3-642-36926-1_16
12. Gagne, D., Trudel, A.: Time-BPMN. IEEE Conf. Commer. Enterp. Comput. **2009**, 361–367 (2009)
13. Habib, G., Lancellotti, P., Antunes, M.J., Bongiorni, M.G., Casalta, J.P., et al.: 2015 ESC guidelines for the management of infective endocarditis. Eur. Heart J. **36**(44), 3075–3128 (2015)
14. Lanz, A., Weber, B., Reichert, M.: Time patterns for process-aware information systems. Requirements Eng. **19**(2), 113–141 (2014)
15. Larsen, J.W., Hager, W.H., Livengood, C.H., Hoyme, U.: Guidelines for the diagnosis, treatment and prevention of postoperative infections. Infect. Dis. Obstet. Gynecol. **11**(1), 65–70 (2003)

16. Leekha, S., Terrell, C.L., Edson, R.S.: General principles of antimicrobial therapy. In: Mayo Clinic Proceedings. vol. 86, pp. 156–167. Elsevier (2011)
17. Object Management Group: Business Process Model and Notation (BPMN), v2.0.2, http://www.omg.org/spec/BPMN/2.0.2/

# Discriminant Chronicles Mining
## Application to Care Pathways Analytics

Yann Dauxais[1(✉)], Thomas Guyet[2], David Gross-Amblard[1],
and André Happe[3]

[1] Rennes University 1/IRISA-UMR6074, Rennes, France
yann.dauxais@irisa.fr
[2] AGROCAMPUS-OUEST/IRISA-UMR6074, Rennes, France
[3] CHRU BREST/EA-7449 REPERES, Brest, France

**Abstract.** Pharmaco-epidemiology (PE) is the study of uses and effects of drugs in well defined populations. As medico-administrative databases cover a large part of the population, they have become very interesting to carry PE studies. Such databases provide longitudinal care pathways in real condition containing timestamped care events, especially drug deliveries. Temporal pattern mining becomes a strategic choice to gain valuable insights about drug uses. In this paper we propose $DCM$, a new discriminant temporal pattern mining algorithm. It extracts chronicle patterns that occur more in a studied population than in a control population. We present results on the identification of possible associations between hospitalizations for seizure and anti-epileptic drug switches in care pathway of epileptic patients.

**Keywords:** Temporal pattern mining · Knowledge discovery · Pharmaco-epidemiology · Medico-administrative databases

## 1 Introduction

Healthcare analytics is the use of data analytics tools (visualization, data abstraction, machine learning algorithms, etc.) applied on healthcare data. The general objective is to support clinicians to discover new insights from the data about health questions.

Care pathways are healthcare data with highly valuable information. A care pathway designates the sequences of interactions of a patient with the healthcare system (medical procedures, biology analysis, drugs deliveries, etc.). Care pathways analytics arises with the medico-administrative databases opening. Healthcare systems collect longitudinal data about patients to manage their reimbursements. Such huge databases, as the SNIIRAM [11] in France, is readily available and has a better population coverage than *ad hoc* cohorts. Compared

This research is supported by the PEPS project funded by the French National Agency for Medicines and Health Products Safety (ANSM).

© Springer International Publishing AG 2017
A. ten Teije et al. (Eds.): AIME 2017, LNAI 10259, pp. 234–244, 2017.
DOI: 10.1007/978-3-319-59758-4_26

to Electronic Medical Records (EMR), such database concerns cares in real life situation over a long period of several years.

Such databases are useful to answer various questions about care quality improvement (*e.g.* care practice analysis), care costs cutting and prediction, or epidemiological studies. Among epidemiological studies, pharmaco-epidemiological (PE) studies answer questions about the uses of health products, drugs or medical devices, on a real population. Medico-administrative databases are of particular interest to relate care pathway events to specific outcomes. For instance, Polard et al. [12] used the SNIIRAM database to assess associations between brand-to-generic anti-epileptic drug substitutions and seizure-related hospitalizations. Dedicated data processing identified drug substitutions and statistical test assessed their relation to seizure-related hospitalization event.

The main drawbacks of such an approach is that (1) epidemiologists often have to provide an hypothesis to assess, (2) they can not handle sequences with more than two events and (3) the temporal dimension of pathways are poorly exploited. Care pathway analytics will help to deeply explore such complex information source. More especially temporal pattern mining algorithms can extract interesting sequences of cares, that would be potential study hypothesis to assess.

Temporal pattern mining is a research field that provides algorithms to extract interesting patterns from temporal data. These methods have been used in various medical data from time series to EMR datasets. Such method can be organized according to the temporal nature of the patterns they extract. *Sequential patterns* only takes into account the order of the events. It has been used in [14] to identify temporal relationships between drugs. These relationships are exploited to predict which medication a prescriber is likely to choose next. *Temporal rules* [2,4], or more complex patterns like *chronicles* [1,8], model inter-event durations based on the event timestamps. Finally, *Time interval patterns* [7,10] extract patterns with timestamps and durations. For large datasets, the number of temporal patterns may be huge and not equally interesting. Extracting patterns related to a particular outcome enables to extract less but more significant patterns. In [9], the temporal patterns are ranked according to their correlation with a particular patient outcome. Nonetheless, their number is not reduced.

The frequency constraint is the limitation of these approaches. While dealing with PE study, pattern discriminancy seems more interesting to identify patterns that occur for some patients of interest but not in the control patients. Only few approaches proposed to mine discriminant temporal patterns [6,13]. Fradkin and Mörchen [6] proposed the BIDE-D algorithm to mine discriminant sequential patterns and Quiniou et al. [13] used inductive logic to extract patterns with quantified inter-event durations.

This article presents a temporal pattern mining algorithm that discovers discriminant chronicle patterns. A chronicle is a set of events linked by quantitative temporal constraints. In constraint satisfaction domain, chronicle is named temporal constraint network [5]. It is discriminant when it occurs more in the set of positive sequences than in the set of negative sequences. To the best of our knowledge, this is the first approach that extracts discriminant pattern with quantitative temporal information. This algorithm is applied to pursue the Polard et *al.* analysis [12]. It aims at identifying possible associations between hospitalizations for seizure and anti-epileptic drug switch from SNIIRAM care pathways of epileptic patients.

## 2   Discriminant Chronicles

This section introduces basic definitions and defines the discriminant chronicle mining task.

### 2.1   Sequences and Chronicles

Let $\mathbb{E}$ be a set of event types and $\mathbb{T}$ be a temporal domain where $\mathbb{T} \subseteq \overline{\mathbb{R}}$. An **event** is a couple $(e, t)$ such that $e \in \mathbb{E}$ and $t \in \mathbb{T}$. We assume that $\mathbb{E}$ is totally ordered by $\leq_{\mathbb{E}}$. A **sequence** is a tuple $\langle SID, \langle (e_1, t_1), (e_2, t_2), \ldots, (e_n, t_n) \rangle, L \rangle$ where $SID$ is the sequence index, $\langle (e_1, t_1), (e_2, t_2), \ldots, (e_n, t_n) \rangle$ a finite event sequence and $L \in \{+, -\}$ a label. Sequence items are ordered by $\prec$ defined as $\forall i, j \in [1, n],\ (e_i, t_i) \prec (e_j, t_j) \Leftrightarrow t_i < t_j \lor (t_i = t_j \land e_i <_{\mathbb{E}} e_j)$.

**Example 1 (sequence set, $\mathcal{S}$).** *Table 1 represents a set of six sequences containing five event types (A, B, C, D and E) and labeled with two different labels.*

A **temporal constraint** is a tuple $(e_1, e_2, t^-, t^+)$, also noted $e_1[t^-, t^+]e_2$, where $e_1, e_2 \in \mathbb{E}$, $e_1 \leq_{\mathbb{E}} e_2$ and $t^-, t^+ \in \mathbb{T}$, $t^- \leq t^+$. A temporal constraint $e_1[t^-, t^+]e_2$ is said satisfied by a couple of events $((e, t), (e', t'))$ iff $e = e_1$, $e' = e_2$ and $t' - t \in [t^-, t^+]$.

A **chronicle** is a couple $(\mathcal{E}, \mathcal{T})$ where $\mathcal{E} = \{\!\{e_1 \ldots e_n\}\!\}$, $e_i \in \mathbb{E}$ and $\forall i, j, 1 \leq i < j \leq n$, $e_i \leq_{\mathbb{E}} e_j$, $\mathcal{T}$ is a temporal constraint set: $\mathcal{T} = \{e[a, b]e' \mid e, e' \in \mathcal{E},\ e \leq_{\mathbb{E}} e'\}$. As the constraint $e[a, b]e'$ is equivalent to $e'[-b, -a]e$, we impose the order on items, $\leq_{\mathbb{E}}$, to decide which one is represented in the chronicle. The set $\mathcal{E}$ is a **multiset**, *i.e.* $\mathcal{E}$ can contain several occurrences of a same event type.

**Table 1.** Set of six sequences labeled with two classes $\{+, -\}$.

| SID | Sequence | Label |
|-----|----------|-------|
| 1 | $(A, 1), (B, 3), (A, 4), (C, 5), (C, 6), (D, 7)$ | + |
| 2 | $(B, 2), (D, 4), (A, 5), (C, 7)$ | + |
| 3 | $(A, 1), (B, 4), (C, 5), (B, 6), (C, 8), (D, 9)$ | + |
| 4 | $(B, 4), (A, 6), (E, 8), (C, 9)$ | − |
| 5 | $(B, 1), (A, 3), (C, 4)$ | − |
| 6 | $(C, 4), (B, 5), (A, 6), (C, 7), (D, 10)$ | − |

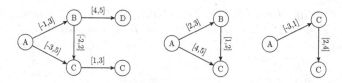

**Fig. 1.** Example of three chronicles occurring in Table 1 (*cf.* Examples 2 and 3). No edge between two events is equivalent to the temporal constraint $[-\infty, \infty]$.

**Example 2.** *Figure 1 illustrates three chronicles represented by graphs. Chronicle* $\mathcal{C} = (\mathcal{E}, \mathcal{T})$ *where* $\mathcal{E} = \{\!\!\{e_1 = A, e_2 = B, e_3 = C, e_4 = C, e_5 = D\}\!\!\}$ *and* $\mathcal{T} = \{e_1[-1,3]e_2,\ e_1[-3,5]e_3,\ e_2[-2,2]e_3, e_2[4,5]e_5, e_3[1,3]e_4\}$ *is illustrated on the left. This graph is not complete. No edge between two events is equivalent to the temporal constraint* $[-\infty, \infty]$, *i.e. there is no constraint.*

## 2.2    Chronicle Support

Let $s = \langle (e_1, t_1), \ldots, (e_n, t_n) \rangle$ be a sequence and $\mathcal{C} = (\mathcal{E} = \{\!\!\{e'_1, \ldots, e'_m\}\!\!\}, \mathcal{T})$ be a chronicle. An **occurrence** of $\mathcal{C}$ in $s$ is a sub-sequence $\tilde{s} = \langle (e_{f(1)}, t_{f(1)}), \ldots, (e_{f(m)}, t_{f(m)}) \rangle$ such that (1) $f : [1, m] \mapsto [1, n]$ is an injective function, (2) $\forall i,\ e'_i = e_{f(i)}$ and (3) $\forall i, j,\ t_{f(j)} - t_{f(i)} \in [a, b]$ where $e'_i[a, b]e'_j \in \mathcal{T}$. It is worth noting that $f$ is not necessarily increasing. In fact, there is a difference between (i) the order of the chronicle multiset defined on items, $\leq_\mathbb{E}$, and (ii) the order on events in sequences, $\prec$, defined on the temporal domain. The chronicle $\mathcal{C}$ **occurs** in $s$, denoted $\mathcal{C} \in s$, iff there is at least one occurrence of $\mathcal{C}$ in $s$. The **support** of a chronicle $\mathcal{C}$ in a sequence set $\mathcal{S}$ is the number of sequences in which $\mathcal{C}$ occurs: $supp(\mathcal{C}, \mathcal{S}) = |\{s \in \mathcal{S} \mid \mathcal{C} \in s\}|$. Given a minimal support threshold $\sigma_{min}$, a chronicle is **frequent** iff $supp(\mathcal{C}, \mathcal{S}) \geq \sigma_{min}$.

**Example 3.** *Chronicle* $\mathcal{C}$, *Fig. 1 on the left, occurs in sequences 1, 3 and 6 of Table 1. We notice there are two occurrences of* $\mathcal{C}$ *in sequence 1. Nonetheless, its support is* $supp(\mathcal{C}, \mathcal{S}) = 3$. *This chronicle is frequent in* $\mathcal{S}$ *for any minimal support threshold* $\sigma_{min}$ *lower or equal to 3. The two other chronicles, so called* $\mathcal{C}_1$ *and* $\mathcal{C}_2$ *from left to right, occur respectively in sequences 1 and 3; and in sequence 6. Their supports are* $supp(\mathcal{C}_1, \mathcal{S}) = 2$ *and* $supp(\mathcal{C}_2, \mathcal{S}) = 1$.

## 2.3    Discriminant Chronicles Mining

Let $\mathcal{S}^+$ and $\mathcal{S}^-$ be two sets of sequences, and $\sigma_{min} \in \mathbb{N}$, $g_{min} \in [1, \infty]$ be two user defined parameters. A chronicle is **discriminant** for $\mathcal{S}^+$ iff $supp(\mathcal{C}, \mathcal{S}^+) \geq \sigma_{min}$ and $supp(\mathcal{C}, \mathcal{S}^+) \geq g_{min} \times supp(\mathcal{C}, \mathcal{S}^-)$. The **growth rate** $g(\mathcal{C}, \mathcal{S})$ of a chronicle is defined by $\frac{supp(\mathcal{C}, \mathcal{S}^+)}{supp(\mathcal{C}, \mathcal{S}^-)}$ if $supp(\mathcal{C}, \mathcal{S}^-) > 0$ and is $+\infty$ otherwise.

**Example 4.** *With chronicle $\mathcal{C}$ of Fig. 1, $supp(\mathcal{C}, \mathcal{S}^+) = 2$, $supp(\mathcal{C}, \mathcal{S}^-) = 1$, where $\mathcal{S}^+$ (resp. $\mathcal{S}^-$) is the sequence set of Table 1 labeled with $+$ (resp. $-$). Considering that $g(\mathcal{C}, \mathcal{S}) = 2$, $\mathcal{C}$ is discriminant if $g_{min} \leq 2$. For chronicles $\mathcal{C}_1$ and $\mathcal{C}_2$, $supp(\mathcal{C}_1, \mathcal{S}^+) = 2$ and $supp(\mathcal{C}_1, \mathcal{S}^-) = 0$ so $g(\mathcal{C}_1, \mathcal{S}) = +\infty$ and $supp(\mathcal{C}_2, \mathcal{S}^+) = 0$ and $supp(\mathcal{C}_2, \mathcal{S}^-) = 1$ so $g(\mathcal{C}_2, \mathcal{S}) = 0$. $\mathcal{C}_2$ is not discriminant, but $\mathcal{C}_1$ is for any $g_{min}$ value.*

The support constraint, using $\sigma_{min}$, prunes the unfrequent, and so insignificant, chronicles. For example, a chronicle like $\mathcal{C}_1$ such that $g(\mathcal{C}_1, \mathcal{S}) = +\infty$ but $supp(\mathcal{C}_1, \mathcal{S}^-) = 0$ is discriminant but not interesting. Pruning can be done efficiently thanks to the anti-monotonicity of frequency, which is also valid for chronicle patterns [1]. More specifically, if a chronicle[1] $(\mathcal{E}, \mathcal{T}_\infty)$ is not frequent, then no chronicle of the form $(\mathcal{E}, \mathcal{T})$ will be frequent. This means that temporal constraints may be extracted only for frequent multisets.

Extracting the complete set of discriminant chronicles is not interesting. Discriminant chronicles with same multiset and similar temporal constraints are numerous and considered as redundant. It is preferable to extract chronicles whose temporal constraints are the most generalized. The approach proposed in the next section efficiently extracts an incomplete set of discriminant chronicles that we want to be meaningful.

## 3  DCM Algorithm

The *DCM* algorithm is given in Algorithm 1. It extracts discriminant chronicles in two steps: the extraction of frequent multisets, and then the specification of temporal constraints for each non-discriminant multiset.

At first, EXTRACTMULTISET extracts $\mathbb{M}$, the frequent multisets in $\mathcal{S}^+$. It applies a regular frequent itemset mining algorithm on a dataset encoding multiple occurrences. An item $a \in \mathbb{E}$ occurring $n$ times in a sequence is encoded by $n$ items: $I_1^a, \dots, I_n^a$. A frequent itemset of size $m$, $(I_{i_k}^{e_k})_{1 \leq k \leq m}$, extracted from this

---

**Algorithm 1.** Algorithm *DCM* for discriminant chronicles mining

---

**Require:** $\mathcal{S}^+, \mathcal{S}^-$ : sequences sets, $\sigma_{min}$ : minimal support threshold, $g_{min}$ : minimal growth threshold

1: $\mathbb{M} \leftarrow$ EXTRACTMULTISET$(\mathcal{S}^+, \sigma_{min})$          ▷ $\mathbb{M}$ is the frequent multisets set
2: $\mathbb{C} \leftarrow \emptyset$                                ▷ $\mathbb{C}$ is the discriminant chronicles set
3: **for all** $ms \in \mathbb{M}$ **do**
4:     **if** $supp\left(\mathcal{S}^+, (ms, \mathcal{T}_\infty)\right) > g_{min} \times supp\left(\mathcal{S}^-, (ms, \mathcal{T}_\infty)\right)$ **then**
5:         $\mathbb{C} \leftarrow \mathbb{C} \cup \{(ms, \mathcal{T}_\infty)\}$ ▷ Discriminant chronicle without temporal constraints
6:     **else**
7:         **for all** $\mathcal{T} \in$ EXTRACTDTC$(\mathcal{S}^+, \mathcal{S}^-, ms, g_{min}, \sigma_{min})$ **do**
8:             $\mathbb{C} \leftarrow \mathbb{C} \cup \{(ms, \mathcal{T})\}$          ▷ Add a new discriminant chronicle
9: **return** $\mathbb{C}$

---

[1] $\mathcal{T}_\infty$ is the set of temporal constraints with all bounds set to $\infty$.

dataset is transformed into the multiset containing, $i_k$ occurrences of the event $e_k$. Itemsets with two items $I_{i_k}^{e_k}$, $I_{i_l}^{e_l}$ such that $e_k = e_l$ and $i_k \neq i_l$ are redundant and thus ignored.

In a second step, lines 3 to 8 extract the discriminant temporal constraints (DTC) of each multiset. The naive approach would be to extract DTC for all frequent multisets. A multiset $\mathcal{E}$ (*i.e.* a chronicle $(\mathcal{E}, \mathcal{T}_\infty)$) which is discriminant may yield numerous similar discriminant chronicles with most specific temporal constraints. We consider them as useless and, as a consequence, line 4 tests whether the multiset $ms$ is discriminant. If so, $(ms, \mathcal{T}_\infty)$ is added to the discriminant patterns set. Otherwise, lines 7–8 generate chronicles from DTC identified by EXTRACTDTC.

## 3.1 Temporal Constraints Mining

The general idea of EXTRACTDTC is to see the extraction of DTC as a classical numerical rule learning task [3]. Let $\mathcal{E} = \{\!\{e_1 \ldots e_n\}\!\}$ be a frequent multiset. A relational dataset, denoted $\mathcal{D}$, is generated with all occurrences of $\mathcal{E}$ in $\mathcal{S}$. The numerical attributes of $\mathcal{D}$ are inter-event durations between each pair $(e_i, e_j)$, denoted $\mathcal{A}_{e_i \to e_j}$. Each occurrence yields one example, labeled by the sequence label ($L \in \{+, -\}$).

A rule learning algorithm induces numerical rules from $\mathcal{D}$. A rule has a label in conclusion and its premise is a conjunction of conditions on attribute values. Conditions are inequalities in the form: $\mathcal{A}_{e_i \to e_j} \geq x \wedge \mathcal{A}_{e_i \to e_j} \leq y$, where $(x, y) \in \mathbb{R}^2$. These rules are then translated as temporal constraints, $e_i[x, y]e_j$, that make the chronicle discriminant.

**Example 5.** *Table 2 is the relational dataset obtained from the occurrences of* $\{\!\{A, B, C\}\!\}$ *in Table 1. The attribute* $\mathcal{A}_{A \to B}$ *denotes the durations between A and B. We observe that several examples can come from the same sequence.*

*The rule* $\mathcal{A}_{A \to B} \leq 5 \wedge \mathcal{A}_{B \to C} \leq 2 \implies +$ *perfectly characterizes the examples labeled by + in Table 2. It is translated by the DTC* $\{A[-\infty, 5]B, B[-\infty, 2]C\}$ *which gives the discriminant chronicle* $\mathcal{C} = (\{\!\{e_1 = A, e_2 = B, e_3 = C\}\!\}, \{e_1[-\infty, 5]e_2, e_2[-\infty, 2]e_3\})$.

**Table 2.** Relational dataset for the multiset $\{\!\{A, B, C\}\!\}$.

| SID | $\mathcal{A}_{A \to B}$ | $\mathcal{A}_{B \to C}$ | $\mathcal{A}_{A \to C}$ | Label |
|-----|-----|-----|-----|-----|
| 1 | 2 | 2 | 4 | + |
| 1 | −1 | 2 | 1 | + |
| 2 | 5 | −2 | 3 | + |
| 3 | 3 | 0 | 3 | + |
| 5 | −1 | 3 | 1 | − |
| 6 | 6 | −1 | 5 | − |

Our DCM implementation uses the Ripper algorithm [3] to induce discriminant rules. Ripper generates discriminant rules using a growth rate computed on $\mathcal{D}$. To take into account multiple occurrences of a multiset in sequences, the growth rate of chronicles is reevaluated a posteriori on sequences datasets.

# 4    Case Study

This section presents the use of DCM to study care pathways of epileptic patients. Recent studies suggested that medication substitutions (so called switches) may be associated with epileptic seizures for patients with long term treatment with anti-epileptic (AE) medication. In [12], the authors did not found significant statistical relationship between brand-to-generic substitution and seizure-related hospitalization. The DCM algorithm is used to extract patterns of drugs deliveries that discriminate occurrences of recent seizure. These patterns may be interesting for further investigations by statistical analysis.

## 4.1    Dataset

Our dataset was obtained from the SNIIRAM database [11] for epileptic patients with stable AE treatment with at least one seizure event. The treatment is said stable when the patient had at least 10 AE drugs deliveries within a year without any seizure. Epileptics seizure have been identified by hospitalization related to an epileptic event, coded G40.x or G41.x with ICD-10[2]. The total number of such patients is 8,379. The care pathway of each patient is the collection of timestamped drugs deliveries from 2009 to 2011. For each drug delivery, the event id is a tuple $\langle m, grp, g \rangle$ where $m$ is the ATC code of the active molecule, $g \in \{0, 1\}$ and $grp$ is the speciality group. The speciality group identifies the drug presentation (international non-proprietary name, strength per unit, number of units per pack and dosage form). Our dataset contains 1,771,220 events of 2,072 different drugs and 20,686 seizure-related hospitalizations.

Since all patient sequences have at least one seizure event, we adapt the case-crossover protocol to apply our $DCM$ algorithm. This protocol, consisting in using a patient as his/her own control, is often used in PE studies. The dataset is made of two sets of 8,379 labeled sequences. A 3-days induction period is defined before the first seizure of each patient. Drugs delivered within the 90 days before inductions yield the positive sequences and those delivered within the 90 days before the positive sequence, $i.e.$ the 90 to 180 days before induction, yield the negative sequences. The dataset contains 127,191 events of 1,716 different drugs.

## 4.2    Results

Set up with $\sigma_{min} = 5.10^{-3}$, $i.e.$ 42 patients[3], and $g_{min} = 1.4$, we generated 777 discriminant chronicles. Chronicles involved 510 different multisets and 128 different event types.

---

[2] ICD-10: International Classification of Diseases 10th Revision.

[3] This number of patients have been initially estimated important by epidemiologists to define a population of patients with similare care sequences associated to seizure.

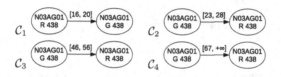

**Fig. 2.** Four discriminant chronicles describing switches between same type of valproic acid (N03AG01) generic (G 438) and brand-named (R 438). $supp(\mathcal{C}_i, \mathcal{S}^+)$ respectively for $i = 1$ to 4 equals 43, 78, 71 and 43 and $supp(\mathcal{C}_i, \mathcal{S}^-)$ equals 23, 53, 39 and 30.

Three types of pattern are of specific interest for clinicians: (1) sequences of AE generic and brand-named drug deliveries, (2) sequences of same AE drug deliveries, (3) sequences with AE drug deliveries and other drug types deliveries. According to these criteria, we selected 55 discriminant chronicles involving 16 different multisets to be discussed with clinicians. For the sake of conciseness, we choose to focus the remaining of this section on chronicles related to *valproic acid* (*N03AG01* ATC code, with different presentations) but our results contain chronicles related to other AE drugs like *levetiracetam* or *lamotrigin*.

**Taking into Account Time in Brand-to-generic Substitution.** We start with patterns representing switches between different presentation of *N03AG01*. Figure 2 illustrates all discriminant patterns that have been extracted. It is noteworthy that all chronicles have temporal constraints, this means that multisets without temporal constraints are not discriminant. This results is consistent with Polard et al. [12] which concluded that brand-to-generic AE drug substitution was not associated with an elevated risk of seizure-related hospitalization. But temporal constraints was not taken into account in the later. The four extracted chronicles suggest that for some small patient groups, drug switches with specific temporal constraints are more likely associated to seizure.

The two first chronicles represent delivery intervals lower than 30 days, respectively from brand-to-generic and generic-to-brand. The third one represents an interval between the two events greater than 30 days but lower than 60 days. The DTC of the last one could be interpreted as [67, 90] because of the bounded duration of the study period (90 days). This chronicle represents a switch occurring more than 60 days but most of the time less than 90 days.

These behaviors may correspond to unstable treatments. In fact, AE drug deliveries have to be renew every months, thus, a regular treatment corresponds to a delay of ≈30 days between two AE drug deliveries.

We next present in Fig. 3 an example of discriminant chronicle that involves three deliveries of *N03AG01* (no chronicle involves more deliveries of this AE drug).

The growth rate of this chronicle is high (2.94). It is moreover simple to understand and, with their DTC, it can be represented on a timeline (see Fig. 3, below). It is noteworthy that the timeline representation loses some constraint information. The first delivery is used as starting point ($t_0$), but it clearly illustrates that last delivery occurs too late after the second one (more 30 days after).

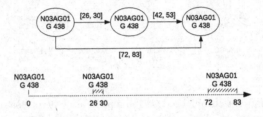

**Fig. 3.** Above, a chronicle describing repetitions of *valproic acid* (N03AG01) generic (G 438) and, below, its timeline representation. The chronicle is more likely related to epileptic seizure: $supp(\mathcal{C}, \mathcal{S}^+) = 50$, $supp(\mathcal{C}, \mathcal{S}^-) = 17$.

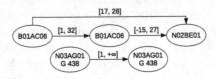

**Fig. 4.** A chronicle describing co-occurrences between anti-thrombosis drugs (*B01AC06*) and *valproic acid* which is more likely associated to seizure: $supp(\mathcal{C}, \mathcal{S}^+) = 42$, $supp(\mathcal{C}, \mathcal{S}^-) = 20$.

As well as previous patterns, this chronicle describes an irregularity in deliveries. More precisely, the irregularity occurs between the second and the third deliveries as described by the DTC [42, 53] and [72, 83].

We conclude from observations about the previous two types of patterns that the precise numerical temporal information discovered by $DCM$ is useful to identify discriminant behaviors. Analyzing pure sequential patterns does not provide enough expression power to associate switch of same AE deliveries with seizure. Chronicles, specifying temporal constraints, allow us to describe the conditions under which a switch of same AE deliveries is discriminant for epileptic seizure.

**Example of a Complex Chronicle.** The chronicle presented in Fig. 4 has been judged interesting by clinicians as a potential adverse drug interaction between an AE drug and a drug non-directly related to epilepsy, more especially aspirin (*B01AC06*), prescribed as an anti-thrombotic treatment. The $DTC$ implies that aspirin and paracetamol (*N02BE01*) are delivered within a short period (less 28 days). There is no temporal relations between these deliveries and the deliveries of *valproic acid*. But their co-occurrence within the period of 90 days is part of the discriminatory factor.

After a deeper analysis of patient care pathways supporting this chronicle, clinicians made the hypothesis that these patients were treated for brain stroke. It is known to seriously exacerbate epilepsy and to increase seizure risk.

## 5    Conclusion

This article presents a new temporal pattern mining task, the extraction of discriminant chronicles, and the $DCM$ algorithm which extracts them from temporal sequences. This new mining task appears to be useful to analyze care pathways in the context of pharmaco-epidemiology studies and we pursue the Polard et al. [12] study on epileptic patients with this new approach. On the one hand, extracted patterns are discriminant. They correspond to care sequences that are more likely associated to a given outcome, i.e. epileptic seizures in our case study. Such patterns are less numerous and more interesting for clinicians. On the other hand, the main contribution of $DCM$ algorithm is the discovery of temporal information that discriminates care sequences. Even if a sequence of care events is not discriminant (e.g. drug switches), the way they temporally occur may witness the outcome (e.g. seizure).

Experimental results on our case study show that $DCM$ extracts a reduced number of patterns. Discriminant patterns have been presented to clinicians who conclude of their potential interestingness by exploring the care pathways of sequences supported by chronicles. At this stage of the work, our main perspective is to integrate $DCM$ in a care pathways analytics tool such that extracted chronicles may easily be contextualized in care pathways and manually modified to progressively build an care sequence of interest.

# References

1. Alvarez, M.R., Felix, P., Carinena, P.: Discovering metric temporal constraint networks on temporal databases. Artif. Intell. Med. **58**, 139–154 (2013)
2. Berlingerio, M., Bonchi, F., Giannotti, F., Turini, F.: Mining clinical data with a temporal dimension: a case study. In: Proceedings of the International Conference on Bioinformatics and Biomedicine, pp. 429–436 (2007)
3. Cohen, W.W.: Fast effective rule induction. In: Proceedings of the International Conference on Machine Learning, pp. 115–123 (1995)
4. Concaro, S., Sacchi, L., Cerra, C., Fratino, P., Bellazzi, R.: Mining healthcare data with temporal association rules: improvements and assessment for a practical use. In: Combi, C., Shahar, Y., Abu-Hanna, A. (eds.) AIME 2009. LNCS, vol. 5651, pp. 16–25. Springer, Heidelberg (2009). doi:10.1007/978-3-642-02976-9_3
5. Dechter, R., Meiri, I., Pearl, J.: Temporal constraint networks. Artif. Intell. **49**, 61–95 (1991)
6. Fradkin, D., Mörchen, F.: Mining sequential patterns for classification. Knowl. Inf. Syst. **45**(3), 731–749 (2015)
7. Guyet, T., Quiniou, R.: Extracting temporal patterns from interval-based sequences. In: Proceedings of the IJCAI, pp. 1306–1311 (2011)
8. Huang, Z., Lu, X., Duan, H.: On mining clinical pathway patterns from medical behaviors. Artif. Intell. Med. **56**, 35–50 (2012)
9. Lakshmanan, G.T., Rozsnyai, S., Wang, F.: Investigating clinical care pathways correlated with outcomes. In: Business Process Management, pp. 323–338 (2013)
10. Moskovitch, R., Shahar, Y.: Fast time intervals mining using the transitivity of temporal relations. Knowl. Inf. Syst. **42**, 21–48 (2015)
11. Moulis, G., Lapeyre-Mestre, M., Palmaro, A., Pugnet, G., Montastruc, J.L., Sailler, L.: French health insurance databases: what interest for medical research? La Revue de Médecine Interne **36**, 411–417 (2015)

12. Polard, E., Nowak, E., Happe, A., Biraben, A., Oger, E.: Brand name to generic substitution of antiepileptic drugs does not lead to seizure-related hospitalization: a population-based case-crossover study. Pharmacoepidemiol. Drug Saf. **24**, 1161–1169 (2015)

13. Quiniou, R., Cordier, M.-O., Carrault, G., Wang, F.: Application of ILP to cardiac arrhythmia characterization for chronicle recognition. In: Rouveirol, C., Sebag, M. (eds.) ILP 2001. LNCS (LNAI), vol. 2157, pp. 220–227. Springer, Heidelberg (2001). doi:10.1007/3-540-44797-0_18

14. Wright, A.P., Wright, A.T., McCoy, A.B., Sittig, D.F.: The use of sequential pattern mining to predict next prescribed medications. J. Biomed. Inform. **53**, 73–80 (2015)

# Is Crowdsourcing Patient-Reported Outcomes the Future of Evidence-Based Medicine? A Case Study of Back Pain

Mor Peleg[1,2(✉)], Tiffany I. Leung[3], Manisha Desai[1],
and Michel Dumontier[1,4]

[1] Stanford Center for Biomedical Informatics Research,
Stanford University, Stanford, CA, USA
[2] Department of Information Systems, University of Haifa, Haifa, Israel
morpeleg@is.haifa.ac.il
[3] Faculty of Health, Medicine and Life Sciences, Maastricht University,
Maastricht, The Netherlands
[4] Institute of Data Science, Maastricht University, Maastricht, The Netherlands

**Abstract.** Evidence is lacking for patient-reported effectiveness of treatments for most medical conditions and specifically for lower back pain. In this paper, we examined a consumer-based social network that collects patients' treatment ratings as a potential source of evidence. Acknowledging the potential biases of this data set, we used propensity score matching and generalized linear regression to account for confounding variables. To evaluate validity, we compared results obtained by analyzing the patient reported data to results of evidence-based studies. Overall, there was agreement on the relationship between back pain and being obese. In addition, there was agreement about which treatments were effective or had no benefit. The patients' ratings also point to new evidence that postural modification treatment is effective and that surgery is harmful to a large proportion of patients.

## 1 Introduction

Lower back pain is a prevalent chronic condition affecting 39% of the population, which causes long-term disability and agony to patients, loss of work days and large healthcare costs [1]. Diagnosis and treatment is complicated by the fact that there is no clear association between pain and abnormalities detected by spine imaging [2]. Hence, many patients who undergo corrective surgery continue to have pain. Treatment options include spine surgery, injections, medications, psychological interventions, exercise, nutritional supplements, and lifestyle change and self-management approaches. Although many treatments exist, very few were shown to have more than moderate effectiveness at long-term pain reduction [3]. Clinical trials often employ small cohorts and cannot point to effective treatments. Some of them even have contradictory results. Furthermore, outcomes, especially patient-reported, are rarely systematically reported in electronic health records (EHRs).

In order to compare treatments by their effectiveness, objective measures need to be complemented with subjective patient-reported outcome measures (PROM). PROM are

© Springer International Publishing AG 2017
A. ten Teije et al. (Eds.): AIME 2017, LNAI 10259, pp. 245–255, 2017.
DOI: 10.1007/978-3-319-59758-4_27

well established indicators of patients' global health [3]. Collecting PROM is a challenging but growing effort, involving clinicians, medical researchers and most importantly, patients/consumers. Efforts by medical care providers focus on collecting from patients, in a standardized way, the changes is their health state, (e.g., level of pain, physical function, anxiety) [3]. On the other hand, collecting PROM from consumer-centric platforms (e.g., PatientsLikeMe, idonine.com, HealthOutcome.org) attract millions of patients and have obtained patients' treatment experiences from ten thousands of patient first-hand, including patients' treatment ratings, which are not collected in provider-centric EHRs.

Healthoutcome.org is a consumer health website that allows patients to report and share their treatment and health outcomes for most common orthopedic injuries and conditions. The site provides aggregated patients treatment outcome ratings as well as access to each patient review that includes patient information, treatment outcome rating and optional free text description. HealthOutcome has over 110,000 treatments ratings from over 15,000 patients, gathered in less than a year. A set of 38 treatment options are offered to lower back pain patients for rating, including a large number (26) of non-invasive/non-pharmacologic options and new treatment options.

Such non-invasive treatments are usually not documented at such granularity in EHR systems. Moreover, information about treatment outcomes is not available directly to patients. The primary limitation of HealthOutcome is that it does not currently collect PROM with a validated item-set; apart from treatment outcome ratings, patients indicate basic information about themselves, including their injury status (cured, in pain, or recovering), as well as their age category, gender, chronicity, and number of weekly hours of physical exercise. A further limitation is that the information entered is not inspected by clinicians to verify validity. Nonetheless, its importance is in providing transparent data about established and new treatments, while allowing treatments comparison by prevalence and crowd-sourced score, which can be filtered according to the characteristics of the reporting patients.

In light of the promise, but recognizing the limitations of such social networks as tools for evidence collection, our main research question is: **How can PROM among patients with low back pain improve our knowledge of effectiveness and harm of available treatments?** To answer this question we address the following objectives:

(1) Characterization of the HealthOutcome dataset' features and (2) its potential biases; (3) validation of associations with treatment and treatment effectiveness known from the literature or evidence-based studies; (4) Demonstration of the types of data analysis that can be done to compare treatments effectiveness; and (5) Reflection on the value and limitations of the crowdsourcing patient reported treatment outcome ratings as a source of evidence, and directions in which it can be improved.

## 2   Background

Different data sources and/or study designs may be used to estimate the effect of treatment on an outcome, with each approach having its own limitations. Randomized controlled studies collect evidence on treatment effectiveness by recruiting a homogenous cohort of patients and then randomly assigning patients to one treatment group or another. Because the group of patients is homogenous and most importantly, because subjects are randomized to treatment assignment, differences in outcomes can be attributed to the treatments, as such a design effectively controls for confounding. Alas, such traditional ways of collecting evidence have important shortcomings. First, most patients do not fit the study's inclusion or exclusion criteria, hence the evidence is not applicable to them. Second, they are expensive and time-consuming to conduct. Usually small cohorts are recruited which limits the validity of evidence that can be generated. Additionally, some studies of intervention have issues with compliance in a randomized controlled setting. Consequently, most studies performed to compare effectiveness of back pain treatments do not provide conclusive evidence [3]. Alternatively, evidence of treatment effectiveness can be collected prospectively from medical records. Such observational designs may be desirable in their inclusiveness of patients and measures, but they pose threats to obtaining a causal effect of treatment due to the presence of confounding. Section 2.1 reviews some statistically-controlled methods to address such evidence collection. When treatment effectiveness cannot be objectively assessed by laboratory tests (i.e., pain medicine), PROM are collected from patients. We review provider- and consumer-based systems for collecting PROM, noting their differences.

## 2.1 Statistically-Controlled Methods to Collect Evidence Prospectively

In recent years, researchers started using electronic medical records as a source of evidence for computing treatment effectiveness. However, Hersh et al. [4] note that "EHR data from clinical settings may be inaccurate, incomplete, transformed in ways that undermine their meaning, unrecoverable for research [e.g., found in textual notes], of insufficient granularity, and incompatible with research protocols" [i.e., treatment recommended as a balance of what is best for patient care and patient preferences]. Moreover, in observational prospective studies, where there is no randomization to intervention, confounding variables, such as demographics, medications at baseline, and medical conditions, may correlate with both the treatment and outcome [5]. Further, in systems that are based on users' decision to report, sampling bias may occur. For example, physicians may under-report adverse events of drugs that are already trusted vs. reporting for new drugs [6], or patients may decide not to rate treatments that they see as less important. Selection bias may also occur because patients had received certain treatment because it was indicated based on their demographic or disease-state, which are also correlated to the outcome being studied, such as adverse drug effects [6, 7] or treatment ratings. One of the most popular methods to address confounding and issues with selection bias is to use propensity score matching [5] to account for confounders.

## 2.2 Provider Systems for Collecting PROM

The National Institutes of Health have assembled a task force on research standards for chronic low back pain [3]. This task force developed a minimal data set of 40 data items, 29 of which were taken from the PROM Information System (PROMIS) instrument. These items were recommended as offering the best trade-off of length with psychometric validity. The full item-set collects medical history, including chronicity, demographics, involvement in worker's compensation, work status, education, comorbidity, and previous treatment. Key self-report domains include pain intensity, pain interference, physical function, depression, sleep disturbance, and catastrophizing (i.e., thinking that pain will never cease). Provider-centric implementations of PROMIS have been Implemented, such as Collaborative Health Outcomes Information Registry (CHOIR, https://choir.stanford.edu/) [8]. All patients with a pain diagnosis who visit clinics that have implemented CHOIR are asked to complete the PROMIS questionnaire prior to each visit. When matched with EHR data, created by clinicians, which record the treatments that patients received, these records can be analyzed together to compare treatment effectiveness on individual and cohort levels.

## 2.3 Using Crowdsourcing to Find Clinical Evidence for Treatment Effectiveness

Unlike provider based medical records, patient social networks introduce sampling bias because not all patients seen by clinicians are active in social networks. In addition, their reports are not validated by clinicians during encounters to assess problems in understanding the semantics of questions asked, correctness and completeness.

Bove et al. [9] validated the multiple sclerosis (MS) rating scale used in PatientsLikeMe.com by asking MS patients from a MS clinic to use the scale to rate the severity of their disease and compared it to the physician-provided scores recorded in their medical records. Having established the validity of the rating scale, they found small nonparametric correlations between BMI and the disease course of MS, adjusting for age, sex, race, disease duration, and disease type.

Nakamura et al. [10] compared clinicians' and patients' perspectives on the effectiveness of treatments for symptoms of amyotrophic lateral sclerosis by comparing data from a traditional survey study of clinicians with data from PatientsLikeMe. The perception of effectiveness for the five symptom-drug pairs that were studied differed. But due to the small number of patients' ratings that were available at the time of the study (20–66), statistical significance could not be evaluated. Nakamura et al. note the difference between the effectiveness provided by patients based on their direct personal experience versus that provided by clinicians, which is indirect, aggregated from their perception of experience of multiple patients but also more systematic as it draws from their clinical knowledge. It is worth noting that the symptoms studied by the authors (sialorrhea, spasticity and stiffness) can also be observed directly by clinicians.

## 3 Methods

### 3.1 Data Collection and Data Set

Patients freely choose whether to post their reviews to HealthOutcome. They may remain anonymous or sign in. The web site is publicized by targeted Facebook ads, sent to adults who have posted content relating to orthopedic problems. The study was approved to review deidentified data by the Stanford University Human Subjects Research and Institutional Review Board (Protocol 40070). Data was obtained for patients with back pain who reported during 12/2008–12/2016. Two comma separated value (csv) files were obtained: one containing 5230 reviews by patients. Columns included: review ID, timestamp, user ID, injury Status (in pain, recovering, cured), age category (18–34, 35–54, 55+), gender, pain chronicity (<6 M, 6–18 M, >18 M), hours of physical activity per week (0–4, 4–8, 8+), repeat injury?, weight, height, location (city, state), #surgeries, #treatments, textual review. From height and weight, we computed body mass index (BMI) category.

The second csv file contained 44,592 treatment ratings provided by patients in their reviews. The columns included review ID, treatment name, treatment rating. There are five possible ratings: worsened, not improved, improved, almost cured, or cured. The two csv files were joined via a script written in Python 3 which extended the first csv file to contain 38 additional columns, one for each possible treatment, recording the treatment ratings provided in a review to each of the possible treatments.

## 3.2  Data Analysis

We tested the following hypotheses (more details are provided in [11]):

(1)  Patients who respond to the website are not meaningfully different from the targeted patients in terms of age and gender; this was determined through a two-sided Chi-squared test comparing frequency distributions for age and gender.

(2)  PROM are internally consistent; We address this hypothesis by evaluating consistency of reporting by comparing reviews among patients who entered multiple reviews. This was determined by manual inspection of a random subset of 5% of these reviews, to see if a patient's demographic data and set of diagnoses and ratings did not change from one report to the next when they were provided within a six-month time period.

(3)  Those with high BMI have greater back pain; determined by linear regression.

(4)  Treatments' effectiveness, as determined in evidence-based studies, will match with ratings by patients; This was determined by comparing the literature-based effective treatments to patient-rated treatments with majority of ratings being improved, almost cured, or cured, which are not harmful; harmful treatments would be those where at least 10% patients ranked them as "worsening".

(5)  We hypothesize that postural modifications (PM) is more effective than spinal fusion surgery (SFS), and we hypothesize that PM is more effective than laminectomy; We addressed these hypotheses through two approaches. The first used generalized linear regression with a logit link (using R's Logit package) adjusting for potential confounders (adjusted model) and the second similarly utilized a logistic regression but with propensity matching (propensity score model). More specifically, for the adjusted model, we included indicators for treatments of interest in the regression: PM, SFS, or laminectomy. In addition, we included potentially confounding demographic variables of age group, gender,

number of treatments, number of physical activity hours a week, chronicity and BMI; these parameters were chosen because they were shown to be predictive of pain status in a decision tree learning analysis, outside the scope of this paper. Regression for predicting treatment showed that spinal stenosis was a potentially confounding variable, hence this diagnosis was added to the demographics confounding variables for propensity score matching (using R's Matchit package). We report the odds ratio and confidence intervals for each analysis.

# 4  Results

## 4.1  Data Characterization

Characteristics of the responding and the targeted patients are listed in Appendix A [11]. Of all responding patients, 43% of the targeted patients were 55 years or older and 80% were women. Responding patients had a larger proportion of the 55+ age group (72.7%, p-value <0.0001) and a smaller percentage of women (78.2%, p-value <0.0001). The representation of older patients is in concordance with the literature. Most of the patients with back pain are in pain (57.9% in HealthOutcome vs. 52.9% of adults 65 or older [12]). Only 5.7% are cured (the rest are recovering). Accordingly, most of the patients' treatment ratings indicate no improvement (52.1%). 37.5% indicate improvement, 2.9% indicate that they are almost cured and only 1.3% say that they are cured. The percentage of reviews of worsening is 6.2%, higher than the total of cured and almost cured patients. This grim picture is consistent with the literature, showing that back pain is most often a chronic condition.

**Missing Values.** The percentage of missing data are: gender 7.4%; age 12.2%; pain chronicity 24.1%; physical activity 22.3%; injury status 24.5%; weight/height 41.7%.

**Data Quality and Consistency.** 1% (27 of 2706) "In Pain" patients inconsistently provided treatment scores of "cured". 32% of patients provided non-anonymous reviews. 8.3% of the reports were by patients who each provided two or more reports. In a manual inspection of a random subset of 5% of these reports, we found that 5.6% of reports were inconsistent with respect to the patient's demographic data while 33% did not report the same set of diagnoses, treatments tried, or treatment ratings.

## 4.2  Consistency of Relationships with Those Described in the Literature

We evaluated whether relationships observed in our data set were consistent with those reported in the literature. Specifically, that high body mass is associated with an increased prevalence of low back pain [13]. Table 1 shows that overweight or obese patients are more in pain than others (linear regression; overweight: p-value = 0.0006; OR = 0.41; obese: p-value = 1.91e-08. OR = 0.204.

Next, we compared patient opinions about treatment effectiveness to evidence based results. We first studied the distribution of treatment ratings. Table 2 shows select results. Complete results are in [11]. Treatments are ordered by prevalence. The

**Table 1.** Patient injury status with different BMI

| BMI status | In pain | Recovering | Cured |
|---|---|---|---|
| Underweight | 25 (1.2%) | 2 (0.5%) | 3 (3.4%) |
| Normal | 378 (18.6%) | 101 (26.9%) | 38 (43.2%) |
| Overweight | 655 (32.2%) | 145 (38.6%) | 27 (30.7%) |
| Obese | 975 (48.0%) | 128 (34.0%) | 20 (22.7%) |
| Total | 2033 | 376 | 88 |

mode is shown in bold. Treatments that worsen the state of at least 10% of patients are circled. Effective treatments (i.e., have $\geq 50\%$ ratings in improved, almost cured, or cured and have <10% ratings of worsened) are shown in capitals. Not shown are treatments that were tried by fewer than 200 patients and ratings for broad classes of treatments –surgery and physical therapy (PT). The individual treatments under these categories are shown (e.g., PT includes TENS, stretching, heat, etc.). The patients who were cured provided 365 treatment ratings of "cured". The treatments that received the highest number of "cured" ratings were strengthening exercises 41/365; postural modifications 37/365; and stretching 30/365. Table 3 compares the benefit of treatments according to their ratings by patient to results of evidence-based studies [14–16].

**Table 2.** Summary of treatment ratings

| Patient ratings: Treatments | Worsened | Not improved | Improved | Almost cured | Cured | #patients tried treatment |
|---|---|---|---|---|---|---|
| NSAIDs | 113 | **1608** | 946 | 49 | 9 | 2725 |
| Cortisone Injection | 151 | **1308** | 778 | 86 | 20 | 2343 |
| **REST** | 71 | 1088 | **1102** | 56 | 12 | 2329 |
| Stretching | 87 | **1084** | 991 | 78 | 40 | 2280 |
| Strengthening Exercises | 142 | **1080** | 868 | 64 | 52 | 2206 |
| Chiropractor | 171 | **901** | 661 | 69 | 39 | 1841 |
| Epidural | 101 | **805** | 492 | 70 | 9 | 1477 |
| **MASSAGE** | 48 | 662 | **666** | 44 | 19 | 1439 |
| Acupuncture | 23 | **414** | 181 | 29 | 4 | 651 |
| **SWIMMING** | 25 | 264 | **299** | 38 | 5 | 631 |
| Spinal Fusion Surgery | (126) | **212** | 167 | 37 | 18 | 560 |
| Oral corticosteroids | 16 | **275** | 184 | 22 | 3 | 500 |
| Laminectomy Surgery | (82) | **171** | 162 | 29 | 24 | 468 |
| **YOGA** | 14 | 132 | **160** | 22 | 17 | 345 |
| **POSTURAL MODIF.** | 8 | 118 | **124** | 22 | 39 | 311 |
| Discectomy Surgery | (33) | 85 | **89** | 12 | 11 | 230 |
| All treatments | 2189 | **20229** | 14571 | 1144 | 527 | 38660 |

## 4.3　Comparing Treatment Effectiveness

As expected, some attenuation in estimates of association were observed across models (Table 4). More specifically, the adjusted model had estimates that were closer to the null than the unadjusted model, and the propensity-score based model had estimates

**Table 3.** Comparison of treatment benefit: evidence-based vs. patient ratings

| | | Evidence from clinical trials | | | |
|---|---|---|---|---|---|
| | | Effective | No benefit | Harmful | No sufficient evidence |
| Patient ratings | Effective | Massage Yoga Exercise (swimming) | | Rest | *Postural modifications* |
| | No Benefit | Acupuncture Spinal manipulation (chiropractor) | Steroid injection | Traction Inversion table | |
| | Harmful | | | | *Spinal Fusion Laminectomy Disectomy* |
| | Not enough data | Functional restoration Interdisciplinary rehab Cognitive-behavioral | Prolotherapy | Home care Topical gel Dithermy | |

**Table 4.** Analysis of association of treatment options and patients having outcome of cured

| Treatment | N (before matching) | Odds ratio (97.5% CI) | p-value |
|---|---|---|---|
| *Traditional unadjusted regression analysis* | | | |
| Postural modifications vs. Spinal fusion surgery | 637 | 7.96 (4.10–17.03) | 7.75e-09 |
| Postural modifications vs. Laminectomy | 637 | 10.02 (4.53–26.61) | 2.09e-07 |
| *Traditional adjusted regression analysis* | | | |
| PM vs. SFS considering demographics + DXs | 355 | 6.61 (1.64–35.35) | 0.01 |
| PM vs. Lam considering demographics + Dxs | 355 | 5.16 (1.30–26.91) | 0.03 |
| *Propensity-score matched analysis* | | | |
| PM vs. SFS considering demographics + Dxs | 224 (236) | 6.52 (1.40–40.83) | 0.025 |
| PM vs. Lam considering demographics + Dxs | 224 (231) | 5.08 (1.01–35.20) | 0.065 |

that were attenuated relative to the adjusted model. All models, however, provided evidence that PM was strongly associated with Cured status relative to SFS (Unadjusted OR = 7.96, p-value <0.001; Adjusted OR = 6.61, p-values = 0.014; and propensity score-based OR = 6.52, p-value = 0.025). Further, both the unadjusted and adjusted models provided evidence of an associated between PM vs Laminectomy and Cured status, whereas the propensity-score based method did not indicate a significant association (Unadjusted OR = 10.03, p-value <0.001; Adjusted OR = 5.16, p-value = 0.029; and propensity score-based OR = 5.08, p-value = 0.065). In addition, results from the propensity score-based model suggested that variables associated with Cured status include spinal stenosis when the diagnoses were considered, or #treatments, when they were not.

## 5    Discussion

Patient crowdsourcing has been shown to provide large quantities of data. The quality of the data collection metrics, and the ability to validate the collected data, could be improved by collecting additional PROM, collecting data from wearable sensors about physical activity, and by linking the patient's reports to provider-based medical records. Even in its current state, the results suggest that patient-reported opinions of treatment effectiveness in a consumer social network are mostly consistent with published medical evidence. The most effective treatments were confirmed to be massage, yoga, and swimming exercises. In a small number of treatments, patients reported lower effectiveness than published literature suggests (acupuncture, spinal manipulation) or higher effectiveness (rest). The data also points to effective treatments that have not been studied in the evidence-based literature, including postural modifications, as well as provides evidence that all forms of surgery are considered as harmful by 14.4–22.5% of patients. These findings are in line with the recommendations of clinical guidelines [17] to delay surgery to later stages, in the absence of neurologic deficits. Surprisingly, 51% and 55% of patients found that stretching and strengthening exercises were not helpful. More details should be collected to evaluate this further. Generalizability to other domains and other designed platforms would also need evaluation.

### 5.1    Regression vs. Propensity Score Matching

Propensity score matching is considered a state of the art approach for handling confounding in observed studies, especially when there are $\leq 7$ events per confounder [18]. But traditional multivariate regression models may also be appropriate in certain settings. Our analysis has shown (1) that the method of adjusting for confounding may produce disparate findings; Importantly, traditional unadjusted regression that does not account for confounding variables shows some results that are not replicated in analyses that accounts for confounding. Namely, that postural modification has better outcomes than laminectomy. In fact, laminectomy seemed to have a higher odds ratio vs. spinal fusion surgery, as compared to PM; and (2) that propensity matching developed to mitigate confounding can result in attenuated estimates even using the same confounders in a traditional regression; traditional adjusted regression (considering demographics with/without diagnoses) showed that PM was superior to both SFS and laminectomy. However, these results were confirmed on the propensity-matched data set only for PM being superior to SFS; regression that considered demographics with diagnoses and was performed on the propensity-matched data did not show that PM had better outcomes than laminectomy.

### 5.2    Limitations and Future Research

**Incompleteness.** The information collected directly from patients, using an easy-to-use interface, allows collection of a large volume of data quickly. However, incomplete data may result from the voluntary nature of reporting; we speculate that the high rate of missing weight and location data may seem too private to share. In addition, HealthOutcome's user interface has changed over time, so items that were added later (e.g., #surgeries) have more missing values.

**Data Quality and Consistency.** Inconsistency in reports by consumers is common. Yet, these are random measurement errors, which mostly just increase the variance. Evaluation with live subjects and corroboration with clinician-recorded data could estimate how well patients understand the items that they rate or indicate as being true. For example, are they aware of their diagnoses? Do they understand that reporting a value of zero or not reporting is not equivalent (e.g., #surgeries)? Do they consider long or short-term relief (e.g., rest)? Our study does not address the potential fake reporting.

**Generalizability.** The large volume of data collected in consumer networks may help address the limitations discussed above, resulting in treatment ratings that could suggest evidence of effectiveness. However, the patient population in the consumer network does not represent all patients with back pain. Specifically, 80% of the users are women, probably reflecting the tendency of women to write posts on personal topics [19]. While the low number of back pain patients reported being cured (5.7%) could be attributed to patients' interest in reporting negative experience or to the severity of this disease, the former seems unlikely, considering that the percentage of patients who reported being cured of plantar fasciitis in HealthOutcome is much higher (27.8%) than back pain. Note however, that sampling bias is also present in provider-based PROM systems such as CHOIR, which collect the more severe patients who visit pain clinics. We thus suggest that the quantitative results would better be used qualitatively, pointing to potentially beneficial treatments. Conversely, the results pertaining to the high ratio of harmful surgeries do not distinguish patients who had the clinical indications for such treatments, who could benefit more from such treatment. In line with this, the fact that in HealthOutcome, spinal fusion surgery and laminectomy were more prevalent than postural modifications may indicate either that patients were referred to surgery before exhausting all non-invasive options or that a sampling bias was present.

### 5.3   Future Work

Healthoutcome are now collecting more detailed temporal information regarding the ranked treatments as well as some PROMIS outcome measures. This would support future research to assess conditions in which particular treatments are successful.

**Acknowledgement.** We thank Ofer Ben-Shachar for supplying the HealthOutcome data and thank him and Tobias Konitzer for the valuable discussions.

## References

1. Hoy, D., Bain, C., William, G., March, L., Brooks, P., Blyth, F., et al.: A systematic review of the global prevalence of low back pain. Arthritis Rheum. **64**, 2028–2037 (2012)
2. Chou, R., Deyo, R.A., Jarvik, J.G.: Appropriate use of lumbar imaging for evaluation of low back pain. Radiol. Clin. North Am. **50**, 569–585 (2012)

3. Deyo, R.A., Dworkin, S.F., Amtmann, D., et al.: Report of the NIH task force on research standards for chronic low back pain. Int. J. Ther. Massage Bodyw. **8**(3), 16–33 (2015)
4. Hersh, W.R., Weiner, M.G., Embi, P.J., et al.: Caveats for the use of operational electronic health record data in comparative effectiveness research. Med. Care **5**, S30–S37 (2014)
5. Tannen, R.L., Weiner, M.G., Xie, D.: Use of primary care electronic medical record database in drug efficacy research on cardiovascular outcomes: comparison of database and randomised controlled trial findings. Br. Med. J. **338**, b81 (2009)
6. Tatonett, N.P., Ye, P.P., Daneshjou, R., Altman, R.B.: Data-driven prediction of drug effects and interactions. Sci. Trans. Med. **4**(125), 1–26 (2013)
7. Harpaz, R., DuMouchel, W., Shah, N.H., et al.: Novel data mining methodologies for adverse drug event discovery and analysis. Clin. Pharmacol. Ther. **91**(6), 1010–1021 (2012)
8. Bhandari, R.P., Feinstein, A.B., Huestis, S.E., et al.: Pediatric-collaborative health outcomes information registry (Peds-CHOIR): a learning health system to guide pediatric pain research and treatment. Pain **157**(9), 2033–2044 (2016)
9. Bove, R., Secor, E., Healy, B., et al.: Evaluation of an online platform for multiple sclerosis research: patient description, validation of severity scale, and exploration of BMI effects on disease course. PLoS ONE **8**(3), e59707 (2013)
10. Nakamura, C., Bromberg, M., Bhargava, S., et al.: Mining online social network data for biomedical research: a comparison of clinicians' and patients' perceptions about amyotrophic lateral sclerosis treatments. J. Med. Internet Res. **14**(3), e90 (2012)
11. Peleg, M.: Appendices (2017). http://mis.haifa.ac.il/~morpeleg/PatientOutcomesAppend.html
12. Morone, N.E., Greco, C.M., Moore, C.G., et al.: A mind-body program for older adults with chronic low back pain: a randomized clinical trial. JAMA Int. Med. **3**, 329–337 (2016)
13. Heuch, I., Hagen, K., Heuch, I., et al.: The impact of body mass index on the prevalence of low back pain: the HUNT study. Spine (Phila Pa 1976) **35**(7), 764–768 (2010)
14. Chou, R., Huffman, L.H.: Nonpharmacologic therapies for acute and chronic low back pain: a review of the evidence for an American Pain Soc. Ann. Int. Med. **147**, 492–504 (2007)
15. Chou, R., Atlas, S.J., Stanos, S.P., Rosenquist, R.W.: Nonsurgical interventional therapies for low back pain: a review of the evidence for an American Pain Soc. Spine **34**, 1078–1093 (2009)
16. Chou, R., Huffman, L.H.: Medications for acute and chronic low back pain: a review of the evidence for an American Pain Soc. Ann. Intern. Med. **147**, 505–514 (2007)
17. Institute for Clinical Systems Improvement. Adult Acute and Subacute Low Back Pain, November 2012
18. Biondi-Zoccai, G., Romagnol, E., Agostoni, P., et al.: Are propensity scores really superior to standard multivariable analysis? Contemp. Clin. Trials. **32**(5), 731–740 (2011)
19. Wang, Y.C., Burke, M., Kraut, R.E.: Gender, topic, and audience response: an analysis of user-generated content on facebook. In: SIGCHI Conference on Human Factors in Computing Systems, pp. 31–34 (2013)

# Multi-level Interactive Medical Process Mining

Luca Canensi[1], Giorgio Leonardi[2], Stefania Montani[2(✉)],
and Paolo Terenziani[2]

[1] Department of Computer Science, Università di Torino, Turin, Italy
[2] DISIT, Computer Science Institute, Università del Piemonte Orientale,
Alessandria, Italy
stefania.montani@uniupo.it

**Abstract.** In this paper, we present a novel process mining approach, specifically tailored to medical applications, which allows the user to build an initial process model from the hospital event log, and then supports further model refinements, by *directly exploiting* her *knowledge-based model evaluation*. In such a way, it supports the *interactive* construction of the *process model at multiple and user-defined levels of abstraction*, ranging from a model which perfectly adheres to the input traces (i.e., all of its paths correspond to at least one trace in the log) to models which increasingly loose precision, but gain generality. Our results in the field of stroke management, reported as a case study in this paper, show that our approach can provide relevant advantages with respect to traditional process mining techniques.

## 1 Introduction

*Process mining* [4] describes a family of a-posteriori analysis techniques that exploit the so-called *event log*, which records information about the sequences (*traces* henceforth) of actions executed at a given organization. The most relevant and widely used process mining technique is *discovery*; process discovery takes as an input the event log and produces in output a process model. *Medical process mining*, in particular, is a research field which is gaining attention in recent years (see, e.g., [6]). Indeed, the complexity of healthcare [5] demands for proper techniques to analyse the *actual* patient management process applied at a given hospital setting, in order to identify bottlenecks and to assess the correct application of clinical guideline directives in the clinical practice. The currently available process mining solutions, including the open-source framework ProM [7], are not specifically tailored to medical applications.

Current process mining approaches mostly operate in three main phases: (i) application of a process mining algorithm to provide a process model; (ii) process model evaluation, "manually" performed by domain experts on the basis of their knowledge; (iii) process model refinement (if required, given the evaluation), by tuning the parameters of the process mining algorithm. Evaluation and refinement are usually enacted as a closed loop. However, they are not directly related:

© Springer International Publishing AG 2017
A. ten Teije et al. (Eds.): AIME 2017, LNAI 10259, pp. 256–260, 2017.
DOI: 10.1007/978-3-319-59758-4_28

if domain experts identify problems in the model (phase (ii)), the miner parameters have to be tuned and the miner itself has to be re-run (phase. (iii)). Even though user-friendly interfaces support these tasks, a *direct* way to *exploit expert's analysis and knowledge* to guide *model refinement* (phase (iii)) is not available.

In this paper, we present a novel process mining approach, specifically tailored to medical applications, which: (1) allows to construct an initial process model (phase (i)), and supports further model refinement (phase (iii)), *directly exploiting the results of the expert's knowledge-based model evaluation* (phase (ii)); (2) in such a way, it supports the construction of the *process model at multiple and user-defined levels of abstraction*, ranging from a model which perfectly adheres to the input traces (i.e., all of its paths correspond to at least one trace in the log, reaching a 100% precision [2]) to models which increasingly loose precision, but gain generality. This approach is indeed particularly useful in medical applications, where physicians typically have a "deep" domain knowledge, whose integration within the refinement procedure can provide critical advances with respect to traditional techniques. Our results in the field of stroke management, reported as a case study in this paper, support this statement.

## 2    Process Mining Approach

Our framework is able to construct an initial process model, by taking in input the event log. Such a process model, called the *log-tree*, perfectly adheres to the input traces, reaching a 100% precision [2], and represents the starting point for further model enhancement/abstraction steps. The log-tree is a data structure whose nodes represent actions, and arcs represent a precedence relation between them. Every node in the model can represent a single action, or a set of actions. In this second situation, the interpretation is that the actions in the set can be executed in *any order*. When multiple arcs exit a node, the node itself represents a XOR splitting point. Details about the log-tree construction are presented in [1].

While the log-tree is already a process model, it is not very "abstract". Indeed, all of its paths correspond to at least one trace in the event log. When such a high precision is not required, medical experts may want to generate a more abstract model. Moreover, further process refinements, dictated by medical knowledge and by the analysis of the currently available model, may be proposed by medical experts themselves. In our framework, this goal can be achieved by means of a *merging facility*. The *merging facility* allows the user to search for subgraphs isomorphic to an input one in the currently available process model, visualize them through the graphical interface, and merge (some of) them, generating a more abstract process model, which looses precision with respect to the previous one, but gains generality and readability. This step can be repeated several times, in an interactive procedure where the user is always allowed to visualize the current output, and to possibly roll back to the previous model, if merging has introduced errors or imprecisions. In this case, the user can look for subgraphs isomorphic to a different input, and/or select only a subset of the identified subgraphs, and then repeat the merging step.

Algorithm 1 illustrates isomorphic graphs search and merging. The function *Search* (lines 1–6) takes in input the process model $M$ and the subgraph pattern *Patt* to be searched for in the model. It executes the function *findIsomorphism* (line 2), which exploits a classical approach to isomorphic subgraph search in large graphs [3], and saves all subgraphs isomorphic to *Patt* in the variable *res*. The user, through the system interface, is then allowed to explicitly indicate which subgraphs in *res* should be merged; to this end, the function *partition* is invoked (line 3). Subgraphs in the same partition $I$ in *res'* will be fused by the function *Merge* (invoked in lines 4–6; details in lines 7–13). As a special case, if the user creates a single partition, and thus *res'*={*res*}, all identified isomorphic graphs will be merged into a single instance in the model.

Function *Merge* (lines 7–13) takes in input the process model $M$ and a partition $I$, containing a (sub)set of isomorphic subgraphs, to be merged through a pairwise process. In particular, the function *redirect* (line 11) merges every subgraph in $I$ (*isom2*) to a single instance, corresponding to the first element of the partition (*isom*), by redirecting all input and output edges of *isom2* to that instance, and then deletes *isom2* itself (line 12).

---

**Algorithm 1.** Search and Merge

---

1 **Search** (M,Patt)
2 $res \leftarrow findIsomorphism(M, Patt)$
3 $res' \leftarrow partition(res)$
4 **foreach** $I \in res'$ **do**
5 | $Merge(M, I)$
6 **end**
7 **Merge** (M,I)
8 $patternToDelete \leftarrow \{\}$
9 $isom \leftarrow \text{first}(I)$
10 **foreach** $isom2 \neq isom \in I$ **do**
11 | $redirect(M, isom2, isom);$
12 | $delete(isom2, M)$
13 **end**

---

## 3   An Interactive Session of Work

In this section, as an example session of work, we simulate the possible operations performed by a stroke care expert, who wants to analyze the modus operandi of her/his health care structure. First, the expert (user of our system) generates a log-tree from the process data. In our example situation, the result of this first step is the log-tree shown on the left of Fig. 1.

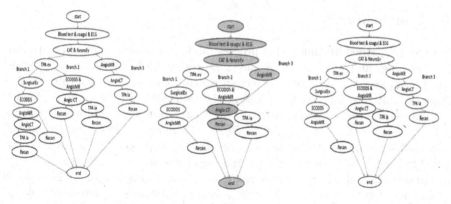

**Fig. 1.** The log-tree mined in the stroke management example (left); the process model obtained by merging all identical subgraphs (center); the process model obtained by selecting the subgraphs to be merged on the basis of medical knowledge (right)

After an observation of the log-tree, the expert can ask the system to obtain a more abstract version. In fact, looking at the log-tree, it is easy to recognize repeated subgraphs. Specifically, the execution of AngioCT (angiographic computerized tomography), followed by the sequence of TPA ia (intra-arterial anti-thrombotic), and then by Recan (recanalization), can be identified at the end of branch 1, branch 2 and branch 3. The user, through the *partition* function, can indicate to the system to merge all the three occurrences together (i.e., a unique partition containing the three occurrences of the subgraph is returned). The application of the *Merge* algorithm to the model and the unique partition merges the three occurrences, returning the model shown in the center of Fig. 1. This model is more compact and readable than the log-tree, but it introduces a path, highlighted in the figure, which is incorrect on the basis of domain knowledge: a recanalization that does not follow any TPA will have the effect of pushing the thrombus even closer to the brain, causing additional damages. This situation can not be accepted by the expert, who can backtrack to the log-tree, and merge only a subset of the retrieved subgraphs. Domain knowledge is explicitly resorted to in this phase, since only an expert is able to identify what subgraphs to merge. In our example, the subgraphs on branches 1 and 2 are actually part of the same therapeutic strategy, performed after the administration of the TPA ev (i.e., endo-venous anti-thrombotic, on patients arriving before 4.5 hours from stroke onset), while branch 3 concerns another type of patients. Therefore, the expert decides to merge only the first two subgraphs, obtaining the model on the right of Fig. 1. This last model, although less compact than the one in the center of Fig. 1, is more abstract than the original one on the left of Fig. 1; it is correct, since it does not introduce any therapeutic path which would be impossible/harmful. The remaining redundancy makes sense, as the different paths refer to patients arriving at the emergency department in different conditions, and who must be treated accordingly. The final process model thus keeps the right balance between generality of representation and correctness.

## 4    Conclusions

We have described an innovative process mining approach that takes in input an event log, and provides in output a process model, which can be progressively evaluated and refined/abstracted, on the basis of the user's needs. Sessions of work are interactive, allowing the user to apply her domain knowledge in the abstraction procedure. Our proposal represents a significant advance with respect to the state of the art in process mining. Indeed, in "standard" approaches, domain experts can analyse the mined process model and, if it is not satisfactory, they can re-run the mining algorithm with different values for its parameters, until a satisfactory model is obtained. However, there is no *direct* connection between the analysis of the experts (a *qualitative* analysis of the current model), and the operations to adjust mining results (a *quantitative* setting of the mining algorithm parameters). Our approach, on the other hand, allows domain experts to *directly* use their *qualitative knowledge to drive the refinement/abstraction process*. This capability promises to be particularly useful in medical applications, where physicians typically have a "deep" domain knowledge, whose integration within the refinement procedure can provide critical advances with respect to traditional techniques. While the initial experiments in the stroke management context have provided satisfactorily results, in the future we plan to conduct a more extensive evaluation in different healthcare settings, to further verify this statement.

## References

1. Bottrighi, A., Canensi, L., Leonardi, G., Montani, S., Terenziani, P.: Trace retrieval for business process operational support. Expert Syst. Appl. **55**, 212–221 (2016)
2. Buijs, J.C.A.M., van Dongen, B.F., van der Aalst, W.M.P.: On the role of fitness, precision, generalization and simplicity in process discovery. In: Meersman, R., et al. (eds.) OTM 2012. LNCS, vol. 7565, pp. 305–322. Springer, Heidelberg (2012). doi:10.1007/978-3-642-33606-5_19
3. Cordella, L.P., Foggia, P., Sansone, C., Vento, M.: A (sub)graph isomorphism algorithm for matching large graphs. IEEE Trans. Pattern Anal. Mach. Intell. **26**(10), 1367–1372 (2004)
4. IEEE Taskforce on Process Mining: Process Mining Manifesto. http://www.win.tue.nl/ieeetfpm. Accessed 4 Nov 2013
5. Mans, R.S., van der Aalst, W.M.P., Vanwersch, R.J.B., Moleman, A.J.: Process mining in healthcare: data challenges when answering frequently posed questions. In: Lenz, R., Miksch, S., Peleg, M., Reichert, M., Riaño, D., ten Teije, A. (eds.) KR4HC/ProHealth -2012. LNCS (LNAI), vol. 7738, pp. 140–153. Springer, Heidelberg (2013). doi:10.1007/978-3-642-36438-9_10
6. Rojas, E., Munoz-Gama, J., Sepulveda, M., Capurro, D.: Process mining in healthcare: a literature review. J. Biomed. Inform. **61**, 224–236 (2016)
7. van Dongen, B.F., de Medeiros, A.K.A., Verbeek, H.M.W., Weijters, A.J.M.M., van der Aalst, W.M.P.: The ProM framework: a new era in process mining tool Support. In: Ciardo, G., Darondeau, P. (eds.) ICATPN 2005. LNCS, vol. 3536, pp. 444–454. Springer, Heidelberg (2005). doi:10.1007/11494744_25

# Declarative Sequential Pattern Mining
# of Care Pathways

Thomas Guyet[1(⊠)], André Happe[2], and Yann Dauxais[3]

[1] AGROCAMPUS-OUEST/IRISA-UMR6074, Rennes, France
`thomas.guyet@irisa.fr`
[2] CHRU Brest/EA-7449 REPERES, Paris, France
[3] Rennes University 1/IRISA-UMR6074, Rennes, France

**Abstract.** Sequential pattern mining algorithms are widely used to explore care pathways database, but they generate a deluge of patterns, mostly redundant or useless. Clinicians need tools to express complex mining queries in order to generate less but more significant patterns. These algorithms are not versatile enough to answer complex clinician queries. This article proposes to apply a declarative pattern mining approach based on Answer Set Programming paradigm. It is exemplified by a pharmaco-epidemiological study investigating the possible association between hospitalization for seizure and antiepileptic drug switch from a french medico-administrative database.

**Keywords:** Answer Set Programming · Epidemiology · Medico-administrative databases · Patient care pathways

## 1 Introduction

Pharmaco-epidemiology applies the methodologies developed in general epidemiology to answer questions about the uses of health products in the population in real condition. In pharmaco-epidemiology studies, people who share common characteristics are recruited. Then, a dataset is built from meaningful data (drug exposures, events or outcomes) collected within a defined period of time. Finally, a statistical analysis highlights the links (or the lack of link) between drug exposures and outcomes (*e.g.* adverse effects).

The data collection of such prospective cohort studies is slow and cumbersome. Medico-administrative databases are readily available and cover a large population. They record, with some level of details, all reimbursed drug deliveries and all medical procedures, for insured people. Such database gives an abstract view on longitudinal care pathways. It has become a credible alternative for pharmaco-epidemiological studies [1]. However, it has been conceived for administrative purposes and their use in epidemiology is complex.

This research is supported by the PEPS project funded by the french agency for health products safety (ANSM) and the SePaDec project funded by Brittany Region.

A. ten Teije et al. (Eds.): AIME 2017, LNAI 10259, pp. 261–266, 2017.
DOI: 10.1007/978-3-319-59758-4_29

Our objective is to propose a versatile pattern mining approach that extracts sequential patterns from care pathways. The flexibility of such new knowledge discovery tools has to enable epidemiologists to easily investigate various types of interesting patterns, *e.g.* frequent, rare, closed or emerging patterns, and possibly new ones. On the other hand, the definition of interesting patterns has to exploit in-depth the semantic richness of care pathways due to complex care event descriptions (*e.g.* units number, strength per unit, drugs and diagnosis taxonomies, etc.). By this mean, we expect to extract less but more significant patterns.

This article presents the application of a declarative pattern mining framework based on Answer Set Programming (ASP) [2] to achieve care pathway analysis answering pharmaco-epidemiological questions.

Answer Set Programming (ASP) is a declarative programming paradigm. It gives a description, in a first-order logic syntax, of what is a problem instead of specifying how to solve it. Semantically, an ASP program induces a collection of so-called *answer sets*. For short, a model assigns a truth value to each propositional atoms of the program. An answer set is a minimal set of true propositional atoms that satisfies all the program rules. ASP problem solving is ensured by efficient solvers. For its computational efficiency, we use *clingo* [3] as a primary tool for designing our encodings. An *ASP program* is a set of rules of the form: $a_0$ :- $a_1, \ldots, a_m$, not $a_{m+1}, \ldots$, not $a_n$, where each $a_i$ is a propositional atom for $0 \leq i \leq n$ and not stands for *default negation*. In the body of the rule, commas denote conjunctions between atoms. If $n = 0$, *i.e.*, the rule body is empty, the rule is called a *fact* and the symbol ":-" may be omitted. Such a rule states that the atom $a_0$ has to be true. If $a_0$ is omitted, *i.e.*, the rule head is empty, the rule represents an integrity constraint meaning that it must not be *true*. *clingo* also includes several extensions to facilitate the practical use of ASP (variables, conditional literals and cardinality constraints).

Recent researches has been focused on the use of declarative paradigms, including ASP, to mine structured datasets, and more especially sequences [2,4]. The principle of declarative pattern mining is closely related to the Inductive Logic Programming (ILP) [5] approach. The principle is to use a declarative language to model the analysis task: supervised learning for ILP and pattern mining for our framework. The encoding benefits from the versatility of declarative approaches and offers natural abilities to represent and reason about knowledge.

## 2 Context, Data and Pharmaco-Epidemiological Question

In this work, we exemplify our declarative pattern mining framework by investigating the possible association between hospitalization for seizure and antiepileptic drug switches, *i.e.*, changes between drugs. The first step was to create a digital cohort of 8,379 patients with a stable treatment for epilepsy (stability criterion detailed in [6] have been used). This cohort has been built from the medico-administrative database, called SNIIRAM [1] which is the database of the french health insurance system. It is made of all outpatient reimbursed health expenditures. Our dataset represents 1,8 M deliveries of 7,693 different drugs and 20,686 seizure-related hospitalizations.

This dataset and background knowledge (ATC drugs taxonomy, ICD-10 taxonomy) are encoded as ASP facts. For each patient $p$, drug deliveries are encoded with $\texttt{deliv}(p, t, d, q)$ atoms meaning that patient $p$ got $q$ deliveries of drug $d$ at date $t$. Dates are day numbers starting from the first event date. We use french CIP, Presentation Identifying Code, as drug identifier. The knowledge base links the CIP to the ATC and other related informations (e.g. speciality group, strength per unit, number of units or volume, generic/brand-named status, etc.). Each diagnosis related to an hospital stay is encoded with $\texttt{disease}(p, t, d)$ meaning that patient $p$ have been diagnosed with $d$ at date $t$. Data, $\mathcal{D}$, and related knowledge base, $\mathcal{K}$, represent a total of 2,010,556 facts.

## 3   Sequential Pattern Mining with ASP

Let $\mathcal{I} = \{i_1, i_2, \ldots, i_{|\mathcal{I}|}\}$ be a set of *items*. A *temporal sequence* $s$, denoted by $\langle (s_j, t_j) \rangle_{j \in [m]}$ is an ordered list of items $s_j \in \mathcal{I}$ timestamped with $t_j \in \mathbb{N}$. Let $p = \langle p_j \rangle_{1 \leq j \leq n}$, where $p_j \in \mathcal{I}$ be a sequential pattern. We denote by $\mathcal{L} = \mathcal{I}^*$ the *pattern search space*. Given the pattern $p$ and the sequence $s$ with $n \leq m$, we say that $s$ *supports* $p$, iff there exists $n$ integers $e_1 < \ldots < e_n$ such that $p_k = s_{e_k}, \forall k \in \{1, \ldots, n\}$. $(e_k)_{k \in [n]}$ is called an *embedding* of pattern $p$ in $s$. $\mathcal{E}_p^s = \{(e_k)_{k \in [n]}\}$ denotes the set of the embeddings of $p$ in $s$. Let $\mathcal{D} = \{s^k\}_{k \in [N]}$, be a dataset of $N$ sequences. We denote by $\mathcal{T}_p$ the sequence set supported by $p$. Given a set of constraints $\mathcal{C}$, the mining of sequential patterns consists in finding out all tuples $\langle p, \mathcal{T}_p, \mathcal{E}_p \rangle$ satisfying $\mathcal{C}$, where $\mathcal{E}_p = \bigcup_{s \in \mathcal{T}_p} \mathcal{E}_p^s$. The most used pattern constraint is the minimal frequency constraint, $c_{f_{min}} : |\mathcal{T}_p| \geq f_{min}$, saying that the pattern support has to be above a given threshold $f_{min}$.

Sequential pattern mining with ASP has been introduced by Guyet et al. [2][1]. It encodes the sequential pattern mining task as an ASP program that process sequential data encoded as ASP facts. A sequential pattern mining task is a tuple $\langle \mathcal{S}, \mathfrak{M}, \mathcal{C} \rangle$, where $\mathcal{S}$ is a set of ASP facts encodings the sequence database, $\mathfrak{M}$ is a set of ASP rules which yields pattern tuples from database, $\mathcal{C}$ is a set of constraints (see [4] for constraint taxonomy). We have $\mathcal{S} \cup \mathfrak{M} \cup \mathcal{C} \models \{\langle p, \mathcal{T}_p, \mathcal{E}_p \rangle\}$.

In our framework, the sequence database is modeled by $\texttt{seq}(s, t, e)$ atoms. Each of these atoms specifies that the event $e \in \mathcal{I}$ occurred at time $t$ in sequence $s$. On the other hand, each answer set holds atoms that encode a pattern tuples. $\texttt{pat}(i, p_i)$ atoms encode the pattern $p = \langle p_i \rangle_i \in [l]$ where $l$ is given by $\texttt{patlen}(l)$, $\texttt{support}(s)$ encodes $\mathcal{T}_s$ and finally $\mathcal{E}_p$ is encoded by $\texttt{occ}(s, i, e_i)$ atoms.

## 4   Declarative Care Pathway Mining

The declarative care pathway mining task can be defined as a tuple of ASP rule sets $\langle \mathcal{D}, \mathfrak{S}, \mathcal{K}, \mathfrak{M}, \mathcal{C} \rangle$ where $\mathcal{D}$ is the raw dataset and $\mathcal{K}$ the knowledge base

---

[1] Original encodings can be found here: https://sites.google.com/site/aspseqmining/.

introduced in Sect. 2; $\mathfrak{M}$ is the encoding of the sequence mining task presented in [2] and $\mathcal{C}$ is a set of constraints. Finally, $\mathfrak{S}$ is a set of rules yielding the sequences database: $\mathfrak{S} \cup \mathcal{D} \cup \mathcal{K} \models \mathcal{S}$. Depending on the study, the expert has to provide $\mathfrak{S}$, a set of rules that specifies which are the events of interest and $\mathcal{C}$, a set of constraints that specifies the patterns the user would like.

In the following of this section, we give examples for $\mathfrak{S}$ and $\mathcal{C}$ to design a new mining tasks inspired from a *case-crossover study* answering our clinical question [6]. For each patient, the $\mathfrak{S}$ rules generate two sequences made of deliveries within respectively the 3 months before the first seizure (positive sequence) and the 3 to 6 months before the first seizure (negative sequence). In this setting the patient serves as its own control. The mining query consists in extracting frequent sequential patterns where a patient is supported by the pattern iff the pattern appears in its positive sequence, but not in its negative sequence. A frequency threshold for this pattern is set up to 20 and we also constraint patterns (1) to have generic and brand-name deliveries and (2) to have exactly one switch from a generic to a brand-name anti-epileptic grugs – AED (or the reverse).

*Defining Sequences to Mine with* $\mathfrak{S}$. Listing above illustrates the sequence generation of deliveries of anti-epileptic drug specialities within the 3 months (90 days) before the first seizure event. It illustrates the use of the knowledge base to express complex sequences generation. In this listing, aed($i, c$) lists the CIP code $i$, which are related to one of the ATC codes for AED ($N03AX09$, $N03AX14$, etc.), and firstseizure($p, t$) is the date, $t$, of the first seizure of patient $p$. A seizure event is a disease event with one of the G40-G41 ICD-10 code. The first seizure is the one without any other seizure event before. ASP enables to use a reified model of sequence where events are functional literals. seq(P,T,deliv(AED,Gr,G)) designates that patient P was delivered at time T with a drug where AED is the ATC code, Gr identify the drug speciality and G indicates whether the speciality is a generic drug or a brand-named one. The same encoding can be adapted for sequences within the 3–6 months before the first seizure event.

```
aed(CIP,AED):-cip_atc(CIP,AED),AED=(n03ax09;n03ax14;n03ax11;n03ag01;n03af01).
firstseizure(P,T) :- disease(P,T,D), is_a(D,g40;g41),
                     #count{Tp: disease(P,Tp,Dp), is_a(Dp,g40;g41), Tp<T}=0.

seq(P,T,deliv(AED,Gr,1)):- deliv(P,T,CIP,Q), aed(CIP,AED), grs(CIP,Gr),
                           generic(CIP), T<Ts, T>Ts-90, firstseizure(P,Ts).

seq(P,T,deliv(AED,Gr,0)):- deliv(P,T,CIP,Q), aed(CIP,AED), grs(CIP,Gr),
                           not generic(CIP), T<Ts, T>Ts-90, firstseizure(P,Ts).
```

*Defining Constraints on Patterns.* On the other side of our framework, $\mathcal{C}$ enables to add constraints on patterns the clinician looks for. Lines 1–3 (see listing above) encode the case-crossover constraints. They select patterns (*i.e.*, answer sets) that are frequently in the 3 months period but not in the 3–6 months period. The frequency threshold is set to 20. Finally, lines 5–6 illustrate a constraint

on the shape of the pattern, that here must contains exactly one switch from a
brand-name to a generic drug (or the reverse).

```
1 discr(T) :- support(T), not neg_support(T).
2 #const th=20.
3 :- { discr(T) } < th.
4
5 change(X) :- pat(X+1,deliv(AEDp,GRSp,Gp)), pat(X,deliv(AED,GRS,G)), Gp!=G.
6 :- #count{X:change(X)}!=1.
```

*Results.* The solver extracts respectively 32 patterns and 21 patterns (against
4359 patterns with a regular sequential pattern mining algorithm). With such
very constrained problem, the solver is very efficient and extracts all patterns in less than 30 s. The following pattern is representative of our results:
$\langle (N03AG01, 438, 1), (N03AG01, 438, 1), (N03AX14, 1023, 0), (N03AX14, 1023, 0) \rangle$ is a
sequence of deliveries showing a change of treatment from a generic drug of the
speciality 438 of valproic acid to the brand-name speciality 1023 of levetiracetam.
According to our mining query, we found more than 20 patients which have this
care sequence within the 3 months before a seizure, but not in the 3 previous
months preceding this period. These new hypothesis of care-sequences are good
candidates for further investigations and possible recommendation about AE
treatments.

## 5   Conclusion

Declarative sequential pattern mining with ASP is an interesting framework to
flexibly design care-pathway mining queries that supports knowledge reasoning
(taxonomy and temporal reasoning). We illustrated the expressive power of this
framework by designing a new mining tasks inspired from case-crossover studies
and shown its utility for care pathway analytics. We strongly believe that our
integrated and flexible framework empowers the clinician to quickly evaluate
various pattern constraints and that it limits tedious pre-processing phases.

## References

1. Martin-Latry, K., Bégaud, B.: Pharmacoepidemiological research using French reim-
   bursement databases: yes we can!. Pharmacoepidemiol. Drug Saf. **19**(3), 256–265
   (2010)
2. Gebser, M., Guyet, T., Quiniou, R., Romero, J., Schaub, T.: Knowledge-based
   sequence mining with ASP. In: Proceedings of IJCAI, pp. 1497–1504 (2016)
3. Gebser, M., Kaminski, R., Kaufmann, B., Ostrowski, M., Schaub, T., Schneider, M.:
   Potassco: the Potsdam answer set solving collection. AI Commun. **24**(2), 107–124
   (2011)
4. Negrevergne, B., Guns, T.: Constraint-based sequence mining using constraint programming. In: Michel, L. (ed.) CPAIOR 2015. LNCS, vol. 9075, pp. 288–305.
   Springer, Cham (2015). doi:10.1007/978-3-319-18008-3_20

5. Quiniou, R., Cordier, M.-O., Carrault, G., Wang, F.: Application of ILP to cardiac arrhythmia characterization for chronicle recognition. In: Rouveirol, C., Sebag, M. (eds.) ILP 2001. LNCS (LNAI), vol. 2157, pp. 220–227. Springer, Heidelberg (2001). doi:10.1007/3-540-44797-0_18
6. Polard, E., Nowak, E., Happe, A., Biraben, A., Oger, E.: Brand name to generic substitution of antiepileptic drugs does not lead to seizure-related hospitalization: a population-based case-crossover study. Pharmacoepidemiol. Drug Saf. 24(11), 1161–1169 (2015)

# Knowledge-Based Trace Abstraction
# for Semantic Process Mining

Stefania Montani[1]([⊠]), Manuel Striani[2], Silvana Quaglini[3], Anna Cavallini[4],
and Giorgio Leonardi[1]

[1] DISIT, Computer Science Institute, Università del Piemonte Orientale,
Alessandria, Italy
stefania.montani@uniupo.it
[2] Department of Computer Science, Università di Torino, Turin, Italy
[3] Department of Electrical, Computer and Biomedical Engineering,
Università di Pavia, Pavia, Italy
[4] I.R.C.C.S. Fondazione "C. Mondino" - on Behalf of the Stroke Unit Network
(SUN) Collaborating Centers, Pavia, Italy

**Abstract.** Many hospital information systems nowadays record data
about the executed medical process instances in the form of *traces* in an
event log. In this paper we present a framework able to convert actions
found in the traces into higher level concepts, on the basis of domain
knowledge. Abstracted traces are then provided as an input to semantic
process mining. The approach has been tested in stroke care, where we
show how the abstraction mechanism allows the user to mine process
models that are easier to interpret, since unnecessary details are hidden,
but key behaviors are clearly visible.

## 1 Introduction

Most commercial information systems, including those adopted by many health
care organizations, record information about the executed process instances in
the form of an *event log* [4]. The event log stores the sequences (*traces* [2] henceforth) of actions that have been executed at the organization, typically together
with key execution parameters, such as times, cost and resources. Event logs can
be provided in input to **process mining** [4] algorithms, a family of a-posteriori
analysis techniques able to extract non-trivial knowledge from these historic
data; within process mining, *process model discovery* algorithms, in particular,
take as input the log traces and build a process model, focusing on its control
flow constructs. Classical process mining algorithms, however, provide a purely
syntactical analysis, where actions in the traces are processed only referring to
their names. Action names are strings without any semantics, so that identical
actions, labeled by synonyms, will be considered as different, or actions that are
special cases of other actions will be processed as unrelated.

Leveraging process mining to the conceptual layer can enhance existing algorithms towards more advanced and reliable approaches. As a matter of fact,
**semantic process mining**, defined as the integration of semantic processing

© Springer International Publishing AG 2017
A. ten Teije et al. (Eds.): AIME 2017, LNAI 10259, pp. 267–271, 2017.
DOI: 10.1007/978-3-319-59758-4_30

capabilities into classical process mining techniques, has been recently proposed in the literature. However, while more work has been done in the field of semantic *conformance checking* (another branch of process mining) [1,3], to the best of our knowledge semantic *process model discovery* needs to be further investigated.

In this paper, we present a **knowledge-based abstraction mechanism**, able to operate on event log traces. In our approach: (i) actions in the log are mapped to the ground terms of an *ontology*; (ii) a *rule base* is exploited, in order to identify which of the multiple ancestors of an action should be considered for abstracting the action itself; *medical knowledge and contextual information* are resorted to in these steps; (iii) when a set of consecutive actions on the trace abstract as the same ancestor, they are merged into the same abstracted *macro-action*, labeled as the common ancestor at hand. This last step requires a proper treatment of delays and/or actions in-between that descend from a different ancestor.

Our abstraction mechanism is then provided as an input to **semantic process mining**. In particular, we rely on classical *process model discovery* algorithms embedded in the open source framework ProM [5], made semantic by the exploitation of domain knowledge in the abstraction phase. We also describe our experimental work in the field of stroke care, where the application of the abstraction mechanism on log traces has allowed us to mine simpler and more understandable process models.

## 2  Knowledge-Based Trace Abstraction

In our framework, trace abstraction has been realized as a multi-step mechanism.

As a first step, every action in the trace to be abstracted is mapped to a ground term of an **ontology**, formalized resorting to domain knowledge. In our current implementation, we have defined an ontology related to the field of stroke management, where ground terms are patient management actions, while abstracted terms represent medical goals. In particular, a set of classes, representing the main goals in stroke management, have been identified, namely: "Administrative Actions", "Brain Damage Reduction", "Causes Identification" "Pathogenetic Mechanism Identification", "Prevention", and "Other". These main goals can be further specialized into subclasses, according to more specific goals (e.g., "Parenchima Examination" is a subgoal of "Pathogenetic Mechanism Identification", while "Early Relapse Prevention" is a subgoal of "Prevention"), down to the ground actions, that will implement the goal itself. Some actions in the ontology can be performed to implement different goals. For instance, a Computer Assisted Tomography (CAT) can be used to check therapy efficacy in "Early Relapse Prevention", or to perform "Parenchima Examination".

As a second step in the trace abstraction mechanism, a **rule base** is exploited to identify which of the multiple ancestors (i.e., goals) of an action in the ontology should be considered for abstracting the action itself. Contextual information (i.e., the actions that have been already executed on the patient at hand, and/or her/his specific clinical conditions) is used to activate the correct rules.

Once the correct ancestor of every action has been identified, trace abstraction can be completed. As a last step, when a set of consecutive actions on the trace abstract as the same ancestor, they have to be merged into the same abstracted **macro-action**, labeled as the common ancestor at hand. This procedure requires a proper treatment of *delays*, and of actions in-between that descend from a different ancestor (*interleaved actions* henceforth). Specifically, the procedure to abstract a trace operates as follows.

For every action $i$ in the trace:

(i) $i$ is abstracted as one of its ancestors (the one identified by the rule based reasoning procedure), at the ontology level chosen by the user; the macro-action $m\_i$, labeled as the identified ancestor, is created;

(ii) for every element $j$ following $i$ in the trace: (ii-a) if $j$ is a delay (or an interleaved action), its length is added to a variable, that stores the total delay duration (or the total interleaved action duration, respectively) accumulated so far during the abstraction of $i$; (ii-b) if $j$ is an action that, according to domain knowledge, abstracts as the same ancestor as $i$, $m\_i$ is extended to include $j$, provided that the total durations mentioned above do not exceed domain-defined thresholds. $j$ is then removed from the actions in the trace that could start a new macro-action, since it has already been incorporated into an existing one;

(iii) the macro-action $m\_i$ is appended to the output abstracted trace which, in the end, will contain the list of all the macro-actions that have been created by the procedure.

## 3 Experimental Results

In this section, we describe the experimental work we have conducted, in the application domain of stroke care. The available event log is composed of more than 15000 traces, collected at the 40 Stroke Unit Network (SUN) collaborating centers of the Lombardia region, Italy. Traces are composed of 13 actions on average. The 40 Stroke Units (SUs) are not all equipped with the same human and instrumental resources: in particular, according to resource availability, they can be divided into 3 classes. Class-3 SUs are top class centers, able to deal with particularly complex stroke cases; class-1 SUs, on the contrary, are the more generalist centers, where only standard cases can be managed. We have tested whether our capability to abstract the event log traces on the basis of their semantic goals allowed to obtained process models where unnecessary details are hidden, but key behaviors are clear. Indeed, if this hypothesis holds, in our application domain it becomes easier to compare process models of different SUs, highlighting the presence/absence of common paths, regardless of minor action changes (e.g., different ground actions that share the same goal) or irrelevant different action ordering or interleaving (e.g., sets of ground actions, all sharing a common goal, that could be executed in any order). Figure 1 compares the process models of two different SUs (SU-A and SU-B), mined by resorting to

Heuristic Miner, operating on ground traces. Figure 2, on the other hand, compares the process models of the same SUs as Fig. 1, again mined by resorting to Heuristic Miner, but operating on traces abstracted according to the goals of the ontology. Generally speaking, a visual inspection of the two graphs in Fig. 1 is very difficult. Indeed, these two ground processes are "spaghetti-like" [2], and the extremely large number of nodes and edges makes it hard to identify commonalities in the two models. The abstract models in Fig. 2, on the other hand, are much more compact, and it is possible for a medical expert to analyze them. In particular, the two graphs in Fig. 2 are not identical, but in both of them it is easy to a identify the macro-actions which correspond to the treatment of a typical stroke patient. However, the model for SU-A exhibits a more complex control flow (with the presence of loops), and shows three additional macro-actions with respect to the model of SU-B, namely "Extracranial Vessel Inspection", "Intracranial Vessel Inspection" and "Recanalization". This finding can be explained, since SU-A is a class-2 SU, where different kinds of patients, including some atypical/more critical ones, can be managed, thanks to the availability of different skills and instrumental resources. These patients

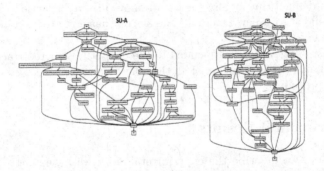

**Fig. 1.** Comparison between two process models, mined operating on ground traces. The figure is not intended to be readable, but only to give an idea of how complex the models can be

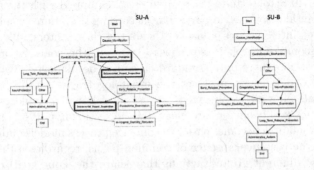

**Fig. 2.** Comparison between the two process models of the same SUs as Fig. 1, mined on abstracted traces. Additional macro-actions executed at SU-A are highlighted in bold

may require the additional macro-actions reported in the model, and/or the repetition of some procedures, in order to better characterize and manage the patient's situation. On the other hand, SU-B is a class-1 SU, i.e., a more generalist one, where very specific human knowledge or technical resources are missing. As a consequence, the overall model control flow is simpler, and some activities are not executed at all. Interestingly, our abstraction mechanism, while hiding irrelevant details, allows to still appreciate these differences.

## 4    Concluding Remarks and Future Work

In this paper, we have presented a framework for knowledge-based abstraction of event log traces. In our approach, abstracted traces are then provided as an input to semantic process mining. Semantic process mining relies on ProM algorithms, but enhances ProM functionalities, which allow one to group multiple events into one class of events - but only syntactically. Experimental results in the field of stroke management have proved that the capability of abstracting the event log traces on the basis of their semantic goal allows to mine clearer process models, where unnecessary details are hidden, but key behaviors are clear. In the future, we plan to substitute/integrate our proof-of-concept ontology with a SNOMED ontology, and to conduct further experiments, by quantitatively comparing different process models (of different SUs) obtained from abstracted traces. We will also extensively test the approach in different application domains.

## References

1. Alves de Medeiros, A.K., van der Aalst, W.M.P., Pedrinaci, C.: Semantic process mining tools: core building blocks. In: Golden, W., Acton, T., Conboy, K., van der Heijden, H., Tuunainen, V.K. (eds.) 16th European Conference on Information Systems, ECIS 2008, Galway, Ireland, pp. 1953–1964 (2008)
2. Van der Aalst, W.: Process Mining. Discovery, Conformance and Enhancement of Business Processes. Springer, Heidelberg (2011)
3. Grando, M.A., Schonenberg, M.H., van der Aalst, W.M.P.: Semantic process mining for the verification of medical recommendations. In: Traver, V., Fred, A.L.N., Filipe, J., Gamboa, H. (eds.) Proceedings of the International Conference on Health Informatics, HEALTHINF 2011, Rome, Italy, 26–29 January 2011, pp. 5–16. SciTePress, Setúbal (2011)
4. IEEE Taskforce on Process Mining: Process Mining Manifesto. http://www.win.tue.nl/ieeetfpm. Accessed 4 Nov 2013
5. van Dongen, B., Alves De Medeiros, A., Verbeek, H., Weijters, A., Van der Aalst, W.: The proM framework: a new era in process mining tool support. In: Ciardo, G., Darondeau, P. (eds.) Knowledge Mangement and its Integrative Elements, pp. 444–454. Springer, Berlin (2005)

# The Spanish Kidney Exchange Model: Study of Computation-Based Alternatives to the Current Procedure

Miquel Bofill[1], Marcos Calderón[3], Francesc Castro[1], Esteve Del Acebo[1(✉)],
Pablo Delgado[2], Marc Garcia[1], Marta García[2], Marc Roig[1],
María O. Valentín[2], and Mateu Villaret[1]

[1] Departament d'Informàtica, Matemàtica Aplicada i Estadística,
Universitat de Girona, Girona, Spain
esteve.acebo@udg.edu
[2] Organización Nacional de Trasplantes, Madrid, Spain
[3] TAISA, Pozuelo de Alarcón, Spain

**Abstract.** The problem of incompatible pairs in living-donor kidney transplant can be solved using paired kidney exchange, i.e., two incompatible patient-donor pairs interchange donors, creating a cycle, which can be extended to three or even more pairs. Finding a set of cycles that maximizes the number of successful transplants is a complex task.

The Organización Nacional de Trasplantes (ONT) is responsible for donation and transplantation processes in Spain. In this paper we compare the current ONT heuristic finding-cycles procedure with an integer programming approach by means of a true-data-based empirical simulation. The obtained results show that, although the two methods provide quite different solutions, they both exhibit weak and strong points.

## 1 Introduction and Previous Work

Living-donor kidney transplant is, nowadays, the treatment of choice for patients with end-stage renal disease. However, in over 30% of the cases, the living donor is incompatible with his/her intended recipient [1]. The nation-wide Kidney Paired Donation (KPD) programs allow incompatible pairs to be added to a pool in which patients can exchange their donors, resulting in transplant cycles. Solving this Kidney Exchange Problem (KEP), i.e., finding, from this pool, an optimal set of disjoint transplant cycles is a computationally hard task [2].

Two important problems arise in KPD programs: (i) Scheduled transplantation cycles do not proceed in all cases because of last-moment-detected compatibility problems, donor backing out, etc.; (ii) it is not easy to find a compatible donor for high-sensitized patients, who tend to remain in the pool for a long period.

A collaboration has started between the Spanish Government-dependent *Organización Nacional de Trasplantes* (ONT) and Computer Science researchers from the University of Girona (UdG). From this collaboration, in this paper we

© Springer International Publishing AG 2017
A. ten Teije et al. (Eds.): AIME 2017, LNAI 10259, pp. 272–277, 2017.
DOI: 10.1007/978-3-319-59758-4_31

compare two different procedures intended to obtain an appropriate set of transplant cycles from a pool of patient-donor pairs. The first procedure is the one currently used by the ONT, a greedy algorithm that prioritizes cycles of length 2 (2-cycles), since they have a lower probability of failure, and repairable cycles of length 3 (3-cycles), i.e. 3-cycles with some 2-cycle embedded. The second procedure is an Integer Programming (IP)-based one which guarantees the optimal solution in the number of proposed transplants, but may have scalability issues. We can find several IP-based approaches in the literature [3–5].

To support this comparison, we provide a long-run simulation of the activity of a living-donor transplant pool, taking into account cycle failures, cycle repairs in case of failures, entrances and releases of patients to/from the pool, etc. This pool is based on data provided by the ONT.

## 2   Problem Definition and Formulation

Traditionally, modeling of the KEP has departed from transplant graphs [2,3] defined by a set of vertices representing donor-recipient incompatible pairs and a set of weighted directed arcs representing compatibilities. The weight of each arc represents the utility of the transplant, defined by a team of medical specialists. A N-way kidney exchange corresponds, then, to a cycle of length N in the transplant graph. The Kidney Exchange Problem (KEP) can be defined as that of finding a maximum weight set of vertex-disjoint cycles having length at most $k$ [3].

The KEP has been proved to be NP-Hard whenever cycles of length greater than two are allowed [2]. This makes it very reasonable to explore the utilization of both exact and heuristic methods to solve it, especially if the possibility of transplant failures need to be taken into account.

## 3   Description of both Approaches

Our simulation compares two approaches to the KEP. The first one, based on Integer Programming, corresponds to the well-known cycle formulation [6], which was also used in [7] to solve the KEP in the UK. The second approach reproduces closely the heuristic method used by the ONT. It is a relatively simple three-step heuristic greedy algorithm which looks for good-enough solutions in an incremental way. Its main features are:

- It gives preference to 2-cycles over 3-cycles, when possible.
- It gives preference to cycles containing high weight transplants.
- It tries to include robust cycles in the solution. Robust cycles are those that have chances to be partially repaired with an embedded cycle if a failure arises during the solution's realization.

The heuristic method input is the set of vertices and the list $C$ of all available cycles built by concatenating the list of all the 3-cycles at the end of the list of all the 2-cycles, both lists being ordered decreasingly by weight. It proceeds by selecting the cycle $f$ in the first position of $C$, adding it to the solution $S$ and removing from the list $C$ all the cycles which are non-vertex-disjoint with $f$, repeating these operations until no cycles remain in $C$. This step of the algorithm gives preference to 2-cycles over 3-cycles (we incorporate them to the solution earlier) and to high-weighted cycles (they appear before in the list). The second step of the algorithm tries to convert 2-cycles to 3-cycles by selecting 2-cycles in the solution and looking for a non-assigned vertex able to be combined with the 2-cycle to form a 3-cycle. The third and last step looks for 3-cycles that can be decomposed into two 2-cycles by using an unassigned vertex.

When implementing the scheduled cycles, it is frequent that the *a priori* realizable transplants turn out to be unfeasible due to unforeseen causes. The ONT copes with the problem by preserving the cycles that do no present any failure and trying to find new cycles for the patients whose cycle is broken, which is done by running again the algorithm considering only the cycles where those patients appear. This reparation policy is also used, for the sake of fairness, in the IP-based approach. Notice that the ONT reparation procedure may take advantage of having chosen repairable 3-cycles (3-cycles containing embedded 2-cycles) in the second step of the heuristic algorithm (see Sect. 3).

## 4   Simulation Description and Results

We have aimed at comparing the ONT heuristic procedure and the IP-based approach. In this sense, We have simulated the evolution of a pool of patients where the best possible cycles are selected and implemented every three months, considering the reparation step described in the previous section. For the sake of realism (i) we allow a number of new patients to enter the pool between each cycle selection, (ii) we consider patient withdrawal (due to several reasons), and (iii) we take into account two failure probabilities:

1. Probability $p_c$ of positive crossmatch, which depends (see [8]) on the patient's PRA and a probability variation $\delta_p$ which is specific for each of the scenarios described later.
2. Probability $p_f$ of failure of patient/related donor.

We have considered the following 3 scenarios in our simulation: (1) $\delta_p = p_f = 0.0$; (2): $\delta_p = 0.1$ and $p_f = 0.05$; (3): $\delta_p = 0.2$ and $p_f = 0.1$. Such scenarios have been chosen according to the guidelines in the work of Bray et al. [8].

In the simulation procedure we have fixed a number of iterations (100 for the results presented below). For each iteration, a number of real-data-based pairs (50) are added to the pool, and next all the 2 and 3-cycles a priori feasible are generated. At this point, a set of cycles for transplantation is computed using each of the two approaches in Sect. 3. Then, transplants in the cycles are simulated and failed transplants are tried to be repaired. Finally, successfully transplanted patients are removed from the pool, and, before the new iteration, some of the remaining patients are also removed in order to simulate patient withdrawn (due to patient death or transplant obtained by another way). The experiment has been performed on 8 GB Intel Xeon E3-1220v2 machines at 3.10 GHz. 10 independent runs have been performed for each approach and scenario, and their results have been averaged to reduce the randomness effect.

First of all we consider the number of cycles and successfully transplanted patients, as well as the size of such cycles. As we can see in Table 1, the number of performed cycles is slightly superior for the ONT method. This is mainly due to the fact that this approach schedules many more 2-cycles (and less 3-cycles) than the IP-based approach (see Table 1). This is a positive aspect for the ONT method in the sense that 2-cycles are less prone to fail.

**Table 1.** Theoretic number of cycles successfully performed and the resulting transplanted patients

| Scenario | Total cycles | | 2-cycles | | 3-cycles | | Transplants | |
|---|---|---|---|---|---|---|---|---|
| | IP | ONT | IP | ONT | IP | ONT | IP | ONT |
| 1 | 360.3 | 370.1 | 84.7 | 158 | 275.6 | 212.1 | 996.2 | 952.3 |
| 2 | 343.9 | 356.7 | 83.1 | 159.4 | 260.8 | 197.3 | 948.6 | 910.7 |
| 3 | 324.9 | 336.5 | 76.3 | 152.1 | 248.6 | 184.4 | 898.4 | 857.4 |

In spite of that, we can also see that the number of transplanted patients in the IP-based approach is slightly superior in comparison with the ONT method. This is caused by the bigger amount of successful 3-cycles. This positive aspect for the IP-based approach is rather foreseeable, since it is just intended to maximize the number of proposed transplants.

We have also considered the stress put on the patients due to last-minute transplant failures. In Table 2 we show the evolution over the iterations of the number of times a patient has been selected for a transplant which at the end has not been performed. We can see in this table that this patient-demoralizing circumstance is noticeably more frequent in the IP-based approach than in the ONT approach, especially in the most pessimistic scenario.

**Table 2.** Evolution over the iterations of the number of times patients have been promised a finally undone transplant

| Scenario | 25 iterations | | 50 iterations | | 75 iterations | | 100 iterations | |
|---|---|---|---|---|---|---|---|---|
| | IP | ONT | IP | ONT | IP | ONT | IP | ONT |
| 1 | 116.1 | 96.8 | 239 | 213.5 | 359.8 | 321.2 | 489.9 | 430.4 |
| 2 | 274.2 | 233.1 | 596.1 | 490.2 | 875.8 | 735.5 | 1170.2 | 1006.7 |
| 3 | 534.8 | 409.6 | 1194.4 | 959.6 | 1800.1 | 1482 | 2411.5 | 1981.5 |

With respect to the average waiting time in the pool, the experiments have shown very similar results for both approaches, with an slight edge in favor of the heuristic algorithm. It is worth to mention that both methods offer an acceptable computing time. In particular, a run involving 750 patients and about 100000 precomputed cycles takes less than one second for both algorithms. It will be interesting in the future to compare their scalability with respect to larger sets of patient/donor pairs and/or higher sized cycles.

## 5    Conclusions and Future Work

We provide an empirical comparison between the ONT algorithm for KEP solving and a classical IP approach. Results show that both of them have strong and weak points. On the one hand, the ONT algorithm performs better in avoiding patient stress due to last-minute failures of scheduled transplants, and in favouring 2-cycles in front of 3-cycles. On the other hand, the IP approach is slightly better regarding the actual number of transplants, while the average waiting time in the pool shows no significant differences between both approaches. The simulation computing time is similar for both methods; nevertheless, we left as future work the scalability comparison of the methods, an issue that could be interesting to study in face of a possible international kidney exchange network.

There are other possible lines of future research. On the one hand, the study of new algorithms that embrace the strong points of both approaches; in this sense, the method proposed in [4] for the UK KPD program could be a good starting point. On the other hand, we aim at extending this algorithm to consider together both cycles and altruistic-donor-based chains.

**Acknowledgements.** Work partially supported by grants TIN2015-66293-R (MINECO/FEDER, UE), TIN2016-75866-C3-3-R, TIN2013-48040-R and MPCU dG2016/055 (UdG).

## References

1. Segev, D.L., Gentry, S.E., Warren, D.S., Reeb, B., Montgomery, R.A.: Kidney paired donation and optimizing the use of live donor organs. JAMA **293**(15), 1883–1890 (2005)
2. Abraham, D.J., Blum, A., Sandholm, T.: Clearing algorithms for barter exchange markets: enabling nationwide kidney exchanges. In: Proceedings of the 8th ACM Conference on Electronic Commerce, pp. 295–304. ACM, New York (2007)
3. Constantino, M., Klimentova, X., Viana, A., Rais, A.: New insights on integer-programming models for the kidney exchange problem. Eur. J. Oper. Res. **231**(1), 57–68 (2013)
4. Manlove, D.F., O'malley, G.: Paired and altruistic kidney donation in the UK: algorithms and experimentation. J. Exp. Algorithmics **19**, 2.6:1–2.6:21 (2015)
5. Klimentova, X., Pedrosa, J., Viana, A.: Maximising expectation of the number of transplants in kidney exchange programmes. Comput. Oper. Res. **73**, 1–11 (2016)
6. Roth, A.E., Sönmez, T., Ünver, M.U.: Efficient kidney exchange: coincidence of wants in markets with compatibility-based preferences. Am. Econ. Rev. **97**(3), 828–851 (2007)

7. Manlove, D.F., O'Malley, G.: Paired and altruistic kidney donation in the UK: algorithms and experimentation. In: Klasing, R. (ed.) SEA 2012. LNCS, vol. 7276, pp. 271–282. Springer, Heidelberg (2012). doi:10.1007/978-3-642-30850-5_24

8. Bray, M., Wang, W., Song, P.X.K., Leichtman, A.B., Rees, M.A., Ashby, V.B., Eikstadt, R., Goulding, A., Kalbfleisch, J.D.: Planning for uncertainty and fallbacks can increase the number of transplants in a kidney-paired donation program. Am. J. Transplant. **15**(10), 2636–2645 (2015)

# A Similarity Measure Based on Care Trajectories as Sequences of Sets

Yann Rivault[1,3,4]($\boxtimes$), Nolwenn Le Meur[1,4], and Olivier Dameron[2,3,4]

[1] EHESP Rennes, Sorbonne Paris Cité, EA 7449 REPERES,
Recherche en Pharmaco-Epidémiologie et Recours aux Soins, Rennes, France
{Yann.Rivault,Nolwenn.LeMeur}@ehesp.fr
[2] Université de Rennes 1, 35000 Rennes, France
Olivier.Dameron@univ-rennesl.fr
[3] IRISA équipe Dyliss, 35042 Rennes, France
[4] PEPS, Pharmacoepidemiology for Health Products Safety, Rennes, France

**Abstract.** Comparing care trajectories helps improve health services. Medico-administrative databases are useful for automatically reconstructing the patients' history of care. Care trajectories can be compared by determining their overlapping parts. This comparison relies on both semantically-rich representation formalism for care trajectories and an adequate similarity measure. The longest common subsequence (LCS) approach could have been appropriate if representing complex care trajectories as simple sequences was expressive enough. Furthermore, by failing to take into account similarities between different but semantically close medical events, the LCS overestimates differences. We propose a generalization of the LCS to a more expressive representation of care trajectories as sequences of sets. A set represents a medical episode composed by one or several medical events, such as diagnosis, drug prescription or medical procedures. Moreover, we propose to take events' semantic similarity into account for comparing medical episodes. To assess our approach, we applied the method on a care trajectories' sample from patients who underwent a surgical act among three kinds of acts. The formalism reduced calculation time, and introducing semantic similarity made the three groups more homogeneous.

**Keywords:** Care trajectories · LCS-based similarity · Semantic similarity

## 1 Introduction

Medico-administrative databases are valuable data source for health research notably because of their large population coverage, as well as of their longitudinal properties [1]. In France, the French national health insurance inter-regime information system (SNIIRAM) records the reimbursements of health care covered by the main insurance funds for workers. These data include ambulatory care data and hospital discharge summaries issued from the French hospital discharge information systems (PMSI). Although their primary goals are essentially financial and managerial, these databases make possible to explore and analyse patients' care trajectories [2]. Understanding and analysing these data are crucial for efficient healthcare planning and fair allocation of health care resources [3]. Moreover, care trajectory analysis can be an asset for

© Springer International Publishing AG 2017
A. ten Teije et al. (Eds.): AIME 2017, LNAI 10259, pp. 278–282, 2017.
DOI: 10.1007/978-3-319-59758-4_32

epidemiology studies by providing statistical indicators to understand and explain care seeking behaviours [4]. Care trajectories' comparison, which is part of their analysis, relies (i) on their representation and (ii) on an adequate comparison method. If we consider medico-administrative data for composing care trajectories, such as diagnoses, medical procedures or drugs prescriptions, an intuitive way of representing it is to write it down as a sequence. However, such formalism could be too simplistic considering the complexity of a care trajectory. The main complexity could be that the temporal order between events is not always known or even meaningful, especially when they occur simultaneously or in a really short time range. The second challenge is to handle the complexity of the large alphabet of trajectories' components. They are medical concepts that belong to detailed taxonomies, and taking into account semantic similarities [5] between these codes could render the method more robust to small variations [6]. The goal of this article is to introduce a representation of care trajectories that better accounts for administrative data complexity and an associated similarity measure for comparing them.

## 2  Materials and Methods

### 2.1  Representing Trajectories as Sequences of Sets

To take into account the uncertainty or simultaneity ignored with the simple sequence formalism, we proposed to group the events as unordered sets of events. Trajectories can then be seen as sequences of sets composed of simultaneous or related events. Similarity measures between sequences [7] could then be generalized to this formalism. To determine the overlapping part between two trajectories, we generalized the principle of the longest common subsequence (LCS) to this formalism [8].

### 2.2  Comparing Sequences of Sets

**Longest Common Subsequence for Sequences of Sets**

*Definition 1: Sequence of sets.*

A sequence of sets is a non-empty sequence composed of sets of elements.

*Definition 2: Size of a sequence of sets.*

Let $X = (x_1, x_2, \ldots, x_m)$ be a sequence of sets. The size of a sequence of set is:

$$|X| = \sum_{i=1}^{m} |x_i| \tag{1}$$

*Definition 3: Subsequence of a sequence of sets.*

Given two sequences of sets $X = (x_1, x_2, \ldots, x_m)$ and $Y = (y_1, y_2, \ldots, y_n)$, with $m \leq n$, $X$ is a subsequence of $Y$ if it exists the indexes $1 \leq j_1 < j_2 < \ldots < j_m \leq n$ such as $x_i \subseteq y_{j_i}$ is true for all $i = 1, 2, \ldots, m$.

*Definition 4: Longest common subsequence of two sequences of sets.*

Given two sequences of sets $X = (x_1, x_2, \ldots, x_m)$ and $Y = (y_1, y_2, \ldots, y_n)$, Z is a LCS of X and Y if $|Z| \geq |Z'|$, for all other common subsequence $Z'$ of X and Y.

**Trajectories Structural Similarity.** To get a similarity measure between sequences, it is intuitive and popular to normalize the size of the LCS by maximal size of the compared sequences.

*Definition 5: Similarity between trajectories.*

We define a similarity measure between X and Y sequences of sets as:

$$sim(X, Y) = \frac{|LCS(X, Y)|}{max(|X|, |Y|)} \tag{2}$$

**Considering Semantic Similarity Between Events.** In order to take into account similarities between elements of two trajectories, we proposed a modification of the $|LCS(X, Y)|$ calculation. Instead of using intersection between sets to determine the common part of two trajectories, we used a similarity measure between sets, which requires a similarity between elements of these sets. These elements are issued from taxonomies, and several semantic similarities based on their hierarchical structures are conceivable [5]. This similarity allows us to compute similarity measure between two sets of concepts.

*Definition 6: Similarity measure between two sets of concepts.*

Given two sets of concepts $X = (x_1, x_2, \ldots, x_m)$ and $Y = (y_1, y_2, \ldots, y_n)$, we note C the set of all matchings between elements of X and Y, and $sem(,)$ a semantic similarity between concepts. The similarity measure between X and Y is defined as:

$$sim(X, Y) = max_{c \in C} \sum_{(x,y) \in c} sem(x, y) \tag{3}$$

Due to this modification, the similarity measure is not anymore based on the longest common subsequence but on what we could call the longest similar subsequence. As for solving many algorithmic text problems, such as sequence alignment, both similarities can be computed using a dynamic programming algorithm [9].

## 2.3    Experimentations

We performed a retrospective analysis using the permanent sample of the SNIIRAM database (EGB). The EGB is a representative cross-sectional sample of the population covered by National Health Insurance. First, we extracted the hospital stay and reimbursed drug prescription information of the patients who underwent an ambulatory care surgery in 2012. All available medical codes, i.e. principal diagnoses, related diagnoses, associated diagnoses, clinical acts and drugs delivered in pharmacies were extracted from three months before to three months after the hospital stay to reconstruct the care trajectories. Next, to experiment the use of our method in performing cluster analysis, we selected three sub-groups of ambulatory surgeries, namely angioplasties, eye surgeries and breast surgeries, which constitute a sample of 287 patients. Elements of the care trajectories were drugs, diagnoses and medical acts, represented respectively by the Anatomical Therapeutic Chemical Classification System (ATC), the International Statistical Classification of Diseases and Related Health Problems – 10th revision (ICD-10), and the Common Classification of Medical Procedures (CCAM). To compute semantic similarities between elements from a same classification, we used the Wu and Palmer's similarity [10] which is based on their hierarchical structure. We then computed Eq. (2) between each pair of the 287 trajectories, with and then without taking into account the semantic similarities. We performed a cluster analysis using the R software (version 3.1.1), with an ascending hierarchical classification and a Ward linkage. Three classes were identified based on the highest drop of inertia between classes. Then we focused on intra-class and inter-class similarity. Because we worked on three sub-groups of patients, we had an *a priori* knowledge of the class a patient belongs to. We computed the ratio between the sum of patient's similarities with its own group and the sum of patient's similarities with the other groups, for both kind of similarities and for each patient.

## 3   Results

Before introducing semantic similarities in the method, the running time was twice faster when considering sequences of sets than sequences of atomic elements. It was no longer the case with their introduction, because it is more complicated to compute semantic similarities between sets of concepts than between atomic concepts.

Before evaluating the relevance of using semantic similarity, we ensured that a cluster analysis based on these similarities led to three distinct clusters associated to the three initial sub-groups. Only three patients were not classified in the correct cluster. Further analysis revealed that all three shared frequent comorbidities with patients from the other groups.

Because including semantic similarity to the care trajectory similarity measure could only increase the final similarity values, we focused on the ratios between intra-class and inter-class similarity. A statistical comparison has shown that they were significantly higher with the semantic similarities' introduction (Wilcoxon signed-rank test, $p = 0.002$). Overall, with this enrichment, the similarity of a patient with the patients of its own group has thus more increased than its similarity with the patients of the other groups, which is desirable in a cluster analysis.

## 4   Discussion and Conclusion

Thanks to its expressivity, the formalism of sequence of sets is appropriate to represent care trajectories based on medico-administrative data, e.g. clinical acts, diagnosis and drug codes. We have proposed a method to compare two care trajectories written as sequences of sets, which relies on a generalization of the LCS problem, in order to identify the homologous parts of two patients' care trajectories. And to make this method less strict, more robust to small variations, we tried to take into account the possible similarity between the care trajectories' components.

Our next objective will be to study the potential of the method in a clinical context, specifically the hospital stays for an angioplasty, to see if the method could be useful for predicting post hospital stay outcomes (e.g. rehospitalisation and adverse events), explaining disease conditions severity or a mode of taking in charge the patients (e.g. ambulatory or inpatient care), and discovering trends in the care consumptions.

We envision enriching the method by taking into account other kinds of similarity between events, such as delay between events or events' durations. It is also our objective to apply this method to patient and guideline comparison, to know the homologous part between a care trajectory and a guideline.

**Funding**
Doctoral fellowship funded by PEPS Research consortium, supported by Agence Nationale de Sécurité des Médicaments et produits de santé (ANSM).

# References

1. Tuppin, P., de Roquefeuil, L., Weill, A., Ricordeau, P., Merlière, Y.: French national health insurance information system and the permanent beneficiaries sample. Rev. Epidemiol. Sante Publique **58**, 286–290 (2010)
2. Moulis, G., Lapeyre-Mestre, M., Palmaro, A., Pugnet, G., Montastruc, J.-L., Sailler, L.: French health insurance databases: what interest for medical research? Rev. Med. Interne **36**, 411–417 (2015)
3. Jay, N., Nuemi, G., Gadreau, M., Quantin, C.: A data mining approach for grouping and analyzing trajectories of care using claim data: the example of breast cancer. BMC Med. Inform. Decis. Mak. **13**, 130 (2013)
4. Le Meur, N., Gao, F., Bayat, S.: Mining care trajectories using health administrative information systems: the use of state sequence analysis to assess disparities in prenatal care consumption. BMC Health Serv. Res. **15**, 200 (2015)
5. Pesquita, C., Faria, D., Falcão, A.O., Lord, P., Couto, F.M.: Semantic similarity in biomedical ontologies. PLoS Comput. Biol. **5**, e1000443 (2009)
6. Girardi, D., Wartner, S., Halmerbauer, G., Ehrenmüller, M., Kosorus, H., Dreiseitl, S.: Using concept hierarchies to improve calculation of patient similarity. J. Biomed. Inform. **63**, 66–73 (2016)
7. Studer, M., Ritschard, G.: A comparative review of sequence dissimilarity measures. LIVES Work. Pap. **2014**, 1–47 (2014)
8. Hirschberg, D.S.: A linear space algorithm for computing maximal common subsequences. Commun. ACM **18**, 341–343 (1975)
9. Bellman, R.: The theory of dynamic programming. Bull. Amer. Math. Soc. **60**, 503–515 (1954)
10. Wu, Z., Palmer, M.: Verbs semantics and lexical selection. In: Presented at the Proceedings of the 32nd Annual Meeting on Association for Computational Linguistics, 27 June 1994

# Machine Learning

# Influence of Data Distribution in Missing Data Imputation

Miriam Seoane Santos[1], Jastin Pompeu Soares[1], Pedro Henriques Abreu[1(✉)],
Hélder Araújo[2], and João Santos[3]

[1] Department of Informatics Engineering, Faculty of Sciences and Technology,
CISUC, University of Coimbra, Coimbra, Portugal
{miriams,jastinps}@student.dei.uc.pt, pha@dei.uc.pt
[2] Department of Electrical and Computer Engineering,
Faculty of Sciences and Technology, ISR, University of Coimbra,
Coimbra, Portugal
helder@isr.uc.pt
[3] IPO-Porto Research Centre (CI-IPOP), Porto, Portugal
joao.santos@ipoporto.min-saude.pt

**Abstract.** Dealing with missing data is a crucial step in the preprocessing stage of most data mining projects. Especially in healthcare contexts, addressing this issue is fundamental, since it may result in keeping or loosing critical patient information that can help physicians in their daily clinical practice. Over the years, many researchers have addressed this problem, basing their approach on the implementation of a set of imputation techniques and evaluating their performance in classification tasks. These classic approaches, however, do not consider some intrinsic data information that could be related to the performance of those algorithms, such as features' distribution. Establishing a correspondence between data distribution and the most proper imputation method avoids the need of repeatedly testing a large set of methods, since it provides a heuristic on the best choice for each feature in the study. The goal of this work is to understand the relationship between data distribution and the performance of well-known imputation techniques, such as Mean, Decision Trees, k-Nearest Neighbours, Self-Organizing Maps and Support Vector Machines imputation. Several publicly available datasets, all complete, were selected attending to several characteristics such as number of distributions, features and instances. Missing values were artificially generated at different percentages and the imputation methods were evaluated in terms of Predictive and Distributional Accuracy. Our findings show that there is a relationship between features' distribution and algorithms' performance, although some factors must be taken into account, such as the number of features per distribution and the missing rate at state.

**Keywords:** Missing data · Machine learning imputation · Data distribution · Healthcare contexts

© Springer International Publishing AG 2017
A. ten Teije et al. (Eds.): AIME 2017, LNAI 10259, pp. 285–294, 2017.
DOI: 10.1007/978-3-319-59758-4_33

# 1   Introduction

In healthcare classification scenarios, the main goal is to provide strong classification results, whereas imputation is considered a necessary pre-processing step to achieve such goal [4]. Therefore, imputation is often evaluated using the classification error (CE): the method that minimizes the CE is considered the best. The use of CE is however controversial, in the sense that the imputation method that minimizes the classification error might produce biased estimates and affect the original data distribution, especially if the same method is used for all different types of features' distributions [11]. Furthermore, using the same method for all features raises two main issues: first, all techniques must be implemented for all features, which increases the number of necessary simulations and consequently computational cost; secondly, imputation is performed based on the assumption that the same technique should perform well for the great majority of features, which could be an over assumption, since different features may benefit the most from different imputation techniques, particularly if different missing rates are taken into account. Studying the influence of data distribution in imputation provides a heuristic on the most appropriate imputation strategy for each feature in the study, avoiding the need of testing a large set of methods.

In this work, we aim to assess which imputation techniques can efficiently reproduce the true, original values in data, without causing a distortion in their distribution, which can be evaluated by Predictive Accuracy (PAC) and Distributional Accuracy (DAC) metrics, respectively. Furthermore, we intend to investigate whether there is a relationship between the imputation methods and a particular distribution. Our study focuses on the best techniques for data imputation across several different distributions, in terms PAC and DAC, rather than CE. To achieve this goal, we have selected several complete healthcare datasets comprising features with different data distributions, and artificially generated missing data in all of them at several rates (5, 10, 15, 20 and 25%). Then the missing values are imputed with the methods most commonly used in related works: Mean imputation, Decision Trees (DT), k-Nearest Neighbours (KNN), Self-Organizing Maps (SOM) and Support Vector Machines (SVM) imputation. Our experiments show that the imputation methods are in fact influenced by data distribution, with the exception of SVM, that does not seem to be affected. Aside for SVM, that achieves the best PAC and DAC results for all distributions, SOM is overall winner in both metrics. However, the choice of the best imputation method depends also on the number of features per distribution and the missing rate at state.

The remainder of the manuscript is organized as follows: Sect. 2 presents some works that studied imputation for classification purposes. Sections 3 and 4 describe the setup used in this work and report on the experimental results, while Sect. 5 presents the conclusions and suggests some directions for future work.

## 2   Related Work

Addressing missing data to increase data quality for classification purposes is a standard procedure in a plethora of contexts, including healthcare. Nanni et al. [8] compared several imputation approaches (including Mean and KNN imputation) by randomly generating missing data at several rates, and used classification-related metrics such as accuracy (1-CE) and Area Under the ROC Curve (AUC) to evaluate the quality of imputation. Kang [7] also generated missing values in complete datasets, at several missing ratios. They compare their approach with other well-known imputation methods (also including Mean imputation and KNN), and the results were evaluated using accuracy. Aisha et al. [1] study the effects of several imputation techniques (including Mean, KNN and SVM imputation) on Bayesian Network classification of datasets with missing data, and evaluate the results also using accuracy. García-Laencina et al. [4] studied the influence of imputation (including KNN and SOM imputation) on the classification accuracy, using synthetic and real datasets. In this work, the authors start by measuring the quality of imputation using PAC (Pearson's coefficient and mean squared error) and DAC (Kolmogorov-Smirnov distance) metrics. However, this analysis in only performed for KNN imputation, and immediately discarded in favor of CE metrics, since the main objective is to solve a classification problem. Rahman and Islam [10] present two imputation techniques based on DT and compare them in terms of their predictive accuracy (PAC), using the Pearson's correlation coefficient, root mean squared error (RMSE) and mean absolute error (MAE) as performance indicators. DAC metrics are, however, completely disregarded. In what concerns healthcare contexts in particular, García-Laencina et al. [3] also compared the performance of standard imputation algorithms (including Mean and KNN imputation) on the survival prediction of breast cancer patients. The results were evaluated in terms of sensitivity, specificity, accuracy and AUC. Rahman and Davis [9] studied the influence of Mean, DT, KNN and SVM imputation on the survival prediction of cardiovascular patients, evaluating the quality of imputation also classification-related metrics (sensitivity, specificity and accuracy). Jerez et al. [6] use imputation (including Mean, KNN and SOM) to predict breast cancer recurrence in a real incomplete dataset, evaluating the results in terms of AUC. In all the previously mentioned works, imputation techniques are frequently evaluated in terms of CE, and the effects they may have in data distribution are ignored. Furthermore, all features are imputed with the same technique, without considering the possibility that some techniques may perform differently for different features. We herein conduct a study on the influence of data distribution in missing data imputation, aiming to assess how different imputation techniques perform across different feature distributions, which to the extent of our knowledge, as never been performed.

## 3    Methodology

This works comprised four main stages: Data Collection, Missing Data Generation, Data Imputation and Evaluation Metrics.

### 3.1    Data Collection

The first stage of this work consisted in choosing several publicly available datasets, all without missing values: Bupa Liver Disorders Dataset (*bupa*), Breast Tissue Dataset (*breast*), Cardiotocography Dataset (*ctg*), Haberman's Survival Dataset (*hsd*), Wisconsin Diagnostic Breast Cancer Dataset (*wdbc*), Parkinsons Dataset (*parkinson*) and Lower Back Pain Symptoms Dataset (*backpain*). All datasets were collected from UCI Machine Learning Repository (http://archive.ics.uci.edu/ml), except for the latter, retrieved from Kaggle Datasets (https://www.kaggle.com/datasets). We have chosen only complete datasets composed exclusively of continuous features so that both the influence of different data distributions and missing rates could more efficiently studied. Table 1 summarizes the datasets' characteristics in what concerns their context, sample size, number of features and number of different distributions comprised in the data. In terms of data distributions, these datasets are somewhat heterogeneous, with the most common distributions being generalized extreme value (all 7 datasets), generalized pareto (6 datasets) and gamma distributions (4 datasets). We have also included the ratio of variables per distribution for each dataset (Ratio). Ratio is estimated as $\frac{\text{No. of features}}{\text{No. of distributions}^2}$, so that a greater weight is given to the number of distributions comprised in the dataset.

### 3.2    Missing Data Generation

Before generating missing values, each dataset's features were fitted against several standard continuous distributions and the distribution of each feature is saved for posterior analysis when assessing the imputation results (Table 1). Missing data was randomly inserted at several rates (5, 10, 15, 20 and 25%) for each feature in the dataset. Therefore, for each of the datasets, 5 different versions exist, one for each considered missing percentage.

### 3.3    Data Imputation

In this section, each imputation technique is briefly explained, with particular emphasis on the implementation details. **Mean imputation** is the most common of imputation techniques [5]. For continuous data, the missing values are replaced with the mean of the observed cases on each respective feature. In **k-Nearest Neighbours** (KNN), the incomplete patterns are imputed according to the values of their $k$ closest neighbours on the missing features: mode for discrete data and the mean or a weighted average for continuous data [6], which is used in this work. Our implementation considers a range of $k$ from 1 to 20

**Table 1.** Summary of datasets' characteristics.

| Dataset | Context | Sample size | No. of features | Ratio | No. of distributions (no. of features) |
|---|---|---|---|---|---|
| bupa | Detect alcoholism problems | 345 | 6 | 0.240 | Generalized extreme value (1) Logistic (1), exponential (1) Loglogistic (2), lognormal (1) |
| breast | Identify breast carcinomas | 106 | 9 | 0.360 | Birnbaum-saunders (2) Generalized extreme value (1) Generalized pareto (2) Rayleigh (1), inverse gaussian (3) |
| ctg | Detect pathologic fetal cardiotocograms | 2126 | 21 | 0.583 | Generalized extreme value (5) Generalized pareto (10), gamma (1) Logistic (1), weibull (3), nakagami (1) |
| hsd | Predict 5-year survivability after breast cancer surgery | 306 | 3 | 0.333 | Generalized extreme value (1) Generalized pareto (1) Nakagami (1) |
| wdbc | Diagnose breast cancer cases | 569 | 30 | 0.469 | Generalized extreme value (16) Generalized pareto (2), gamma (1) Birnbaum-saunders (2), exponential (1) Inverse gaussian (1), loglogistic (1) Lognormal (4) |
| parkinson | Diagnose cases of parkinson disease | 195 | 22 | 0.344 | Generalized extreme value (9) Generalized pareto (5), gamma (1) Beta (1), inverse gaussian (1) Normal (1), weibull (1), lognormal (3) |
| backpain | Detect abnormal back pain | 310 | 12 | 0.245 | Generalized extreme value (2) Generalized pareto (4), gamma (2) Beta (1), birnbaum-saunders (1) Logistic (1), rayleigh (1) |

closest neighbours and the Heterogeneous Euclidean-Overlap Metric (HEOM) as distance measure between patterns [12]. In **DT imputation**, each incomplete feature must be used as target: the remaining features are used as training data, to fit the model, and missing values are determined as if they were class labels. For this work, only regression trees are constructed, given the nature of all our features. In **Self-Organizing Maps** (SOM), each incomplete pattern is imputed according to its Best Matching Unit (BMU), its most similar unit in the SOM map. Several map configurations were tested: from 10 to 100 nodes. **Support Vector Machines** (SVM) are currently the state-of-the-art algorithms in pattern recognition, due to their good trade-off between the model's complexity, generalization and quality of fitting the training data, and have proven to perform well for missing data imputation [4]. In this work, only regression SVMs were used for imputation: in particular, we have implemented several Radial Basis Function (RBF) SVMs, with different values of $C$ and $\gamma$ (both from $1e^{-5}$ to $1e^5$, increasing by a factor of 10).

## 3.4    Evaluation Metrics for Missing Data Imputation

The metrics used in this work concern mainly two aspects: Predictive Accuracy (PAC) and Distributional Accuracy (DAC) [2]. PAC relates to the efficiency of an imputation technique to retrieve the true values in data, while DAC represents the technique's ability to preserve the distribution of those true values. For PAC assessment, two measures were used: Pearson Correlation Coefficient (Pearson's $r$) and Mean-Squared Error (MSE). For DAC assessment, the Kolmogorov-Smirnov distance ($D_{KS}$) was implemented. Considering a complete feature $x$, and its imputed version $\hat{x}$, Pearson's $r$ provides a measure of the correlation between the two, and is given by $r = \frac{\sum_{i=1}^{n}(x_i - \bar{x}_i)(\hat{x}_i - \bar{\hat{x}}_i)}{\sqrt{\sum_{i=1}^{n}(x_i - \bar{x}_i)^2 \sum_{i=1}^{n}(\hat{x}_i - \bar{\hat{x}}_i)^2}}$, where an efficient imputation technique should have a value close to 1. MSE is traduced by $\frac{1}{n}\sum_{i=1}^{n}(x_i - \hat{x}_i)^2$ and measures the difference between the imputed and original values of a given feature $j$, the average square deviation of $\hat{x}_i$ from the true values $x_i$, for all $n$ values of a feature $j$. In this case, values closer to 0 traduce a better imputation. Finally, $D_{KS}$ is given by $\max(\|F_x - F_{\hat{x}}\|)$, where $F_x$ and $F_{\hat{x}}$ are the empirical cumulative distribution functions of $x$ and $\hat{x}$, respectively. Smaller distance values represent better imputations.

## 4    Experimental Results and Discussion

Considering all five imputation methods (Mean, DT, KNN, SOM and SVM), the results clearly show that SVM is the winning method for all distributions (see Total and Total SVM in Table 2). For all metrics, SVM outperforms the remaining methods, with a maximum total mean MSE, Pearson's $r$ and $D_{KS}$ of 0.014, 0.993 and 0.01, respectively, versus the 0.039, 0.98 and 0.13 achieved by the remaining methods. Moreover, SVM does not seem to be affected by data distribution, with good performance indicators across all distributions. However, a preliminary analysis of our simulation results suggested that this was not the case for the remaining methods, which lead us to investigate them more closely, and further divide our analysis in particular ranges of missing data. Therefore, Table 2 also presents the winning methods with respective means and standard-deviations in several missing rate scenarios (5/10, 15/20 and 25%), and summarizes the number of victories and draws of each method. Note that for 25% missing rate, some methods do not show a mean/standard deviation, which happens in distributions included in only two datasets and where the methods tie (each wins in one dataset, and the presented value refers to the result achieved for that dataset). In what concerns PAC results, although DT and KNN may outperform or match SOM's results for low percentages of missing data (5–10%), SOM is generally the best approach for percentages above 10%. In terms of DAC, KNN and SOM have similar results for missing percentages between 5% and 20%. Nevertheless, for percentages higher that 20%, SOM is the method that better preserves the original data distribution. Due to space constraints, it is not possible to show the results for each dataset and distribution, but we provide a more detailed discussion for certain distributions in

**Table 2.** Simulation results by distribution: means and standard-deviations are shown for the winning methods regarding each distribution, metric and missing percentage (n.a - not applicable).

**Distributions**

| Metric | Scenario | beta | birnbaumsaunders | generalized extreme value | gamma | rayleigh | inverse gaussian |
|---|---|---|---|---|---|---|---|
| MSE | 5%-10% | SOM[0.054 ± 0.049] | SOM[0.015±0.017] | SOM[0.06±0.007] | SOM[0.023±0.012] | SOM[0.022±0.018] | DT[0.008±0.006]/KNN[0.008±0.004] |
| | 15%-20% | SOM[0.143 ± 0.011] | SOM[0.049±0.029] | SOM[0.086±0.059] | DT[0.063±0.017/SOM[0.087±0.027] | DT[0.068±0.021] | SOM[0.037±0.023] |
| | 25% | KNN[0.021/SOM[0.091]] | SOM[0.091±0.063] | SOM[0.151±0.106] | SOM[0.068±0.027] | SOM[0.132±0.085] | SOM[0.062±0.047] |
| | Total | KNN[0.141±0.099]/SOM[0.099±0.059] | SOM[0.02±0.04] | SOM[0.085±0.077] | SOM[0.093±0.04] | SOM[0.061±0.008] | SOM[0.039±0.033] |
| | Total SVM | SVM[0.07±0.084] | SVM[0.02±0.044] | SVM[0.03±0.078] | SVM[0.03±0.078] | SVM[0.044±0.047] | SVM[0.014±0.028] |
| Pearson | 5%-10% | KNN[0.982±0.004]/SOM[0.97±0.036] | SOM[0.993±0.008] | DT[0.989±0.00]/SOM[0.996±0.038] | SOM[0.088±0.006] | SOM[0.988±0.0] | DT[0.995±0.003]/KNN[0.996±0.002] |
| | 15%-20% | KNN[0.974±0.028]/SOM[0.971±0.015] | SOM[0.976±0.015] | SOM[0.985±0.044] | DT[0.968±0.009]/SOM[0.966±0.014] | DT[0.666±0.011] | SOM[0.982±0.011] |
| | 25% | KNN[0.881/SOM[0.923] | SOM[0.954±0.028] | SOM[0.912±0.00] | SOM[0.945±0.014] | SOM[0.932±0.046] | SOM[0.968±0.025] |
| | Total | KNN[0.939±0.051]/SOM[0.943±0.03] | SOM[0.978±0.021] | SOM[0.956±0.042] | SOM[0.965±0.021] | SOM[0.985±0.017] | SOM[0.98±0.017] |
| | Total SVM | SVM[0.964±0.028] | SVM[0.985±0.017] | SVM[0.974±0.042] | SVM[0.984±0.013] | SVM[0.975±0.024] | SVM[0.993±0.013] |
| DKS | 5%-10% | KNN[0.011±0.003] | SOM[0.011±0.005] | KNN[0.015±0.009] | SOM[0.013±0.008] | SOM[0.014±0.007] | DT[0.014±0.007] |
| | 15%-20% | KNN[0.038±0.005] | SOM[0.027±0.012] | KNN[0.018±0.013] | SOM[0.021±0.01] | SOM[0.018±0.01] | DT[0.018±0.004]/SOM[0.023±0.008] |
| | 25% | SOM[0.053±0.006] | SOM[0.037±0.013] | KNN[0.028±0.004] | SOM[0.027±0.003] | DT[0.039]/SOM[0.047] | SOM[0.03±0.007] |
| | Total | KNN[0.027±0.014]/SOM[0.039±0.023] | SOM[0.023±0.01] | KNN[0.025±0.014] | SOM[0.019±0.007] | SOM[0.025±0.015] | DT[0.018±0.009] |
| | Total SVM | SVM[0.024±0.011] | | SVM[0.015±0.009] | SVM[0.018±0.007] | SVM[0.015±0.009] | SVM[0.018±0.009] |

| Metric | Scenario | generalised pareto | logistic | exponential | loglogistic | lognormal | nakagami |
|---|---|---|---|---|---|---|---|
| MSE | 5%-10% | DT[0.027±0.023]/SOM[0.032±0.036] | KNN[0.033±0.028]/SOM[0.031±0.0001] | SOM[0.035±0.01] | DT[0.064±0.001/SOM[0.119±0.015] | DT[0.015±0.012]/KNN[0.018±0.014]/SOM[0.082±0.04] | KNN[0.031±0.037] |
| | 15%-20% | SOM[0.04±0.0] | SOM[0.099±0.052] | SOM[0.13±0.1] | | | KNN[0.177±0.032]/SOM[0.014±0.013] |
| | 25% | SOM[0.071±0.059] | SOM[0.072±0.047] | KNN[0.218]/SOM[0.038] | SOM[0.15±0.058] | | KNN[0.191±SOM[0.082] |
| | Total | SOM[0.044±0.005] | SOM[0.06±0.093] | SOM[0.085±0.078] | SOM[0.101±0.086] | | KNN[0.121±0.086]/SOM[0.048±0.013] |
| | Total SVM | SVM[0.046±0.057] | | SVM[0.019±0.025] | SVM[0.06±0.044] | SVM[0.031±0.026] | SVM[0.072±0.081] |
| Pearson | 5%-10% | DT[0.083±0.016]/SOM[0.084±0.018] | DT[0.973±0.028]/KNN[0.984±0.014]/SOM[0.985±0.001] | | DT[0.982±0.006]/SOM[0.968±0.03] | SOM[0.974±0.018] | KNN[0.084±0.019] |
| | 15%-20% | SOM[0.969±0.031] | SOM[0.964±0.038] | | KNN[0.025±0.01] | SOM[0.953±0.032] | KNN[0.908±0.018]/SOM[0.98±0.006] |
| | 25% | SOM[0.958±0.034] | SOM[0.953±0.024] | | SOM[0.925±0.01] | SOM[0.92±0.04] | MEAN[0.890/SOM[0.968] |
| | Total | SOM[0.977±0.029] | SOM[0.969±0.033] | | SOM[0.969±0.023] | SOM[0.946±0.046]/SOM[0.376±0.006] | DT[0.004/KNN[0.007]/SOM[0.015±0.009] |
| | Total SVM | SVM[0.977±0.029] | | | SVM[0.06±0.012] | SVM[0.031±0.016] | KNN[0.016±0.01]/SOM[0.026±0.01] |
| DKS | 5%-10% | KNN[0.026±0.011]/SOM[0.024±0.007] | SOM[0.009±0.004] | | | SOM[0.025±0.011] | SVM[0.011±0.012] |
| | 15%-20% | KNN[0.038±0.013] | SOM[0.014±0.008] | | | SOM[0.034±0.007] | |
| | 25% | KNN[0.02±0.012]/SOM[0.022±0.011] | SOM[0.023±0.009] | | | SOM[0.024±0.011] | |
| | Total | | SOM[0.018±0.008] | | | SOM[0.031±0.009] | |
| | Total SVM | SVM[0.019±0.009] | SVM[0.012±0.009] | | | | |

**Distributions**

| Metric | Scenario | weibull | normal | No. of victories and draws per algorithm | | |
|---|---|---|---|---|---|---|
| | | | | SOM | KNN | DT |
| MSE | 5%-10% | DT[0.014±0.012] | KNN[0.0114] * SOM[0.036] | 6/4 | 1/4 | 2/4 |
| | 15%-20% | SOM[0.041±0.013] | DT[0.0883]/SOM[0.163] | 9/4 | 0/2 | 0/3 |
| | 25% | SOM[0.037±0.002] | SOM[0.124] | 11/3 | 0/3 | 0 |
| | Total | SOM[0.039±0.015] | | 12/2 | 0/2 | 0 |
| | Total SVM | SVM[0.024±0.023] | SVM[0.072±0.066] | n.a | n.a | n.a |
| Pearson | 5%-10% | SOM[0.993±0.0] | KNN[0.954]/SOM[0.982] | 5/5 | 1/4 | 2/4 |
| | 15%-20% | SOM[0.979±0.007] | DT[0.951/SOM[0.915] | 9/4 | 0/2 | 1/2 |
| | 25% | SOM[0.971±0.001] | SOM[0.944] | 12/2 | 0/1 | 0 |
| | Total | SOM[0.98±0.008] | SOM[0.963±0.034] | 12/2 | 0/2 | 0 |
| | Total SVM | SVM[0.98±0.012] | | n.a | n.a | n.a |
| DKS | 5%-10% | KNN[0.005±0.002]/SOM[0.013±0.004] | KNN[0.01]/SOM[0.015] | 3/5 | 4/5 | 2/2 |
| | 15%-20% | SOM[0.021±0.005] | DT[0.031]/KNN[0.031]/SOM[0.041] | 4/5 | 4/4 | 1/2 |
| | 25% | KNN[0.02/SOM[0.024] | KNN[0.029±0.004] | 7/5 | 2/3 | 0/1 |
| | Total | KNN[0.013±0.0]/SOM[0.018±0.006] | | 6/5 | 2/5 | 1/0 |
| | Total SVM | SVM[0.01±0.006] | SVM[0.019±0.007] | n.a | n.a | n.a |

what follows. For birnbaum-saunders datasets, SOM was always chosen as the best approach regarding all metrics. For datasets with a considerable number of features following the generalized extreme value distribution (*wdbc*: 16, *parkinson*: 9 and *ctg*: 5) and considering the range 5–10% of missing data, DT achieves the best results for in terms of PAC, although KNN achieves better results in terms of DAC. When the missing percentage increases, SOM is then considered the best approach in both metrics. Nevertheless, for datasets where only one variable of this type exists (*hsd* and *breast*), KNN outperforms or match SOM's results in both PAC and DAC metrics, for all missing rates. Dataset *bupa*, also with one variable of this type, seems to be an exception, with SOM achieving better results in all metrics, except when the missing rate increases (25%), where KNN is considered the best approach. Datasets *backpain*, *ctg* and *bupa* have one variable following the logistic distribution, where for small percentages of missing data, DT and KNN are feasible approaches. As the missing rate increases, only *bupa* includes KNN as best approach, while the remaining are better imputed with SOM. Dataset *bupa* seems to be a special case, where results are somewhat variable with increasing rates of missing values. This fact could be due to the ratio of features per distribution of this dataset (see Table 1). In fact, in a total of 6 features, *bupa* includes 5 different distributions, which causes it to have the lowest feature per distribution ratio (0.240). Intrigued by these results of *bupa*, we have further compared the overall MSE, Pearson's $r$ and $D_{KS}$ results for datasets with the lowest (*bupa* and *backpain*) and highest (*wdbc* and *ctg*) ratio of features per distribution, where a particular distribution is present in only one feature: exponential and logistic distributions (see Table 1). In the case of logistic distribution, PAC results of *backpain* and *bupa* differ from *ctg*: a mean MSE of 0.1/0.12 versus 0.04 and a mean Pearson's $r$ of 0.95/0.94 versus the 0.98, respectively. Regarding DAC, all datasets are similar (maximum difference of 0.01). For the exponential distribution, the results follow the same trend: a mean MSE of 0.025/0.147 and Pearson's $r$ of 0.99/0.92 for *wdbc*/*bupa*. DAC results are practically the same, with a difference of 0.005. This suggests that, when a particular distribution is present in only one feature, datasets with a low ratio of features per distribution (*backpain*: 0.245, *bupa*: 0.240), are more challenging than datasets with a higher ratio (*wdbc*: 0.469, *ctg*: 0.583), in what concerns retrieving the true values in data. However, imputation algorithms are able to considerably preserve the data distribution in both cases.

## 5   Conclusions and Future Work

Our results show that SVM is the winning method for all distributions in both PAC and DAC metrics. Aside for SVM, SOM is generally the best approach in terms of PAC when the missing rates increases above 10%, although for DAC its superiority its only noticeable for percentages higher that 20%. Regarding particular distributions, SOM was the best approach for birnbaum-saunders distributions in all considered missing percentages. In datasets with a great number of features following a generalized extreme value distribution, DT and SOM are the best approaches in terms of PAC, in 5–10% and 15–20% ranges of missing data, respectively. Furthermore, PAC metrics seem to be affected by the ratio of features per distribution, when a particular distribution is present in only one feature. Lower ratios generally achieve worst PAC results, although the data distribution is not significantly affected (DAC results are similar for both cases). There are several directions for future work the authors would like to address. To the extent of authors' knowledge, this approach has never been applied in imputation studies for healthcare contexts in particular or other subjects in general. Therefore, its application for other contexts and other data distributions is yet to be addressed. The extension of this methodology for discrete features, fitting discrete distributions and investigating how the studied imputation techniques perform in this case, could also be a possibility for future work. An ongoing work is the evaluation of the proposed approach in more extreme setups, where missing values are not generated completely at random, but rather affecting specific areas of features' probability density functions. Finally, from a classification perspective, it would also be interesting to study whether the best imputation techniques regarding PAC and DAC metrics also achieve reasonable results in terms of classification error.

**Acknowledgments.** This article is a result of the project NORTE-01-0145-FEDER-000027, supported by Norte Portugal Regional Operational Programme (NORTE 2020), under the PORTUGAL 2020 Partnership Agreement, through the European Regional Development Fund (ERDF).

# References

1. Aisha, N., Adam, M.B., Shohaimi, S.: Effect of missing value methods on Bayesian network classification of hepatitis data. Int. J. Comput. Sci. Telecommun. **4**(6), 8–12 (2013)
2. Chambers, R.: Evaluation Criteria for Statistical Editing and Imputation. National Statistics Methodological Series No. 28. University of Southampton, Southampton (2001)
3. García-Laencina, P.J., Abreu, P.H., Abreu, M.H., Afonso, N.: Missing data imputation on the 5-year survival prediction of breast cancer patients with unknown discrete values. Comput. Biol. Med. **59**(2015), 125–133 (2015)
4. García-Laencina, P.J., Sancho-Gómez, J.L., Figueiras-Vidal, A.R.: Pattern classification with missing data: a review. Neural Comput. Appl. **19**(2), 263–282 (2010)
5. García-Laencina, P.J., Sancho-Gómez, J.L., Figueiras-Vidal, A.R.: Classifying patterns with missing values using multi-task learning perceptrons. Expert Syst. Appl. **40**(4), 1333–1341 (2013)
6. Jerez, J.M., Molina, I., García-Laencina, P.J., Alba, E., Ribelles, N.: Missing data imputation using statistical and machine learning methods in a real breast cancer problem. Artif. Intell. Med. **50**(2), 105–115 (2010)
7. Kang, P.: Locally linear reconstruction based missing value imputation for supervised learning. Neurocomputing **118**, 65–78 (2013)
8. Nanni, L., Lumini, A., Brahnam, S.: A classifier ensemble approach for the missing feature problem. Artif. Intell. Med. **55**(1), 37–50 (2012)

9. Rahman, M.M., Davis, D.N.: Fuzzy unordered rules induction algorithm used as missing value imputation methods for K-mean clustering on real cardiovascular data. In: Proceedings of the World Congress on Engineering, vol. 1, pp. 391–395 (2012)

10. Rahman, M.G., Islam, M.Z.: Missing value imputation using decision trees and decision forests by splitting and merging records: two novel techniques. Knowl.-Based Syst. **53**, 51–65 (2013)

11. Van Buuren, S.: Flexible Imputation of Missing Data. CRC Press, Boca Raton (2012)

12. Wilson, D.R., Martinez, T.R.: Improved heterogeneous distance functions. J. Artif. Intell. Res. **6**, 1–34 (1997)

# Detecting Mental Fatigue from Eye-Tracking Data Gathered While Watching Video

Yasunori Yamada$^{(\boxtimes)}$ and Masatomo Kobayashi

IBM Research - Tokyo, Tokyo, Japan
{ysnr,mstm}@jp.ibm.com

**Abstract.** Monitoring mental fatigue is of increasing importance for improving cognitive performance and health outcomes. Previous models using eye-tracking data allow inference of fatigue in cognitive tasks, such as driving, but they require us to engage in a specific cognitive task. A model capable of estimating fatigue from eye-tracking data in natural-viewing situations when an individual is not performing cognitive tasks has many potential applications. Here, we collected eye-tracking data from 18 adults as they watched video clips (simulating the situation of watching TV programs) before and after performing cognitive tasks. Using this data, we built a fatigue-detection model including novel feature sets and an automated feature selection method. With eye-tracking data of individuals watching only 30-seconds worth of video, our model could determine whether that person was fatigued with 91.0% accuracy in 10-fold cross-validation (chance 50%). Through a comparison with a model incorporating the feature sets used in previous studies, we showed that our model improved the detection accuracy by up to 13.9% (from 77.1 to 91.0%).

**Keywords:** Mental fatigue · Cognitive fatigue · Feature selection · Natural viewing · Free viewing · Visual attention model

## 1 Introduction

Health monitoring in a smart environment such as adaptive workplaces and smart houses has been increasingly recognized for its importance in improving health outcomes [1]. Especially for supporting a rapidly aging population, research has been focused on developing technologies that help mitigate critical situations and enable individuals to manage their own health, with the dual aim of increasing quality of lives and reducing healthcare costs [2].

One area of an individual's daily health status that has yet to be utilized is mental fatigue, which refers to the feeling people might experience during or after cognitive activities [3]. Mental fatigue is a common problem in modern everyday life and comes at a huge public health cost [4]. It is a warning sign of harmful accumulation of stress that can have a detrimental effect on one's health [5] and an important symptom in general practice due to its association with a large number of chronic medical conditions such as cancer, Alzheimer's disease, and Parkinson's disease [6].

© Springer International Publishing AG 2017
A. ten Teije et al. (Eds.): AIME 2017, LNAI 10259, pp. 295–304, 2017.
DOI: 10.1007/978-3-319-59758-4_34

Previous studies for monitoring mental fatigue have primarily focused on detecting fatigue during cognitive tasks such as driving [6–8]. Unobtrusive methods, such as those that use remote- or webcam-based eye tracking, typically monitor changes in an individual's pupil response, blinking behavior, and eye movement to determine levels of fatigue during cognitive tasks [9]. Although these methods have shown the usefulness for monitoring fatigue during a specific cognitive task that requires visual processing, no study has yet developed a model that enables us to infer mental fatigue from eye-tracking data in natural-viewing situations when the individual is not performing cognitive tasks. Such a system would enable people to monitor mental fatigue in a condition close to everyday life. Moreover, it would be used to infer fatigue induced by not only specific cognitive visual tasks, but also various factors such as cognitive auditory tasks, multiple cognitive tasks, or poor health [6].

In contrast, recent studies have attempted to discriminate clinical populations from eye-tracking data in natural viewing conditions [10,11]. For example, Crabb and colleagues demonstrated that patients with neuro-degenerative eye diseases can be separated from healthy controls by using eye-tracking data collected while the patients freely watched TV movies [11]. However, there has been no investigation on the associations of eye-tracking data in natural viewing conditions with mental fatigue.

In this paper, we present a novel model that detects mental fatigue in natural viewing situations in which people watched video clips such as a TV program. More specifically, we collected eye-tracking data from 18 adults as they watched video clips before and after they performed auditory cognitive tasks. From this data, we extracted 181 quantitative features, categorized into six feature sets related to oculomotor-based metrics, blinking behavior, pupil measurements, gaze allocation, eye-movement directions, and saliency-based metrics using a saliency model (a computational model of visual attention). Although the last three feature sets have already been used for characterizing eye movements, especially in natural viewing situations [10,11], they have not been used for inferring mental fatigue. Using these features and an automated feature selection method, we built a two-class classifier for detecting mental fatigue. With eye-tracking data of individuals watching only 30 s worth of video, our model could determine whether that person was fatigued or not with 91.0% accuracy in 10-fold cross-validation (chance 50%). To make a comparison with a model based on the existing work, we also built a model using the three feature sets related to oculomotor-based metrics, blinking behavior, and pupil measurements used in a previous study, where the detection accuracy was 77.1%.

## 2    Data Collection

To build a model for inferring mental fatigue from eye-tracking data in a natural viewing situation, we collected data while participants watched video clips (simulating the situation of watching a TV program) before and after performing an auditory cognitive task.

## 2.1   Participants

We collected data from 20 participants (8 females, 12 males; 24–76 years; mean $\pm$ SD age 47.5 $\pm$ 20.5 years). All participants were well-rested and in good health, as measured by self-reports, and they had normal or corrected-to-normal vision. They were unaware of the purpose of the experiment. Written informed consent was obtained prior to the study. Eye-tracking data from two participants (one female, one male) were excluded from our analysis because of problems calibrating the eye tracker. Thus, our sample size was $N = 18$.

## 2.2   Experimental Design and Procedure

The experimental procedure is summarized in Fig. 1A. Participants performed a 17-minute mental calculation task designed to induce mental fatigue two times (Fig. 1B). They were asked to take questionnaires and watch video clips prior to and following each mental calculation task. Prior to the experiment, all participants were given oral instructions about the experiments and allowed to practice the mental calculation task.

In regard to the questionnaires, we used numerical rating scales to measure the participant's current ("right now, at this moment") perceived intensity of feelings regarding mental and physical fatigue, sleepiness, and motivation. The intensity was scaled from 0 to 10, with zero indicating an absence of those feelings and 10 indicating the strongest feeling ever experienced.

To collect eye-tracking data, the participants were asked to watch video clips approximately five minutes in length during each phase. As in previous studies [10], they were instructed to simply "watch and enjoy the videos."

As an auditory cognitive task to induce mental fatigue, we used a modified version of the Paced Auditory Serial Attention Test (mPASAT) [12] (Fig. 1B). Participants listened to a series of numbers ranging from one to nine. They were asked to add the number they had just heard to the number they had heard before and then to press a button whenever the sum of the two consecutive numbers equaled ten. One phase consisted of five 3-minute on-periods and four 30-second off-periods for a total 17 min. Each number was presented every 1.5 s. Participants were also asked to visually focus on three numbers on the display, which randomly changed every 0.5 s. These visual numbers were intended to distract and interfere with the primary auditory task, thereby increasing the complexity and attentional demands of the task in order to induce further mental fatigue.

## 2.3   Stimuli and Eye-Tracking Data Acquisition

To simulate the situation of watching a TV program, we used video clips made in the same manner as previous studies that investigated how neurodevelopmental and neurodegenerative disorders affect eye movements in natural viewing situations [10,13] (Fig. 1C).

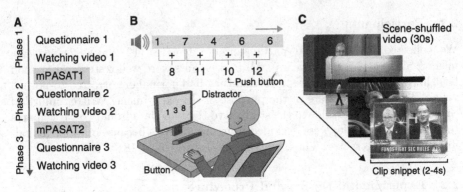

**Fig. 1.** Experimental setup: (A) overall procedure, (B) mental calculation task (mPASAT), (C) examples of scene-shuffled video clips.

Each 5-minute phase consisted of nine scene-shuffled videos (SVs), approximately 30 s each. Between the SVs, there were 5-second off-periods for rest. The SVs were made by assembling randomly extracted snippets from video clips. The lengths of the snippets were determined so that they were within the range of typical television programs [14,15]. Specifically, the lengths of the snippets were uniformly distributed between two and four seconds, so each SV consisted of nine to eleven snippets with no temporal gaps in between. For the original video clips, we utilized two datasets: CRCNS-ORIG [16] and DIEM [17], consisting of heterogeneous sources with different styles of programs that are commonly watched on a daily basis.

The participants' eye movements and pupil data were recorded using a noninvasive infrared EMR ACTUS eye-tracking device at a sample rate of 60 Hz (nac Image Technology Inc.; spatial resolution for eye movements and pupil diameter less than $0.5°$ and 0.1 mm, respectively). The eye tracker was calibrated using 9-point calibration at the beginning of each recording phase.

## 3    Mental Fatigue Detection Model

Our model uses 30 s worth of eye-tracking data in each SV to make a decision whether a participant is in a fatigued or non-fatigued state at that time. Thus, we obtained 9 samples for each 5-min phase of video watching.

First, we extracted 181 features categorized into six feature sets from the eye-tracking data that may change according to an individual's state of mental fatigue. Next, we built a two-class classifier for inferring mental fatigue using a subset of the features selected by a feature selection method through recursive evaluation and selection to avoid over-fitting (Fig. 2).

### 3.1    Data Preprocessing and Features

The raw eye-position data were segmented into blink, saccade, and fixation (or smooth-pursuit) periods. First, we extracted blink periods by using eyelid

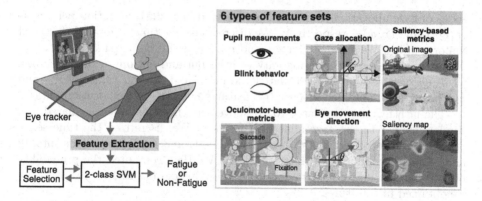

**Fig. 2.** Overview of our fatigue-detection model.

occlusion of both eyes. Apart from the blink periods, artifacts detected by the eye tracker were removed by using a linear interpolation algorithm. Finally, we used the mean-shift clustering method in the spatio-temporal domain to identify saccade and fixation periods [18].

We extracted six feature sets that we hypothesized would be differentially influenced by mental fatigue from each 30-second-long SV (Fig. 2). The first three feature sets related to oculomotor-based metrics, blinking behavior, and pupil measurements were used in previous studies on mental fatigue during cognitive tasks [6–8]. The other three feature sets were related to gaze allocation, eye movement directions, and saliency-based metrics. Although these feature sets have been used for characterizing eye movements in natural viewing conditions as well as inferring neurodevelopmental and neurodegenerative disorders [10,11,19], they have not been used for inferring mental fatigue.

The oculomotor-based features consisted of nine features: saccade amplitude, saccade duration, saccade rate, inter-saccade interval (mean, standard deviation, and coefficient of variance), saccadic mean velocity (mean and median), and fixation duration. We calculated seven features related to blinking behavior: blink duration, blink rate, blink duration per minute (the total time of all durations), and inter-blink interval (mean, standard deviation, and coefficient of variance). The pupil measurements were subdivided into six features related to pupil diameter, constriction velocity, and amplitude of each eye, and nine features related to the coordination of the pupil diameters of both eyes. Of these nine features, one was computed using Pearson's correlation coefficient. The other eight features were extracted using the phase locking value [20], which can identify transient synchrony over shorter time scales than Pearson's correlations. We used the mean and maximum values of the phase locking values with four different time windows (5, 10, 30, 60 frames).

The fourth feature set was calculated from a time-series of gaze allocation. We first converted gaze allocation values into radius and angle $(r, \phi)$ in a polar coordinate system situated at the center of the display. We then defined two time series of gaze allocations during all periods and only during fixation periods as $(r, \phi)_{\text{all}}$ and $(r, \phi)_{\text{fx}}$, respectively. We discretized each time series with $k$ bins of uniform width. We set $k = 8$ for $r_{\text{all}}$ and $r_{\text{fx}}$, $k = 36$ for $\phi_{\text{all}}$, and $k = 12$ for $\phi_{\text{fx}}$. As features, we used the probability of each bin and entropy estimated using these histograms and also calculated the mean and median values of $r_{\text{all}}$ and $r_{\text{fx}}$. In total, we obtained seventy-two features from the gaze-allocation data.

The fifth feature set related to eye-movement directions was calculated in a similar manner to the gaze allocation features. We discretized the time series of eye-movement directions $\theta$ during all periods and saccades periods into 12 and 36 bins of uniform width, respectively. We then computed the probability of each bin and entropy estimated using these histograms as features. In total, we obtained fifty features.

The sixth and final set consisted of features using a saliency model. The saliency model was proposed as a biologically-inspired computational model of human attention [21,22]. The saliency model computes a topographic map of conspicuity for every location in each video frame, highlighting locations that may attract attention in a stimulus-driven manner. We used the graph-based visual saliency model, where conspicuity maps of six low-level features (intensity contrast, color contrast, intensity variance, oriented edges, temporal flicker, and motion contrast) are linearly combined and normalized to form a saliency map [23]. Using both the saliency map and the six conspicuity maps, we obtained $4 \times 7 = 28$ saliency-based features in total. For more details about how to calculate saliency-based features, please see the original papers [10,23].

### 3.2  Classification and Feature Selection

For a two-class classification model for detecting mental fatigue, we used support vector machine (SVM) models [24,25] with a radial basis function kernel as follows: $K(x_i, x_j) = \exp(-\gamma ||x_i - x_j||^2)$. We set $\gamma = b_{\text{SVM}}/n_{\text{f}}$, where $n_{\text{f}}$ is the number of features and $b_{\text{SVM}}$ is a hyper-parameter. We used the algorithm for SVM implemented in MATLAB (MathWorks Inc., Natick, MA) and LIBSVM toolbox [25].

To identify useful features and avoid over-fitting of the model, we performed a feature selection through recursive evaluation and selection. One of the well-known methods is support vector machine recursive feature elimination (SVM-RFE) in a wrapper approach [26]. However, when the candidate feature set has highly correlated features, the ranking criterion of SVM-RFE tends to be biased, which would have a negative effect on the results. Our feature set contained highly correlated features such as features about saccade duration, amplitude, and velocity. We then used an improved SVM-RFE algorithm with a correlation bias reduction strategy in the feature elimination procedure [27].

## 4  Results

We first determined whether or not the tasks succeeded in inducing mental fatigue in the participants by using subjective ratings and objective measurements of mental fatigue that have been used in previous studies. We then evaluated our mental fatigue detection model in terms of their average scores after 20 iterations of 10-fold cross-validation.

## 4.1  Mental Fatigue After the Cognitive Tasks

We first investigated the participants' reported mental fatigue before and after performing the mPASAT, with subjective ratings from 0 to 10 (Fig. 3A). Compared with the subjective ratings in phase 1, i.e., before engaging in the cognitive task, 12 and 14 out of 18 participants in phases 2 and 3, respectively, reported increased ratings of mental fatigue. We performed a repeated-measures Friedman non-parametric ANOVA followed by Dunn's multiple comparisons and found a significant increase in mental fatigue from phases 1 to 3 ($p < .05$), but no significant difference between phases 1 and 2 or between phases 2 and 3.

We also tried to determine whether objective measurements of mental fatigue changed after the participant performed cognitive tasks. We used pupil diameters and blink behaviors, which are widely used as fatigue-related biomarkers [6–8] as the objective measurements. One-way repeated measures ANOVA with post hoc Bonferroni multiple comparisons were used to calculate the statistical significance over the phases. In this analysis, we computed these measurements by taking averages during each 5-minute phase. As a result, we found a significant decrease of pupil diameters from phase 1 to 2 and 3 ($p < .05, p < .005$, respectively) for the left eye and from phase 1 to 3 ($p < .005$, Fig. 3B; from phase 1 to 2, $p = .14$) for the right eye. In regard to the blink behaviors, we also found significant changes indicative of increased mental fatigue in duration, blink rate, and blink duration per minute over the phases. Among them, the blink duration per minute showed the biggest difference ($\eta_p^2 = .348$; from phase 1 to 2, $p < .05$; from phase 1 to 3, $p < .001$; Fig. 3C).

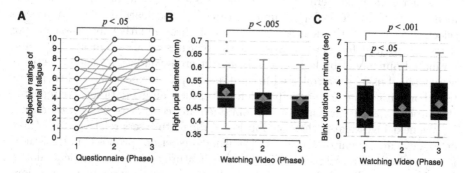

**Fig. 3.** Changes in subjective and objective measurements for mental fatigue after performing mPASAT. (A) Subjective ratings for mental fatigue on an 11-point numerical rating scale from 0 to 10. (B), (C) Right pupil diameter and blink duration per min. Boxes show the median, 25th, and 75th percentiles, filled symbols show outliers, and squares represent mean values.

**Table 1.** Fatigue-detection-model performance in 10-fold cross validation. $F_{pre}$: three feature sets related to oculomotor, blinks, and pupil measurements used in the previous studies, $F_{sal}$: saliency-based features, $F_{emd}$: features related to eye movement directions, and $F_{ga}$: features related to gaze allocation.

| Model | Detection performance (%) | | | |
|---|---|---|---|---|
| | Accuracy | Precision | Recall | F-measure |
| $F_{pre}$ | 77.1 | 78.6 | 72.9 | 75.6 |
| $F_{pre} + F_{sal}$ | 80.7 | 79.4 | 83.0 | 81.0 |
| $F_{pre} + F_{emd}$ | 82.9 | 83.2 | 82.4 | 82.7 |
| $F_{pre} + F_{ga}$ | 84.7 | 84.6 | 84.9 | 84.7 |
| $F_{pre} + F_{sal} + F_{emd} + F_{ga}$ | **91.0** | **91.4** | **90.3** | **90.8** |

Through these analyses, the results regarding the subjective and objective measurements indicate that the participants experienced increased mental fatigue after engaging in the cognitive tasks two times, i.e., in phase 3. We thus regarded phase 1 as a non-fatigued circumstance and phase 3 as a fatigued circumstance and proceeded to build a model that classifies the eye-tracking data of phases 1 and 3.

## 4.2    Model Performance

We built a fatigue detection model to differentiate eye-tracking data before and after performing the cognitive tasks. We used eye-tracking data of 18 participants in phases 1 and 3. In our model, features were extracted from each 30-second SV trial, and each phase consisted of nine SVs. Thus, the number of samples was $18 \times 9 \times 2 = 324$.

As a result of 20 iterations of 10-fold cross-validation, our model detected mental fatigue with 91.0% accuracy (Table 1). The feature selection process selected 55 of the 181 features as the most discriminative for classifiers for detecting mental fatigue and selected the features of all six groups. We also evaluated our model by leave-one-subject-out cross-validation, where classifiers were trained using data collected from all participants expect one and then were tested on data of the one participant left out of the training data set. We repeated this process for all participants, and obtained an accuracy of 88.5% accuracy.

We next investigated the contribution of the feature sets related to gaze allocation, eye movement directions, and saliency predictions proposed in this study. First, we built a model using only the three feature sets related to oculomotor-based metrics, blink behavior, and pupil measurements used in the previous studies. We did the feature selection and hyper-parameter optimization in the same way as our model. The model performance was 77.1% accuracy in 10-fold cross validation. Next, we separately added each feature set to this model. As a result, the model accuracies increased to 84.7%, 82.9%, and 80.7% as a result of adding gaze-location features, eye-movement direction features, and saliency-based features, respectively (Table 1). Therefore, we found that the novel use of three feature sets each improved the model's performance, and when taken together improved the model's performance by up to 13.9% (from 77.1 to 91.0%).

## 5 Conclusion

In contrast to previous studies focusing on detecting mental fatigue during cognitive tasks, we aimed to develop a system enabling us to infer mental fatigue in natural-viewing situations when an individual is not performing cognitive tasks. To this end, we devised a fatigue-detection model including novel feature sets and an automated feature selection method. Through experimentation with 18 adults, we showed that our model could detect mental fatigue with an accuracy of 91.0% in 10-fold cross-validation. One of the limitations in this study is that the study took place in a lab setting. We need to investigate whether our model can infer mental fatigue induced by everyday tasks. In addition, there is a possibility that the controlled setting might influence the way people watch video clips. Thus, future work will include an in-situ study to test our model in more realistic situations.

**Acknowledgments.** This research was partially supported by the Japan Science and Technology Agency (JST) under the Strategic Promotion of Innovative Research and Development Program.

## References

1. Alemdar, H., Ersoy, C.: Wireless sensor networks for healthcare: a survey. Comput. Netw. **54**(15), 2688–2710 (2010)
2. Favela, J., Castro, L.A.: Technology and aging. In: García-Peña, C., Gutiérrez-Robledo, L.M., Pérez-Zepeda, M.U. (eds.) Aging Research-Methodological Issues, pp. 121–135. Springer, Cham (2015)
3. Boksem, M.A., Tops, M.: Mental fatigue: costs and benefits. Brain Res. Rev. **59**(1), 125–139 (2008)
4. Avlund, K.: Fatigue in older adults: an early indicator of the aging process? Aging Clin. Exp. Res. **22**(2), 100–115 (2010)
5. Maghout-Juratli, S., Janisse, J., Schwartz, K., Arnetz, B.B.: The causal role of fatigue in the stress-perceived health relationship: a MetroNet study. J. Am. Board Family Med. **23**(2), 212–219 (2010)
6. Hopstaken, J.F., Linden, D., Bakker, A.B., Kompier, M.A.: A multifaceted investigation of the link between mental fatigue and task disengagement. Psychophysiology **52**(3), 305–315 (2015)
7. Schleicher, R., Galley, N., Briest, S., Galley, L.: Blinks and saccades as indicators of fatigue in sleepiness warnings: looking tired? Ergonomics **51**(7), 982–1010 (2008)
8. Di Stasi, L.L., Renner, R., Catena, A., Cañas, J.J., Velichkovsky, B.M., Pannasch, S.: Towards a driver fatigue test based on the saccadic main sequence: a partial validation by subjective report data. Transp. Res. Part C: Emerg. Technol. **21**(1), 122–133 (2012)

9. Dawson, D., Searle, A.K., Paterson, J.L.: Look before you (s)leep: evaluating the use of fatigue detection technologies within a fatigue risk management system for the road transport industry. Sleep Med. Rev. **18**(2), 141–152 (2014)

10. Tseng, P.H., Cameron, I.G., Pari, G., Reynolds, J.N., Munoz, D.P., Itti, L.: High-throughput classification of clinical populations from natural viewing eye movements. J. Neurol. **260**(1), 275–284 (2013)

11. Crabb, D.P., Smith, N.D., Zhu, H.: What's on TV? Detecting age-related neurodegenerative eye disease using eye movement scanpaths. Frontiers in Aging Neurosci. **6**, 312 (2014)

12. Cook, D.B., O'Connor, P.J., Lange, G., Steffener, J.: Functional neuroimaging correlates of mental fatigue induced by cognition among chronic fatigue syndrome patients and controls. Neuroimage **36**(1), 108–122 (2007)

13. Carmi, R., Itti, L.: The role of memory in guiding attention during natural vision. J. Vis. **6**(9), 4 (2006)

14. Cutting, J.E., DeLong, J.E., Brunick, K.L.: Visual activity in Hollywood film: 1935 to 2005 and beyond. Psychol. Aesthet. Creat. Arts **5**(2), 115 (2011)

15. Bordwell, D.: Intensified continuity visual style in contemporary American film. Film Q. **55**(3), 16–28 (2002)

16. Itti, L., Carmi, R.: Eye-tracking data from human volunteers watching complex video stimuli (2009)

17. Mital, P.K., Smith, T.J., Hill, R.L., Henderson, J.M.: Clustering of gaze during dynamic scene viewing is predicted by motion. Cogn. Comput. **3**(1), 5–24 (2011)

18. Xu, P., Ehinger, K.A., Zhang, Y., Finkelstein, A., Kulkarni, S.R., Xiao, J.: TurkerGaze: crowdsourcing saliency with webcam based eye tracking. arXiv preprint arXiv:1504.06755 (2015)

19. Zhang, Y., Wilcockson, T., Kim, K.I., Crawford, T., Gellersen, H., Sawyer, P.: Monitoring dementia with automatic eye movements analysis. In: Czarnowski, I., Caballero, A.M., Howlett, R.J., Jain, L.C. (eds.) Intelligent Decision Technologies 2016. SIST, vol. 57, pp. 299–309. Springer, Cham (2016). doi:10.1007/978-3-319-39627-9_26

20. Tass, P., Rosenblum, M., Weule, J., Kurths, J., Pikovsky, A., Volkmann, J., Schnitzler, A., Freund, H.J.: Detection of n: m phase locking from noisy data: application to magnetoencephalography. Phys. Rev. Lett. **81**(15), 3291 (1998)

21. Treisman, A.M., Gelade, G.: A feature-integration theory of attention. Cogn. Psychol. **12**(1), 97–136 (1980)

22. Itti, L., Koch, C., Niebur, E.: A model of saliency-based visual attention for rapid scene analysis. IEEE Trans. Pattern Anal. Mach. Intell. **11**, 1254–1259 (1998)

23. Harel, J., Koch, C., Perona, P.: Graph-based visual saliency. In: Advances in Neural Information Processing Systems, pp. 545–552 (2006)

24. Boser, B.E., Guyon, I.M., Vapnik, V.N.: A training algorithm for optimal margin classifiers. In: Proceedings of the Fifth Annual Workshop on Computational Learning Theory, pp. 144–152. ACM (1992)

25. Chang, C.C., Lin, C.J.: LIBSVM: a library for support vector machines. ACM Trans. Intell. Syst. Technol. (TIST) **2**(3), 27 (2011)

26. Guyon, I., Weston, J., Barnhill, S., Vapnik, V.: Gene selection for cancer classification using support vector machines. Mach. Learn. **46**(1–3), 389–422 (2002)

27. Yan, K., Zhang, D.: Feature selection and analysis on correlated gas sensor data with recursive feature elimination. Sens. Actuators B: Chem. **212**, 353–363 (2015)

# Automatic Identification of Intraretinal Cystoid Regions in Optical Coherence Tomography

Joaquim de Moura, Jorge Novo[✉], José Rouco, Manuel G. Penedo, and Marcos Ortega

Department of Computer Science, University of A Coruña, A Coruña, Spain
{joaquim.demoura,jnovo,jrouco,mgpenedo,mortega}@udc.es

**Abstract.** Optical Coherence Tomography (OCT) is, nowadays, one of the most referred ophthalmological imaging techniques. OCT imaging offers a window to the eye fundus in a non-invasive way, permitting the inspection of the retinal layers in a cross sectional visualization. For that reason, OCT images are frequently used in the analysis of relevant diseases such as hypertension or diabetes. Among other pathological structures, a correct identification of cystoid regions is a crucial task to achieve an adequate clinical analysis and characterization, as in the case of the analysis of the exudative macular disease.

This paper proposes a new methodology for the automatic identification of intraretinal cystoid fluid regions in OCT images. Firstly, the method identifies the Inner Limitant Membrane (ILM) and Retinal Pigment Epithelium (RPE) layers that delimit the region of interest where the intraretinal cystoid regions are placed. Inside these limits, the method analyzes windows of a given size and determine the hypothetical presence of cysts. For that purpose, a large and heterogeneous set of features were defined to characterize the analyzed regions including intensity and texture-based features. These features serve as input for representative classifiers that were included in the analysis.

The proposed methodology was tested using a set of 50 OCT images. 502 and 539 samples of regions with and without the presence of cysts were selected from the images, respectively. The best results were provided by the LDC classifier that, using a window size of $61 \times 61$ and 40 features, achieved satisfactory results with an accuracy of 0.9461.

**Keywords:** Computer-aided diagnosis · Retinal imaging · Optical Coherence Tomography · Intraretinal cystoid regions

## 1 Introduction and Previous Work

The analysis of the retina is crucial for the diagnosis of different relevant pathologies. For that reason, the identification of the main retinal structures as the optic disc [1] or the vascular tree [2] can provide evidences for an appropriate characterization of diseases like hypertension or diabetes. Among the different image modalities, Optical Coherence Tomography (OCT) have spread their use over

© Springer International Publishing AG 2017
A. ten Teije et al. (Eds.): AIME 2017, LNAI 10259, pp. 305–315, 2017.
DOI: 10.1007/978-3-319-59758-4_35

the years as they offer a cross-sectional view of the retina and sub-retinal layers with microscopic resolution in a non-invasive and contactless way, providing a more detailed source of information than other modalities such as retinographies. This more detailed set of information about the retinal layers can help the specialists to perform more accurate analysis of relevant diseases as age-related macular degeneration (AMD) or glaucoma [3].

AMD can lead to exudative macular disease, one of the main causes of blindness in developed countries. The intraretinal cystoid fluid is directly related with exudative macular disease, originated by abnormal vasculature growing that leaks fluid, deriving in a progressive retinal architecture degeneration and the corresponding vision loss (Fig. 1). For that reason, an appropriate identification and characterization of the cystoid regions is a crucial task as it represents a measurement of the disease severity, helping clinicians to produce more accurate diagnosis and treatments [4].

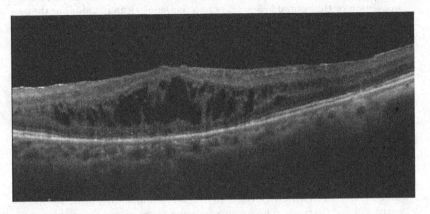

**Fig. 1.** Example of OCT image with the presence of intraretinal cystoid regions.

In recent years, some works have been proposed facing the issue of cyst extraction. The proposals frequently used an initial denoising stage to minimize the impact of the typical speckle noise that normally appears in OCT imaging. Most of the approaches addressed directly the problem by the segmentation of cyst candidates followed by a morphological and intensity analysis and/or a post-processing stage to reduce the false positive (FP) detections and return the final results. Following this strategy, Wilkins et al. [5] faced the cystoid macular edema identification by an initial thresholding of dark structures to identify the cyst candidate contours. They posteriorly applied a couple of rules to reduce the FP detections and produce the final identifications. Roychowdhury et al. [6] also segments dark regions in a bright neighborhood after identifying the 6 main retinal layers. This cyst candidate set is posteriorly analyzed in terms of solidity, mean and maximum intensities to produce the final cyst identifications. In the case of Wieclawek et al. [7], a combination of image processing techniques were applied to extract the candidate segmentations. Redundant regions are posteriorly removed to produce the final cyst extractions. González *et al.* [8] used watershed to produce the initial candidates segmentation. The extracted regions are posteriorly grouped by connectivity and similarity in terms of intensity. Given the large amount of FPs that this step produces, the method posteriorly filters the candidates using discarding rules combined with a learning strategy to reduce the FPs set. Esmaeili *et al.* [9] used a K-SVD dictionary learning in curvelet transform to help with the speckle noise reduction and facilitate the posterior thresholding strategy that the authors proposed. Miss-extractions are posteriorly removed in a post-processing stage. A combined strategy for cysts segmentation was proposed by Wang *et al.* [10], using a fuzzy level set that integrates fuzzy C-Means and level sets. The method extracts the fluid regions by intensity, thanks to the fuzzy C-Means, with a adequate contour segmentation, incorporated by the level set method. The work of Xu *et al.* [11] defined a layer-dependent stratified sampling to produce symptomatic exudate-associated derangements segmentations using voxel classification. In the case of Lang *et al.* [12] a pixel classification system was also designed, but limited to the domain of micro-cystic segmentations.

This strategy, that was followed by most of the approaches, presents as main limitation the high dependency in the candidate segmentation stage. A poor segmentation technique may produce initial large candidate sets, hardening the posterior refinement to remove the detected FPs. Large sets of FPs can provoke the necessity of strong reduction stages that may carry the elimination of real cysts. Additionally, incorrect cyst candidate segmentations may also alter the candidate characteristics, deriving in confusions in the posterior morphological/intensity analysis, penalizing the final cyst identification results.

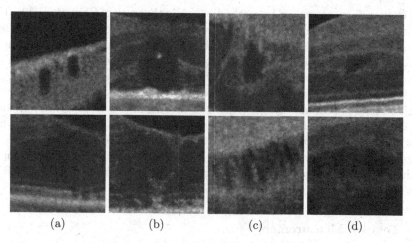

|   (a)   |   (b)   |   (c)   |   (d)   |

**Fig. 2.** Examples of cysts with different levels of complexity.

Many times, this segmentation dependency can be overcome as fluid regions and cyst contours can be acceptably obtained, as illustrated with the examples of Fig. 2, $1^{st}$ row. However, many other times, the cyst contours cannot be clearly identified as there is no enough intensity contrast in the entire cyst region (Fig. 2, second row, (a) and (b)). Other times, cysts appear in nearby groups, making extremely complicated the identification and delimitation of all of them, even for the human eye of the experts (Fig. 2, second row, (c) and (d)).

In this work, we propose a new methodology that faces the issue of cyst identification with a novel strategy. Instead of classical cyst candidate segmentations and FPs removal, we identify intraretinal cystoid regions, that is, regions of the OCT images that contain cysts. The method uses a window size for the analysis, extracts a set of image characteristics and determines the presence of cysts inside the analyzed regions.

## 2    Methodology

The proposed system firstly identify the retinal layer limits that contain the intraretinal cystoid regions. Then, inside this region of interest, the method analyze windows of a defined size to identify the cyst presence including: feature measurement, feature selection and classification.

### 2.1    Retinal Layer Segmentation

As cysts appear inside the retinal layers, we can reduce the search space identifying this region of the OCT images. The region of interest is delimited between the Inner Limitant Membrane (ILM), first intraretinal layer, and the Retinal Pigment Epithelium (RPE), formed by pigmented cells at the external part of the retina.

For that purpose, we used a method based on the work of Chiu et al. [13]. This method uses graph theory to represent each image as a graph of nodes. Then, the optimum connected paths from both sides of the image are obtained using dynamic programming. In this case, dark-to-light gradient images are firstly calculated as these gradients identify the limits of adjacent layers. These gradients are used to generate weights for the layer segmentations. The minimum weighted paths are found by the Dijkstra's algorithm [14] to progressively identify the main layers of the retina. Despite this approach was designed to find eight different layers, we aimed in this work for the ILM and RPE layers, as they constitute the limits of the retinal layers, sufficient for the delimitation of the intraretinal cysts search space. Figure 3 shows an example of ILM and RPE layer identification for a particular OCT image.

### 2.2    Feature Measurement

In order to characterize each analyzed region and identify the presence of cysts, a complete set of 189 features was defined. This feature set includes intensity and texture-based properties that help to maximize the discrimination power of cysts identification with respect to other structures and patterns of the retina.

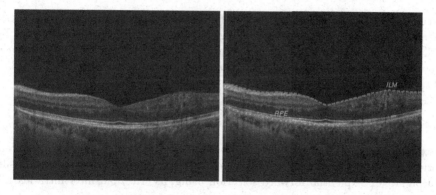

**Fig. 3.** Example of ILM and RPE retinal layer segmentation.

**Intensity Statistics.** We measured 13 global characteristics of the analyzed region, including: *maximum, minimum, mean, median, standard deviation, variance, 25<sup>th</sup> and 75<sup>th</sup> percentile, skewness* and *maximum likelihood estimates for a normal distribution.*

**Gray-Level Intensity Histogram (GLIH).** Using the intensity histogram of the analyzed region, the method derives the following measurements: *obliquity, kurtosis, energy* and *entropy.*

**Eigenvalues.** Eigenvalues can be useful to capture regions with intensity changes in different directions, as happens with the presence of cysts. We calculated the eigenvalues of the analyzed region and selected the 4 highest ($\lambda_{max_i}$) and the 2 lowest ($\lambda_{min_i}$) values. Additionally, several ratios among them were also included in the feature set.

**Histogram of Oriented Gradients (HOG).** The orientation of the gradients can be useful in this issue as cysts typically present closed/oval contours. Instead of that, non-cystoid regions present a more uniform shape, with lower levels of gradients and, if present, they usually appear with a parallel horizontal pattern (due to the presence of the retinal layers) or vertical and tubular patterns (due to the shadows of vessels or other structures). HOG features [15] can help to capture these patterns, presenting some invariance to scale, rotation or translation changes, properties also useful in this issue. 9 HOG windows per bound box and 9 histogram bins were analyzed, adding a total of 81 characteristics.

**Local Binary Patterns (LBP).** LBPs [16] can help to identify local patterns that may appear with and without the presence of cysts. LBPs presents a low complexity and a low sensitivity to changes in illumination, common conditions in OCT imaging due to the variability of capture machines and configuration parameters. A wide range was analyzed, calculating a total of 64 features that were added to the feature set.

**Gray-Level Co-Ocurrence Matrix (GLCM).** These second order statistics measure the simultaneous ocurrence of gray levels in pairs of pixels, separated by a displacement vector. Based on the proposal of Haralick *et al.* [17], we performed the analysis at a distance of 2 pixels and 4 directions: 0°, 45°, 90° and 135°, obtaining a total of 16 features.

### 2.3  Feature Selection and Classification

Next step involves feature selection to avoid irrelevant and redundant characteristics by selecting the most useful ones and, therefore, facilitating the classification stage. Sequential forward selection using, as criterion, inter-intra feature distance was used.

Finally, representative classifiers, frequently used in medical imaging solutions, were trained and tested using the extracted feature set: Linear Discriminant Analysis (LDA), $k-$nearest neighbors (kNN) and Support Vector Machines (SVM) were analyzed. In the case of kNN, 3 configurations were tested using $k = [3, 5, 7]$ whereas 3 configurations were also defined for the case of the SVM, using an exponential kernel with values of $\theta = [1, 2, 3]$:

$$k(x, y) = exp\left(-\frac{\| x - y \|}{\theta}\right) \tag{1}$$

The classification stage was done using a constructed dataset that was randomly divided in two smaller datasets with the same size (each one with the 50% of all the samples). The first dataset is used for the training stage whereas the second one is reserved for testing the trained classifiers. This process of dataset random division, training and testing was repeated 50 times, calculating the mean accuracy in order to obtain a global performance measurement.

## 3  Results and Discussion

The proposed method was validated using a set of 50 OCT histological images. These images were captured by a confocal scanning laser ophtalmoscope, a CIRRUS™HD-OCT–Carl Zeiss Meditec. The images correspond to scans centered in the macula, from both left and right eyes of different patients, and with resolutions that vary from 924 × 616 to 1200 × 800. Several intensity and contrast configurations are also present in the image set. No preprocessing was applied to the images.

The images were labeled by an expert clinician, identifying the location of any present cyst. Using this information, we constructed a dataset by the selection of 502 and 539 samples of regions with and without the presence of cysts, using a window size of 61 × 61. As said, this dataset was randomly divided in two smaller datasets with the same size, one for training and other for testing, using 50 repetitions for each configuration to obtain, for each case, a global measurement of its performance.

**Table 1.** Accuracy results that were obtained with the tested classifiers using different feature set sizes.

| N. Features | 1 | 5 | 10 | 30 | 50 |
|---|---|---|---|---|---|
| LDC | 0.7747 | 0.8721 | 0.9016 | 0.9394 | 0.9456 |
| 3-kNN | 0.7140 | 0.8568 | 0.8870 | 0.9088 | 0.9110 |
| 5-kNN | 0.7348 | 0.8626 | 0.8986 | 0.9099 | 0.9142 |
| 7-kNN | 0.7453 | 0.8617 | 0.8920 | 0.9067 | 0.9049 |
| 1-SVM | 0.7692 | 0.8698 | 0.9043 | 0.9270 | 0.9262 |
| 2-SVM | 0.7742 | 0.8652 | 0.9014 | 0.9196 | 0.9219 |
| 3-SVM | 0.7747 | 0.8638 | 0.8986 | 0.9157 | 0.9165 |

Regarding the selected features, the majority of them were taken from HOG and also from LBP feature sets as they include a high potential in the differentiation of common layer patterns and other structures with respect to the cyst presence. Global intensity statistics, as *minimum*, were also selected in the first positions as the cyst presence typically implies a depression in intensity values and changes in intensity profiles of the analyzed window.

Table 1 presents the accuracy results that were achieved by the different classifier configurations using progressive larger feature sets. A maximum of 50 features was set as no further improvements were obtained from that point, achieving the best performance of each case with smaller feature sets, as Table 2 details. Generally, the obtained accuracy results in all the cases are significantly high, being the best results obtained with the LDC classifier and 40 features, returning a performance of 0.9461. In the case of the SVM, it offers better results than the kNN, being the lowest value the case of the 7-kNN, with a performance of 0.9099. The results of the first degree SVM are significantly high, 0.9299, but at a distance of the LDC classifier.

**Table 2.** Best accuracy obtained by each tested classifier, indicating the number of needed features.

| Classifier | LDC | 3-kNN | 5-kNN | 7-kNN | 1-SVM | 2-SVM | 3-SVM |
|---|---|---|---|---|---|---|---|
| N. Features | 40 | 42 | 44 | 43 | 44 | 44 | 28 |
| Accuracy | 0.9461 | 0.9136 | 0.9151 | 0.9099 | 0.9299 | 0.9240 | 0.9202 |

Figure 4 shows some examples of cystoid and non-cystoid regions from the testing dataset that were correctly classified. Regarding non-cystoid regions, the method is capable to identify the tissue and layer patterns with different levels of intensity and contrast (Fig. 4(a) and (b)) but also to discard dark patterns that are derived from shadows of vessels and other artifacts (Fig. 4(a), (c) and (d)). In the case of the cystoid regions, we can see that the method is capable to detect the simple cases (Fig. 4(a)) where cyst regions and contours are clearly delimited but also other complex cases of cysts or groups with low contrast and imperfect contour definition (Fig. 4(b), (c) and (d)).

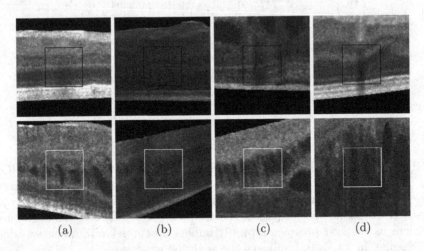

(a)                    (b)                    (c)                    (d)

**Fig. 4.** Examples of testing samples correctly classified. $1^{st}$ row, non-cystoid regions. $2^{nd}$ row, cystoid regions.

Figure 5 includes some representative incorrect classifications. Many misclassified cystoid regions are omitted due to extremely poor contrast and too fuzzy contours (Fig. 5(a) and (b)) being extremely complicated to detect its presence by the system. The common mistakes in non-cystoid regions are due to the presence of other structures (Fig. 5(c) and (d)) that create artificial high-to-low intensity regions that may be confused with the typical patterns that appear with cysts.

We also tested the performance of the system using progressive window sizes. In the dataset construction, we built the corresponding datasets using the same central points in all the cases but using progressive lower window sizes. Table 3

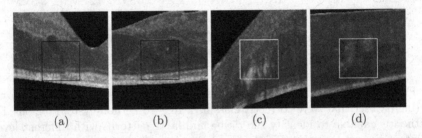

(a)                    (b)                    (c)                    (d)

**Fig. 5.** Examples of testing samples incorrectly classified. (a) and (b), cystoid regions classified as non-cystoid. (c) and (d), non-cystoid regions classified as cystoid.

**Table 3.** Best accuracy obtained by each tested classifier using different window sizes.

| Window size | $11 \times 11$ | $15 \times 15$ | $21 \times 21$ | $31 \times 31$ | $61 \times 61$ |
|---|---|---|---|---|---|
| LDC | 0.8599 | 0.8906 | 0.9118 | 0.9339 | 0.9461 |
| 5-kNN | 0.8450 | 0.8581 | 0.8872 | 0.9081 | 0.9151 |
| 1-SVM | 0.8654 | 0.8915 | 0.9103 | 0.9247 | 0.9299 |

details the best performances that were achieved with each window size and the best configurations of the SVM, KNN and LDC classifiers. As we can see, the performance is progressively penalized with smaller windows, as they do not offer the same information to identify the presence of cysts in the samples, specially in the cases of windows that include large cysts. The final window size should be selected as a balance between detail in the detections and accuracy in the performance.

## 4    Conclusions

The extraction and analysis of intraretinal cystoid fluid regions is a relevant issue for the diagnosis and treatment of relevant pathologies as can represent the exudative macular disease, one of the main causes of blindness in developed countries. Most of the existing approaches faced this issue by an initial segmentation of cyst candidates, segmentations that are posteriorly analyzed using intensity or morphological properties to discard wrong detections. These strategies present a high dependency in the performance of the candidates segmentation stage, given the complex conditions that are frequently present in OCT images. Imperfect or wrong segmentations may carry, in many cases, the removal of existing cysts or the preservation of FPs by the posterior analysis and refinement, penalizing the performance of any proposed system.

In this work, we propose a novel methodology for the identification of intraretinal cystoid fluid regions in OCT images. We face the issue with a different strategy, by the analysis of regions inside the retinal layers and the determination of the presence of cysts. Hence, the dependency of the segmentation stage is omitted, detecting directly the cystoid regions inside the OCT images.

The system defined a set of 189 features, being selected the ones with higher power of discrimination. A set of representative classifiers were studied, including 3 SVM and 3 kNN configurations as well as the LDC classifier. The method was validated with a set of 1041 samples, including 502 and 539 samples of cystoid and non-cystoid regions, respectively. Satisfactory results were obtained, being the best performance achieved by the LDC classifier that, using 40 features and a window size of $61 \times 61$, reported an accuracy of 0.9461.

As future work, a further analysis and inclusion of suitable characteristics should be done as well as the use of wrapped based feature selection methods. Moreover, a wider range of classifiers, like artificial neural networks, should be tested in the classification stage.

**Acknowledgments.** This work is supported by the Instituto de Salud Carlos III, Government of Spain and FEDER funds of the European Union through the PI14/02161 and the DTS15/00153 research projects and by the Ministerio de Economía y Competitividad, Government of Spain through the DPI2015-69948-R research project.

# References

1. Novo, J., Penedo, M.G., Santos, J.: Optic disc segmentation by means of GA-optimized topological active nets. In: Campilho, A., Kamel, M. (eds.) ICIAR 2008. LNCS, vol. 5112, pp. 807–816. Springer, Heidelberg (2008). doi:10.1007/978-3-540-69812-8_80
2. Moura, J., Novo, J., Ortega, M., Charlón, P.: 3D Retinal vessel tree segmentation and reconstruction with OCT images. In: Campilho, A., Karray, F. (eds.) ICIAR 2016. LNCS, vol. 9730, pp. 716–726. Springer, Cham (2016). doi:10.1007/978-3-319-41501-7_80
3. Geitzenauer, W., Hitzenberger, C.K., Schmidt-Erfurth, U.M.: Retinal optical coherence tomography: past, present and future perspectives. Br. J. Ophthalmol. **95**(2), 171–177 (2011)
4. Bogunovic, H., Abramoff, M.D., Zhang, L., Sonka, M.: Prediction of treatment response from retinal OCT in patients with exudative age-related macular degeneration. In: Ophthalmic Medical Image Analysis Workshop, MICCAI 2014, pp. 129–136 (2014)
5. Wilkins, G.R., Houghton, O.M., Oldenburg, A.L.: Automated segmentation of intraretinal cystoid fluid in optical coherence tomography. IEEE Trans. Biomed. Eng. **59**(4), 1109–1114 (2012)
6. Roychowdhury, S., Koozekanani, D.D., Radwan, S., Parhi, K.K.: Automated localization of cysts in diabetic macular edema using optical coherence tomography images. In: International Conference of the IEEE Engineering in Medicine and Biology Society, pp. 1426–1429 (2013)
7. Wieclawek, W.: Automatic cysts detection in optical coherence tomography images. International Conference on Mixed Design of Integrated Circuits and Systems, pp. 79–82 (2015)
8. González, A., Remeseiro, B., Ortega, M., Penedo, M.G., Charlón, P.: Automatic cyst detection in OCT retinal images combining region flooding and texture analysis. IEEE International Symposium on Computer-Based Medical Systems, pp. 397–400 (2013)
9. Esmaeili, M., Dehnavi, A.M., Rabbani, H., Hajizadeh, F.: Three-dimensional segmentation of retinal cysts from spectral-domain optical coherence tomography images by the use of three-dimensional curvelet based K-SVD. J. Med. Signals Sens. **6**(3), 166–171 (2016)
10. Wang, J., Zhang, M., Pechauer, A.D., Liu, L., Hwang, T.S., Wilson, D., Li, D.J., Jia, Y.: Automated volumetric segmentation of retinal fluid on optical coherence tomography. Biomed. Opt. Express **7**(4), 1577–1589 (2016)
11. Xu, X., Lee, K., Zhang, L., Sonka, M., Abràmoff, M.D.: Stratified sampling voxel classification for segmentation of intraretinal and subretinal fluid in longitudinal clinical OCT data. IEEE Trans. Med. Imaging **34**(7), 1616–1623 (2015)
12. Lang, A., Carass, A., Swingle, E.K., Al-Louzi, O., Bhargava, P., Saidha, S., Ying, H.S., Calabresi, P.A., Prince, J.L.: Automatic segmentation of microcystic macular edema in OCT. Biomed. Opt. Express **6**(1), 155–169 (2014)

13. Chiu, S.J., Li, X.T., Nicholas, P., Toth, C.A., Izatt, J.A., Farsiu, S.: Automatic segmentation of seven retinal layers in SDOCT images congruent with expert manual segmentation. Opt. Expr. **10**(10), 19413–19428 (2010)
14. Dijkstra, E.W.: A note on two problems in connexion with graphs. Numer. Math. **1**(1), 269–271 (1959)
15. Dalal, N., Triggs, B.: Histograms of oriented gradients for human detection. In: Computer Vision and Pattern Recognition, CVPR 2005, pp. 886–893 (2005)
16. Ojala, T., Pietikainen, M., Maenpaa, T.: Multiresolution gray-scale and rotation invariant texture classification with local binary patterns. IEEE Trans. Pattern Anal. Mach. Intell. **24**(7), 971 (2002)
17. Haralick, R.M., Shanmugam, K., Dinstein, I.H.: Textural features for image classification. IEEE Trans. Syst Man Cybern. **3**(6), 610–621 (1973)

# Convolutional Neural Networks for the Identification of Regions of Interest in PET Scans: A Study of Representation Learning for Diagnosing Alzheimer's Disease

Andreas Karwath⊙, Markus Hubrich⊙, Stefan Kramer$^{(\boxtimes)}$⊙,
and the Alzheimer's Disease Neuroimaging Initiative

Institut für Informatik, Johannes Gutenberg-Universität, Mainz, Germany
kramer@informatik.uni-mainz.de

**Abstract.** When diagnosing patients suffering from dementia based on imaging data like PET scans, the identification of suitable predictive regions of interest (ROIs) is of great importance. We present a case study of 3-D Convolutional Neural Networks (CNNs) for the detection of ROIs in this context, just using voxel data, without any knowledge given a priori. Our results on data from the Alzheimer's Disease Neuroimaging Initiative (ADNI) suggest that the predictive performance of the method is on par with that of state-of-the-art methods, with the additional benefit of potential insights into affected brain regions.

**Keywords:** PET Scans · Alzheimer's disease · Dementia · Deep learning · Convolutional Neural Networks · Regions of interest · Representation learning

## 1 Introduction

Alzheimer's disease is a progressive, degenerative and incurable disease of the brain and a common cause of dementia. One of the imaging modalities often used in the Computer-Aided Diagnosis (CAD) of Alzheimer's disease is positron emission tomography (PET). Recent years have seen a number of machine learning based approaches applied to PET scans to distinguish between Alzheimer's disease (AD), mild cognitive impairment (MCI) and normal control (NC), amongst others [1,2]. Deep learning approaches have attracted particular interest in attempts to transfer the progress made in computer vision to medical image analysis.[1] Besides classification, the identification of regions of interest (ROIs) is of major concern in medical research, i.e. regions in 2-D or 3-D medical images that are of particular importance for assessing the functioning of organs, for diagnosis and for understanding disease progression. In this paper, we continue

---

[1] For a good overview see the recent paper by Vieira *et al.* [2]. Apparently many more papers can be found in online archives than papers that have appeared already.

© Springer International Publishing AG 2017
A. ten Teije et al. (Eds.): AIME 2017, LNAI 10259, pp. 316–321, 2017.
DOI: 10.1007/978-3-319-59758-4_36

our previous work [1] and report first results towards the identification of ROIs in PET scans, in the context of AD diagnosis, using deep learning methods. While deep learning on PET scans is currently an active research topic and ROIs have been investigated in other contexts (e.g. investigating different tracers), the combination of deep learning and ROIs has not received much attention yet. The approach employs a binary convolutional neural network (CNN) based classifier to generate importance maps of PET voxel data.

## 2 Methodology

The main aim is to establish ROIs with respect to AD without medical background knowledge in an automated manner. In the following, we describe how to apply a binary classifier to generate importance maps for extracting ROIs.

### 2.1 Data Acquisition and Preprocessing

To train the binary classifier used to extract importance maps (or ROIs), we employed the Alzheimer's Disease Neuroimaging Initiative[2] (ADNI) data set. The initiative has a large pool of PET images (co-registered, averaged), which have been acquired on various scanners using different imaging parameters. In the work presented here, we chose the $[^{18}F]$ fluorodeoxyglucose PET scans from enrolled normal control (NC), mild cognitive impairment (MCI) and Alzheimer's disease (AD) subjects. Not all scans from a single patient possess the same class, as the disease might occur at a later stage of a patient's life time. To avoid possible information leakage during the validation of our approach, we always employed a $k$-fold validation based on patients and not on scans. Our aim is to use this data to compare the predictive performance of the identified ROIs to a smaller data set used in a previous study by Li *et al.* [1]. However, as this data also originates from the same source, we excluded 84 subjects from the ADNI data set (see Table 1). Overall, we used 1258 subjects and 2630 scans.

The ADNI data set contains globally normalized scans [3], which are not suitable for our purposes as this kind of normalization does not easily allow the

**Table 1.** The two different data sets used in this work originating from the ADNI project, $\mathcal{D}_C$ the ADNI data set without the patients used in Li *et al.* ($\mathcal{D}_{\mathcal{ROI}}$).

| Data Set | Name | Class Label | No.Patients (male:female) | Age | No.Scans |
|---|---|---|---|---|---|
| | | NC | 396 (186:210) | 74 ($\pm 6$) | 749 |
| $\mathcal{D}_C$ | ADNI \ Li *et al.* | MCI | 662 (376:286) | 73 ($\pm 7$) | 1292 |
| | | AD | 330 (199:131) | 76 ($\pm 8$) | 589 |
| | | NC | 30 (21:9) | 74 ($\pm 5$) | 30 |
| $\mathcal{D}_{\mathcal{ROI}}$ | Li *et al.* | MCI | 29 (23:6) | 74 ($\pm 6$) | 29 |
| | | AD | 25 (15:10) | 72 ($\pm 6$) | 25 |

---

[2] URL: http://adni.loni.usc.edu/ (visited on Jan. 12, 2016).

comparison of PET images to each other. Therefore, and to allow comparison to the works of Li *et al.*, we used a different type of normalization. We employed the same kind of spatial normalization as Li *et al.* based on the Automated Anatomical Labeling (AAL) digital brain atlas: The normalization yields scans in a $91 \times 109 \times 91$ with $2\,mm^3$ voxel format. Next, each scan is intensity normalized with respect to its cerebral global mean (also called grand mean normalization). Finally, all scans are smoothed by an $8\,mm$ FWHM Gaussian kernel.

## 2.2   Generation of Binary Classifiers

As mentioned above, the approach taken is to train a binary classifier in order to extract informative regions of interest (ROIs). To train the binary classifier a 3-D CNN is employed, expected to be able to incorporate local voxel relations as well as to extract complex spatial patterns. Training a 3-D CNN on input volumes with a size of $91 \times 109 \times 91$ can be computationally expensive. Therefore, we down-sampled all scans reducing the number of operations. First, all non-brain voxels are removed to eliminate noise caused by the PET scanners. As a consequence, we trim all scans to their minimum bounding box, which yields a new size of $88 \times 108 \times 88$ voxels. The first and last ten sagittal slices contain very little brain volume, but a considerably large ratio of the skull. Hence, we discard these slices and obtain cubic volumes of dimension $88 \times 88 \times 88$. Next, the scans are down-sampled using "average pooling", i.e. the input is divided into cubic pooling regions of a particular size and the output consists of the average of each region. We chose a pooling size of 4, resulting in scans of size $22 \times 22 \times 22$.

For the binary classifier, we chose the following deep architecture: Overall seven convolutional layers are followed by three fully connected layers, where all of them use rectified linear units as activation function. The input layer corresponds to the $22^3$ down-sampled scan size and the output layer consists of a single sigmoid output node (see Fig. 1). All feature maps use a stride of 1. We employed 10-fold cross validation to estimate the optimal parameters using $\mathcal{D_C}$. To cater for variations, we augmented the data set by mirroring the original scans on the center coronal plane. Furthermore, we used a mini-batch size of 128 and trained the final classifier $\mathcal{C}$ on $\mathcal{D_C}$ with the parameters found in the 10-fold cross validation.

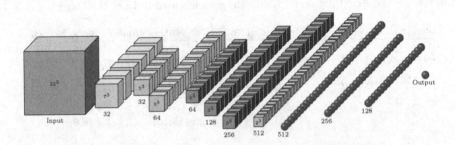

**Fig. 1.** The architecture for the CNN classifier.

## 2.3   ROI Discovery and Ranking

Given the binary classifier $\mathcal{C}$ from the previous section and a scan $X$, one can assign each voxel of $X$ a value of its importance towards the classification of $X$. We assume $X : \mathbb{N}^{n \times n \times n} \to \mathbb{R}_{\geq 0}$. For $1 \leq x, y, z \leq n$, we denote with $v_{xyz}$ the voxel at position $(x, y, z)$. We can exclude the information of $v_{xyz}$ from $X$ by setting $X \setminus v_{xyz}(x, y, z) = X(x, y, z) = \gamma^3$. We denote $Conf_{\mathcal{C}}(X) \in [0, 1]$ to be the confidence for correctly classifying the example $X$ by $\mathcal{C}$. Using this information, the influence of $v_{xyz} \in X$ can be assessed as $\Delta Conf(\mathcal{C}, X, v_{xyz}) = Conf_{\mathcal{C}}(X) - Conf_{\mathcal{C}}(X \setminus v_{xyz})$. If $\Delta Conf(\mathcal{C}, X, v_{xyz}) = 0$, $v_{xyz}$ can be regarded as redundant and can be omitted. However, if $\Delta Conf(\mathcal{C}, X, v_{xyz}) > 0$, the voxel $v_{xyz}$ can be considered as important. Finally, $\Delta Conf(\mathcal{C}, X, v_{xyz}) < 0$ indicates a decrease in the classification performance. Let $b_{\tilde{x}\tilde{y}\tilde{z}}^k$ be a cubic block of voxels with center $(\tilde{x}, \tilde{y}, \tilde{z})$ and size $2k - 1$. Analogously to the exclusion of single voxels, we can exclude the information contained in a block from $X$ as follows:

$$X \setminus b_{\tilde{x}\tilde{y}\tilde{z}}^k(x, y, z) = \begin{cases} \gamma, & \text{if } v_{xyz} \in b_{\tilde{x}\tilde{y}\tilde{z}}^k \\ X(x, y, z), & \text{otherwise.} \end{cases} \tag{1}$$

Using this, we can can generate a map of voxel importance for the exclusion of blocks with length $2k - 1$ as $I_{\mathcal{C},X}^k(x, y, z) = \frac{1}{|b_{xyz}^k|} \Delta Conf(\mathcal{C}, X, b_{xyz}^k)$. To accommodate for blocks in the border areas, we normalize $I_{\mathcal{C},X}^k$ using the total amount of voxels in a particular block, i.e. $|b_{xyz}^k|$. To investigate the varying size of potential patterns, we employed different block sizes $k$ from $\mathcal{K} \subset \{1, 2, \ldots, \frac{n+1}{2}\}$.

---

**Workflow 1:** General workflow for the extraction of ROIs

**Input:** $\mathcal{D}_{\mathcal{C}}$: a data set to train a binary classifier, $\mathcal{D}_{ROI}$: a data set to extract ROIs, $\delta$: a threshold value, $\mathcal{K}$: the set of block sizes

**Output:** an importance map of voxels $(\geq \delta)$

$\mathcal{C} = Train_{\mathcal{C}}(\mathcal{D}_{\mathcal{C}})$ (train binary classifier $\mathcal{C}$ using $\mathcal{D}_{\mathcal{C}}$)

$\forall(x, y, z) \in \mathbb{N}^{n \times n \times n}, \forall X \in \mathcal{D}_{ROI}$ calculate $I_{\mathcal{C},X}^{\mathcal{K}}(x, y, z)$

with $I_{\mathcal{C},X}^{\mathcal{K}}(x, y, z) = \frac{1}{|\mathcal{K}|} \sum_{k \in \mathcal{K}} I_{\mathcal{C},X}^k(x, y, z)$, $\mathcal{K} \subset \{1, 2, \ldots, \frac{n+1}{2}\}$

**return** $sort(filter(\forall(x, y, z) \in \mathbb{N}^{n \times n \times n}, \forall X \in \mathcal{D}_{ROI}\ I_{\mathcal{C},X}^{\mathcal{K}}(x, y, z) \geq \delta))$

---

Parts of the process of generating $I_{\mathcal{C},X}^{\mathcal{K}}$ (definition in Workflow 1) can exemplarily seen in Fig. 2.

# 3   Experiments and Results

First, we trained the initial binary classifier $\mathcal{C}$ as described above on $\mathcal{D}_{\mathcal{C}}$ using all scans with label NC and AD. The accuracy of a 10-fold cross validation was 89%, with $TPr = 0.85$, $TNr = 0.91$ and $AUC = 0.95$. To evaluate the benefit of using importance maps, we used two different settings on $\mathcal{D}_{ROI}$ using a $10 \times 10$-fold

---

[3] Different $\gamma$s are possible. Here, we used $\gamma_1 = \bar{v}_{xyz}$ (the average voxel of all scans in the data set) as well as $\gamma_2 = round\left(\frac{v_{xyz}}{max(X)}\right)$.

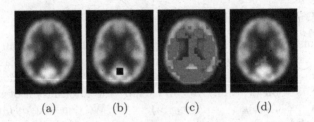

|        (a)         |        (b)         |        (c)         |        (d)         |

**Fig. 2.** The original scan $X$ is shown in (a). In (b), a block of voxels is excluded from $X$. Subfigure (c) shows all voxels of $\mathcal{I}_{\mathcal{C},X}^{\mathcal{K}}$ for a threshold of $\delta \geq 10^{-6}$. Lastly, (d) shows some of the 50 most informative voxels obtained by $\mathcal{I}_{\mathcal{C},X}^{\mathcal{K}}$.

**Table 2.** Performance measure obtained from different settings.

| Performance Measure | Li *et al.* | $S_{all}$ | $S_{\mathcal{ROI}}$ | | | |
|---|---|---|---|---|---|---|
| | | | $\delta = 0.001$ ($\#v = 3251$) | $\delta = 0.05$ ($\#v = 198$) | $\delta = 0.10$ ($\#v = 42$) | $\delta = 0.15$ ($\#v = 16$) |
| *Acc* (%) | 88.1 | $87.1_{\pm 2.5}$ | $87.0_{\pm 1.8}$ | $91.6_{\pm 2.9}$ | $85.6_{\pm 3.8}$ | $79.3_{\pm 2.7}$ |
| *TPr* | 0.91 | $0.84_{\pm 0.06}$ | $0.81_{\pm 0.05}$ | $0.88_{\pm 0.05}$ | $0.82_{\pm 0.06}$ | $0.73_{\pm 0.06}$ |
| *TNr* | 0.83 | $0.90_{\pm 0.02}$ | $0.92_{\pm 0.02}$ | $0.95_{\pm 0.02}$ | $0.89_{\pm 0.05}$ | $0.85_{\pm 0.04}$ |
| *AUC* | 0.97 | $0.94_{\pm 0.02}$ | $0.95_{\pm 0.02}$ | $0.99_{\pm 0.01}$ | $0.94_{\pm 0.02}$ | $0.92_{\pm 0.03}$ |

cross validation. Setting $S_{all}$ uses all available voxels, while $S_{\mathcal{ROI}}$ only employs voxels considered important according to varying $\delta$s. To only evaluate the use of individual voxels and not their spatial neighbours, we train linear support vector machines (SVM) based on individual voxels. A simple grid search for $C$ was performed to find optimal parameters. We used eight folds for training, one as validation set, and the final one to evaluate the performance using the optimal $C$. Furthermore, we employed block sizes $\mathcal{K} = \{1, 3, 5\}$ and $\gamma_2$. The averaged performance of $S_{all}$ and $S_{\mathcal{ROI}}$ for different $\delta$s is given in Table 2. The comparison to the works by Li *et al.* in the table is given as indication of state-of-the-art methods using the same data set $\mathcal{D}_{\mathcal{ROI}}$.

## 4   Conclusions and Future Work

We have presented an approach to extract regions of interest from PET scans based on binary classifiers. The extracted regions, or importance maps, can be employed as a voxel subset to differentiate between scans labeled NC and AD. The binary classifier employed here is based on a deep CNN architecture for the extraction of complex spatial patterns. In contrast to many deep learning approaches just aiming for the classification, our approach explicitly aims for ROIs, but is, at this point, evaluated only quantitatively. As a next step, we are going to evaluate the ROIs qualitatively together with our collaboration partners.

**Acknowledgments.** This work was partially supported by the Carl-Zeiss-Foundation Competence Center for High Performance Computing.

Data used in preparation of this article were obtained from the Alzheimer's Disease Neuroimaging Initiative (ADNI) database (adni.loni.usc.edu). As such, the investigators within the ADNI contributed to the design and implementation of ADNI and/or provided data but did not participate in analysis or writing of this report. A complete listing of ADNI investigators can be found at: http://adni.loni.usc.edu/wp-content/uploads/how_to_apply/ADNI_Acknowledgement_List.pdf.

# References

1. Li, R., Perneczky, R., Drzezga, A., Kramer, S.: Gaussian mixture models and model selection for [18F] fluorodeoxyglucose positron emission tomography classification in Alzheimer's disease. PLoS ONE **10**(4), e0122731 (2015)
2. Vieira, S., Pinaya, W.H., Mechelli, A.: Using deep learning to investigate the neuroimaging correlates of psychiatric and neurological disorders: Methods and applications. Neurosci. Biobehav. Rev. **74**, (Part A), 58–75 (2017)
3. Jagust, W.J., Landau, S.M., Koeppe, R.A., Reiman, E.M., Chen, K., Mathis, C.A., Price, J.C., Foster, N.L., Wang, A.Y.: The ADNI PET core: 2015. Alzheimers Dement. **11**(7), 757–771 (2015)

# Skin Hair Removal in Dermoscopic Images Using Soft Color Morphology

Pedro Bibiloni[✉], Manuel González-Hidalgo, and Sebastia Massanet

SCOPIA, Department of Mathematics and Computer Science,
University of the Balearic Islands, 07122 Palma, Spain
{p.bibiloni,manuel.gonzalez,s.massanet}@uib.es

**Abstract.** Dermoscopic images are useful tools towards the diagnosis and classification of skin lesions. One of the first steps to automatically study them is the reduction of noise, which includes bubbles caused by the immersion fluid and skin hair. In this work we provide an effective hair removal algorithm for dermoscopic imagery employing soft color morphology operators able to cope with color images. Our hair removal filter is essentially composed of a morphological curvilinear object detector and a morphological-based inpainting algorithm. Our work is aimed at fulfilling two goals. First, to provide a successful yet efficient hair removal algorithm using the soft color morphology operators. Second, to compare it with other state-of-the-art algorithms and exhibit the good results of our approach, which maintains lesion's features.

**Keywords:** Dermoscopy · Hair removal · Soft color morphology · Black top-hat · Curvilinear objects · Inpainting

## 1 Introduction

Malignant melanoma, or simply melanoma, is the most dangerous form of skin cancer. Although it only represents 4% of all skin cancers, it causes the 80% of deaths related to skin cancer. In Spain, approximately 5000 new patients are diagnosed each year, which is a quantity increasing around 7% yearly [5].

Dermoscopic imagery is a non-invasive and effective tool towards the clinical diagnosis of skin lesions, both for superficial spreading ones but also in the case of vertically growing melanomas [1]. Such images are acquired with a dermatoscope, which is a camera specially designed to capture small regions and to avoid the reflection of light on the skin surface. The former is achieved with appropriate optics, whereas the latter is typically attained with polarized or non-polarized light and an immersion fluid (e.g. alcohols or mineral oil). Computer-aided image analysis technologies can be leveraged to design diagnosis support tools and thus help practitioners. In this way, a variety of techniques have been proposed to remove hair form dermoscopic images. One of such algorithms is Dullrazor [8] that uses top hat transforms and bilinear interpolation to remove hair. Another state-of-the-art algorithm is presented in [11] in which Canny edge detector

© Springer International Publishing AG 2017
A. ten Teije et al. (Eds.): AIME 2017, LNAI 10259, pp. 322–326, 2017.
DOI: 10.1007/978-3-319-59758-4_37

jointly with coherence transport inpainting are used. In this paper, we present an effective hair removal algorithm employing soft color morphology operators. This novel algorithm is compared with the aforementioned two algorithms exhibiting good results, removing hair effectively while maintaining lesion's features.

## 2    Soft Color Morphology

Mathematical morphology has been widely used since its introduction by Serra and Matheron due to their satisfactory trade-off among expressive power and efficiency [10]. Initially designed to deal with binary images, it was promptly extended to grayscale images. Along with the CIELab color space, the soft color morphology operators are designed to be coherent with human color perception and to avoid generating colors whose chroma is different to that of any color already present in the image. Due to its nature, these operators process the image as a whole rather than as a number of independent channels.

Our operators are designed using fuzzy logic operators. In particular, they employ conjunctions [3] and fuzzy implication functions [2]. They also use a structuring element $B$, which can be represented as a grayscale image. In this work, we always consider that, in the origin $B(0)$, its value is always 1.

We can now formally introduce the basic operators of the *Soft Color Morphology*. They are the soft color dilation and the soft color erosion:

**Definition 1.** *Let $C$ be a conjunction, let $I$ be a fuzzy implication function, let $A$ be a multivariate image and let $B$ be a structuring element. Then, the* soft color dilation *of $A$ by $B$, $\mathcal{D}_C(A, B)$, is*

$$\mathcal{D}_C(A, B)(y) = \Big( C\big(B(x - y), A_1(x)\big), A_2(x), \ldots, A_m(x) \Big) \ s.t.$$
$$x \in d_A \cap T_y(d_B) \ and \ C\big(B(x - y), A_1(x)\big) \ is \ maximum, \quad (1)$$

*and the* soft color erosion *of $A$ by $B$, $\mathcal{E}_I(A, B)$, is*

$$\mathcal{E}_I(A, B)(y) = \Big( I\big(B(x - y), A_1(x)\big), A_2(x), \ldots, A_m(x) \Big) \ s.t.$$
$$x \in d_A \cap T_y(d_B) \ and \ I\big(B(x - y), A_1(x)\big) \ is \ minimum. \quad (2)$$

*The maximum or minimum may not be unique, and so ties are resolved by choosing the candidate with the nearest location (i.e. the $x$ such that the Euclidean distance $d(x, y)$ is minimum), and resolve any further ties with the lexicographical order.*

Given these two definitions, we define the closing, opening and black top-hat with the straightforward generalization of the same operators for binary images [10], where the difference has been generalized to the Euclidean distance between colors.

These mathematical morphology operators generalize the ones of the fuzzy mathematical morphology [7]. Besides, these operators preserve the chromatic components of L*a*b*-encoded images, and also preserve colors in any color space when, essentially, the structuring element is a binary image [4].

## 3    Hair Removal Algorithm

In this section, we introduce the hair removal algorithm and its main steps: a curvilinear object detector for color images and an inpainting algorithm also for color images.

The curvilinear object detector is based on a combination of soft color top-hat transforms. We leverage the appearance of hairs in dermoscopic images by fine-tuning the orientations and size of the structuring elements. Hairs appear as thin, elongated regions with clearly differentiated photometric features than those of its local background, which is non-uniform. More specifically, they are darker than the background, with possible brighter surroundings due to noise in acquisition and to lossy image formats. Besides, their width usually ranges between 4 and 9 pixels and the ratio of hair pixels can range from 0% to almost 40%. The black top-hat, or top-hat by closing, extracts exactly those objects.

The detector $\mathcal{CD}$, visually depicted in Fig. 1, is defined as:

$$\mathcal{CD}(A) = \max_{\alpha} \left\{ \mathcal{BTH}_{C,I}(A, B_\alpha) \right\} - \min_{\alpha} \left\{ \mathcal{BTH}_{C,I}(A, B_\alpha) \right\} \qquad (3)$$

where the parameters used are the minimum operator as conjunction $C(x, y) = \min(x, y)$; and its residuated implication, the Gödel implication $I = I_{\mathbf{GD}}$ [2], and $\alpha \in \{0°, 22.5°, 45° \ldots, 157.5°\}$ ranges over different orientations of the structuring element $B_\alpha$. All of the structuring elements $B_\alpha$ are bar-like shapes with a Gaussian decayment enclosed in a squared region. We remark that the maximum of the transforms

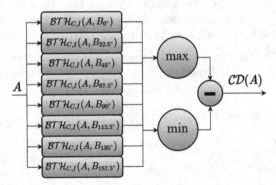

**Fig. 1.** Workflow of the curvilinear object detector.

recovers both isolated regions and thin regions, whereas the minimum only recovers isolated areas. We consider structuring elements at 8 different orientations enclosed in $9 \times 9$ regions.

The inpainting method for color images based on the soft color morphology operators is inspired in the grayscale filter presented in [6]. The inpainting procedure rewrites the pixels that are *missing* (denoted by the symbol $\perp$) as soon as enough information is available, while always maintains pixels whose value is already known. More formally, the iterative inpainting of an image $A_1$ is defined as the limiting case of the following series:

$$A_{n+1}(x) = \begin{cases} A_n(x), & \text{if } A_n(x) \neq \perp, \\ (\mathcal{O}_{T_{\mathbf{M}}, I_{\mathbf{GD}}}(A_n, B_5) + \mathcal{C}_{T_{\mathbf{M}}, I_{\mathbf{GD}}}(A_n, B_5))/2, & \text{otherwise,} \end{cases}$$

where $B_5$ is a $5 \times 5$, flat, rounded structuring element. We also remark that the dilation and erosion from Definition 1 are slightly modified to handle images with *missing* pixels: they ignore them and output *missing* if there is possible candidate.

Once the two main building blocks have been introduced, the complete hair removal algorithm is presented. The course of the algorithm, composed by a series of sequential operations, is presented as follows. The image is converted into L*a*b* and is divided by 100. The channel L* is preprocessed with the CLAHE [12], which increases the contrast with a histogram-based equalization. Then, it is used as input for the curvilinear detector. To postprocess it, the curvilinear mask is smoothed with a $9 \times 9$ median filter, then binarized by a fixed threshold ($t = 0.1$), and finally enlarged two pixels in each direction. In the L*a*b* image, we replace by $\perp$ the colors in the locations indicated by the mask. Missing pixels are inpainted, providing the hairless image.

It is designed for middle-sized dermoscopic images (approximately $600 \times 600$ pixels) with different pathologies, like the ones in the PH$^2$ dataset [9].

## 4   Experiments and Conclusions

In Fig. 2 we compare the results of the Dullrazor® algorithm [8], the algorithm by Toossi *et al.* [11] and the proposed algorithm.

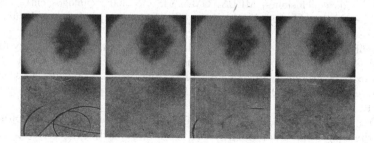

**Fig. 2.** Original (left), Dullrazor® by Lee *et al.* [8] (center left), Toossi *et al.* [11] (center right) and our algorithm (right) of sample IMD047 (top) and a detail of sample IMD101 (bottom) of the PH$^2$ dataset [9].

Our algorithm effectively detects and inpaints almost all hair, while hairless images remain almost unchanged, and it maintains the features of the lesion. The shape and size are never affected due to the behaviour of the closing and opening when inpainting. The color is not affected either: colors with new chroma can not appear due to the nature of the soft color morphological operations. The texture of the lesion, on the other hand, is maintained in the majority of situations. In contrast with other procedures that employ smoothing filters or averages, our inpainting procedure does not tend to blur lesions: it recovers uniform regions correctly, and creates smooth but small transitions when inpainting the missing contours of different regions. The method by Toossi *et al.* leaves a considerable amount of hair and does not respect textures in general. Dullrazor®, on the other side, removes a fair amount of hair. However, it fails at recognizing hair within the lesion, and does not always preserve textures.

**Acknowledgments.** The Spanish grants TIN 2016-75404-P AEI/FEDER, UE and TIN 2013-42795-P partially supported this work. P. Bibiloni also benefited from the fellowship FPI/1645/2014 of the *Conselleria d'Educació, Cultura i Universitats* of the *Govern de les Illes Balears* under an operational program co-financed by the European Social Fund.

# References

1. Argenziano, G., Longo, C., Cameron, A., Cavicchini, S., et al.: Blue-black rule: a simple dermoscopic clue to recognize pigmented nodular melanoma. Br. J. Dermatol. **165**(6), 1251–1255 (2011)
2. Baczyński, M., Jayaram, B.: Fuzzy Implications. Studies in Fuzziness and Soft Computing, vol. 231. Springer, Heidelberg (2008)
3. Beliakov, G., Pradera, A., Calvo, T.: Aggregation Functions: A Guide for Practitioners, vol. 221. Springer, Heidelberg (2007)
4. Bibiloni, P., González-Hidalgo, M., Massanet, S.: Soft color morphology. In: Submitted to IEEE International Conference on Fuzzy Systems (FUZZ-IEEE 2017) (2017)
5. Informe de conclusiones. MELANOMA VISIÓN 360°: Diálogos entre pacientes y profesionales. Madrid (2015). http://fundacionmasqueideas.org/documentos/. Accessed 20 July 2016
6. González-Hidalgo, M., Massanet, S., Mir, A., Ruiz-Aguilera, D.: A fuzzy filter for high-density salt and pepper noise removal. In: Bielza, C., Salmerón, A., Alonso-Betanzos, A., Hidalgo, J.I., Martínez, L., Troncoso, A., Corchado, E., Corchado, J.M. (eds.) CAEPIA 2013. LNCS, vol. 8109, pp. 70–79. Springer, Heidelberg (2013). doi:10.1007/978-3-642-40643-0_8
7. Kerre, E.E., Nachtegael, M.: Fuzzy Techniques in Image Processing. Studies in Fuzziness and Soft Computing, vol. 52. Physica, Heidelberg (2013)
8. Lee, T., Ng, V., Gallagher, R., Coldman, A., McLean, D.: Dullrazor®: a software approach to hair removal from images. Comput. Biol. Med. **27**(6), 533–543 (1997)
9. Mendonça, T., Ferreira, P.M., Marques, J.S., Marcal, A.R., Rozeira, J.: PH 2 - a dermoscopic image database for research and benchmarking. In: 2013 35th Annual International Conference of the IEEE Engineering in Medicine and Biology Society (EMBC), pp. 5437–5440. IEEE (2013)
10. Serra, J.: Image Analysis and Mathematical Morphology, vol. 1. Academic Press, Cambridge (1982)
11. Toossi, M.T.B., Pourreza, H.R., Zare, H., Sigari, M.H., et al.: An effective hair removal algorithm for dermoscopy images. Skin Res. Technol. **19**(3), 230–235 (2013)
12. Zuiderveld, K.: Contrast limited adaptive histogram equalization. In: Graphics GEMS IV, pp. 474–485. Academic Press Professional, Inc. (1994)

# Risk Mediation in Association Rules

## The Case of Decision Support in Medication Review

Michiel C. Meulendijk[1]([⊠]), Marco R. Spruit[2],
and Sjaak Brinkkemper[2]

[1] Department of Public Health and Primary Care,
Leiden University Medical Center, Leiden, The Netherlands
m.c.meulendijk@lumc.nl
[2] Department of Information and Computing Sciences, Utrecht University,
Utrecht, The Netherlands
{m.r.spruit,s.brinkkemper}@uu.nl

**Abstract.** We propose a model for the incorporation of risk in association rule application. We validate this model using data gathered in a randomized controlled trial from a recommender system for medication reviews in primary care. The model's outcomes are found to have predictive value when tested against decisions made by physicians on 261 patients' health records.

**Keywords:** Association rules · Decision support · Risk management

## 1 Introduction

Association rule mining remains one of the most prominent knowledge discovery methods in existence. Association rules have been implemented in countless software applications in a wide variety of domains, including the domain of primary care [1].

The typical example of association rule implementation, the 'market basket', reveals one of its weaknesses: grocery items that are frequently purchased together form an association rule, which is used for personalized adverts. In a safe, insensitive domain such as shopping this poses no problems, but in a precarious domain with vulnerable datasets, where association rules can have potentially far-reaching implications, these kinds of associative suggestions are risky.

Therefore, in this paper we propose a model for the incorporation of risk in association rules. We implement and validate it in a primary care setting.

## 2 Background

### 2.1 Recommender Systems in Precarious Domains

The impact association rules may have depends on the sensitivity of the dataset on which they are applied. This danger involved in applying discovered association rules to sensitive datasets in precarious domains has not received substantial attention. For the remainder of this paper, we define *precarious domains* as 'domains in which

© Springer International Publishing AG 2017
A. ten Teije et al. (Eds.): AIME 2017, LNAI 10259, pp. 327–331, 2017.
DOI: 10.1007/978-3-319-59758-4_38

association rules' consequences have potentially major impact on its datasets'. An example of a precarious domain is that of primary care.

## 2.2   Risk as a Post-mining Metric

Risk management involves a set of coordinated activities to direct and control an organization with regard to risk. Risk is typically understood as the probability of loss in any given situation, or, in a more generic definition, the effect of uncertainty on objectives. In risk management, risk is usually incorporated as a function of these uncertain effects' impact and their likelihood of occurring.

Association rules and risk share the concept of probability; association rules are expressed with a degree of confidence, while risk incorporates the likelihood of dangerous consequences. While association rules' impact is not a standardized concept, the implications their consequences have imply its risk's severity (Table 1).

**Table 1.**  Relations between association rules' characteristics and risk.

*Risk Probability*

|  |  | High | Low |  |  |
|---|---|---|---|---|---|
| *Rule* | High | ✗ | ? | High | *Risk* |
| *Impact* | Low | ? | ✓ | Low | *Severity* |
|  |  | Low | High |  |  |

*Rule Confidence*

## 3   Risk Model Formulation

Recommender systems, whether powered by explicit knowledge bases or implicit content-based or collaborative-based filtering, depend on inference rules. Inference rules are logical functions which analyze premises and, based on their syntax, return conclusions. Recommender systems depend on datasets containing all relevant items to trigger their inference rules. In propositional logic inference rules can be written as $x \rightarrow y$, with a dataset $D = \{d_1, \ldots, d_n\}$ and $x \in D$. Thus, for a specific rule dataset $D$ contains its premises, along with other items, but never its consequence.

The risk associated with a rule is a function of its unwanted consequences and their likelihood of occurring. The formula to determine the risk of an inference rule $x \rightarrow y$ reads:

$$risk(x \rightarrow y) = (1 - probability(x \rightarrow y)) * \sum_{i=D,y} severity(i) \qquad (1)$$

The probability of the consequences being unacceptable is one minus the confidence with which the rule is accepted. In the case of association rules that have been discovered using common algorithms, the probability in the formula is equal to its confidence.

The severity of a rule's risk is the sum of the impact of the objects associated with it. This comprises not just the danger associated with the inference's consequence $y$, but also detrimental characteristics of items in dataset $D$; it may be riskier to perform a certain action on a vulnerable dataset than on a safer one. The danger associated with these items can be estimated through risk formulas as well. As such, when implemented in a domain, the final formula results in the sum of its associated objects' risks, multiplied with its rule's inverse probability.

## 4 Implementation

The above-mentioned formulae are demonstrated and validated by applying them to the STRIP Assistant (STRIPA), a recommender system for medication reviews in primary care [1]. STRIPA's hybrid rule base consists of guidelines and inference rules acquired through association rule mining, and has been shown to be effective [2].

### 4.1 Health Records' Risk

The dataset in STRIPA consists of a patient's health record. The items in this dataset, such as diseases or drugs, serve as premises for the system's inference rules. All inference rules modify one or more drugs by prescribing new medicines, removing existing ones, or adjusting their dosages. A dataset $D$ comprising a certain patient's diseases, drugs, contra-indications, allergies, and measurements can be described as a set:

$$D = \left\{ \begin{array}{c} disease_1, \ldots, disease_k; \; drug_1, \ldots, drug_l; \\ contraindication_1, \ldots, contraindication_m; \\ measurement_1, \ldots, measurement_n; \\ allergy_1, \ldots, allergy_p \end{array} \right\} \tag{2}$$

Following $x \in D$ and the fact that all inference rules adjust drugs, the implemented risk formula reads:

$$risk(x \rightarrow drug) = (1 - probability(x \rightarrow drug)) * (severity(D) + severity(drug)) \tag{3}$$

### 4.2 Health Records' Severity

Determining the risk associated with a patient's health record, or dataset, involves taking into account a multitude of domain-dependent variables, such as his or her age, frailty, physical properties, and cognitive state. This results in the severity of a dataset $D$ being the sum of its domain-relevant, patient-specific, risk factors:

$$severity(D) = \sum_{riskFactor \in D} riskFactor \tag{4}$$

The Dutch multidisciplinary guideline for polypharmacy in elderly people[1] proposes seven factors that increase patients' risk of harm due to inappropriate drug use: age (over 65 years old), polypharmacy (using four or more drugs simultaneously), impaired renal function, impaired cognition, frequent falling, decreased therapy adherence, and living in a nursing home. In this study's dataset, four of these risk factors were available: age, polypharmacy, impaired renal function, and impaired cognition.

## 4.3   Drug Severity

A drug's severity can be expressed as a function of its dose, or toxicity, and its adverse effects, or harm:

$$severity(drug) = toxicity(drug) * harm(drug) \qquad (5)$$

An adverse effect is a response to a drug that is noxious and unintended and occurs at normal doses. These effects are usually classified in terms of their likelihood of occurring. Their frequency and potential impact are used to determine drugs' safety for prescribing. As such, the function of the number of adverse effects a drug has and their frequency can be useful to determine a substance's potential harm:

$$harm(drug) = \sum_{e \in E} e.frequency \qquad (6)$$

The sets of adverse effects, defined per active substance, were retrieved from a database maintained by the Royal Dutch Pharmacists Association[2]. Adverse effects are classified due to their likelihood of occurring: often (over 30% of patients or more), sometimes (10–30% of patients), rarely (1–10% of patients), and very rarely (less than 1% of patients). As such, each drug's adverse effects can be described as a set $E = \{e_1, \ldots, e_n\}$, where $e_i = (id, freqency)$.

The definition provided for adverse effects also takes into account the prescribed dosage, something not accounted for in the formula above. Research has shown that the probability of adverse effects generally increases with higher dosages being used. For each drug, the World Health Organization has defined an average strength with which it is typically prescribed. This Defined Daily Dose[3] (DDD) is the assumed average maintenance dose per day for a drug used for its main indication in adults. Dividing a patient's actual daily dosage of a drug by the DDD provides a factor that can be used to relativize the drug's risk. Its toxicity can thus be calculated as such:

$$toxicity(drug) = \frac{prescribedDailyDose(drug)}{definedDailyDose(drug)} \qquad (7)$$

---

[1] https://www.nhg.org/themas/publicaties/multidisciplinaire-richtlijn-polyfarmacie-bij-ouderen.

[2] http://www.apotheek.nl/medicijnen.

[3] http://www.whocc.no/ddd/definition_and_general_considera/.

# 5   Validation

The risk model was validated by comparing its predictions with actual actions taken by experts. STRIPA was used on real patient cases by dedicated teams of GPs and pharmacists for the duration of a year as part of a randomized controlled trial, which was performed in 25 general practices in Amsterdam, the Netherlands and included 500 patients [3]. For the 261 of these patients that were placed in the intervention arm, four teams consisting of one GP and one pharmacist each used the software to optimize their medical records. The users would respond to patient-specific advice, recommending them to prescribe new drugs for particular diseases. Their responses to advices were gathered; each time a suggestion was heeded or ignored, the instance, along with relevant patient case information, was logged. A total of 776 responses to advices, of which 311 were heeded, has been gathered and will be used to validate the risk model.

Assuming that users will strive for the minimization of risk, we hypothesize that users will have chosen the least risky option whenever possible. Based on the assumption that riskier patients – i.e. patients who have multiple risk factors in their dataset (or health record) – are best served with as little change to their drug regimen as possible, it was hypothesized that the higher an action's risk was, the least likely it was to be performed by users. To test this, the risk factors of each generated recommendation were calculated; its proposed drug's risk and the relevant patient's risk factors were summed according to the introduced model. An independent t-test affirmed the hypothesis, showing a statistical difference in the risk associated with proposed actions which were followed (M = 2.42, SD = 0.57) and the risk of proposed actions which were not followed (M = 2.57, SD = 0.60); t(623) = 3.040, p = .002.

# 6   Conclusion

In this study, we explored the potential usefulness of the concept of risk in association rules. Validation shows that the risk model has predictive power in the domain of medication reviews. Application examples can be found in a technical report [4].

We would like to thank Pieter Meulendijk for his contributions to the conceptual model following his expertise in risk management.

# References

1. Meulendijk, M., et al.: STRIPA: A Rule-Based Decision Support System for Medication Reviews in Primary Care. In: Proceedings of ECIS 2015, Münster, Germany (2015)
2. Meulendijk, M., et al.: Computerized decision support improves medication review effectiveness: an experiment evaluating the STRIP Assistant's usability. Drugs Aging 32(6), 495–503 (2015)
3. Meulendijk, M., et al.: Efficiency of clinical decision support systems improves with experience. J. Med. Syst. 40(4), 76–82 (2016)
4. Meulendijk, M., Spruit, M., Brinkkemper, S.: Risk Mediation in Association Rules: Application Examples, Utrecht University, UU-CS-2017-004 (2017)

# Identifying Parkinson's Patients:
# A Functional Gradient Boosting Approach

Devendra Singh Dhami[1(✉)], Ameet Soni[2], David Page[3],
and Sriraam Natarajan[1]

[1] Indiana University Bloomington, Bloomington, USA
ddhami@indiana.edu
[2] Swarthmore College, Swarthmore, USA
[3] University of Wisconsin-Madison, Madison, USA

**Abstract.** Parkinson's, a progressive neural disorder, is difficult to identify due to the hidden nature of the symptoms associated. We present a
machine learning approach that uses a definite set of features obtained
from the Parkinson's Progression Markers Initiative (PPMI) study as
input and classifies them into one of two classes: PD (Parkinson's disease) and HC (Healthy Control). As far as we know this is the first
work in applying machine learning algorithms for classifying patients
with Parkinson's disease with the involvement of domain expert during the feature selection process. We evaluate our approach on 1194
patients acquired from Parkinson's Progression Markers Initiative and
show that it achieves a state-of-the-art performance with minimal feature engineering.

**Keywords:** Functional gradient boosting · Parkinson's · Human advice

## 1 Introduction

We consider the problem of predicting the incidence of Parkison's disease, a progressive neural disorder. Specifically, we consider the data collected as part of the
Parkisons Progression Marker Initiative (PPMI) and aim to predict if a subject
has Parkinson's based on clinical data - particularly, motor assessments (motor
functions) and non-motor assessments (neurobehavioral and neuropsychological
tests). One of the most important challenges for this task is that there seem to
be no real strong indicator that explains the progression clearly [1].

Our hypothesis, which we verify empirically is that instead of considering a
small set of strongly influencing risk factors, it might be more effective to consider
a large set of weakly influencing risk factors. To this effect, we adapt the recently
successful gradient-boosting algorithm [2] for this task. We exploit the use of a
domain expert in identifying the right set of features and consider learning from
the longitudinal data. Unlike standard methods that require projecting the data
to a feature vector format (using what are called aggregation or propositional-
ization methods), our proposed approach models the data faithfully using time

A. ten Teije et al. (Eds.): AIME 2017, LNAI 10259, pp. 332–337, 2017.
DOI: 10.1007/978-3-319-59758-4_39

as a parameter of a logical representation. We evaluate our proposed approach on ≈1200 patients from the PPMI study and demonstrate the effectiveness and efficiency of the proposed approach.

We make the following key contributions: we consider the challenging task of predicting Parkinson's from 37 different features. These features were chosen by interacting with a domain expert. The key advantage of this approach is that it models the underlying data faithfully by utilizing logic based framework. Evaluation on the PPMI dataset shows that our approach is superior to standard classification approaches.

## 2  Background

Parkinson's Progression Markers Initiative (PPMI) is an observational study with the main aim of identifying features or biomarkers that impact Parkinson's disease progression [1]. The collected data can be divided broadly into four distinct categories: Imaging data, Clinical data, Biospecimens and Subject demographic data. We focus primarily on the clinical data which mainly consists of motor assessments and non-motor assessments. Since Parkinson's mainly affect the motor system (i.e. the part of the nervous system associated with movement) and the initial symptoms are mostly movement related, using motor assessment data seems a natural approach. A subset of the clinical data is selected based on the expert input after which a set of 37 features are obtained which can be broadly defined in these categories: (1) Motor-UPDRS (consists of 34 features) [4], (2) Montreal Cognitive Assessment (MoCA), a non-motor assessment that contains the MoCA score of a patient for a single visit and the difference in MoCA scores of the patient from the last visit, (3) total UPDRS score.

MoCA consists of series of questions assessing various parameters of a subject such as the visual capability, capacity of recognizing objects, the attention span and memorizing words to name a few. Each of the questions are scored with the total score being 30. A subject with score of ≥26 is considered to be normal. Unified Parkinson Disease Rating Scale (UPDRS) is a rating scale used for determining the extent of Parkinson's disease progression in a subject and each assessment ranges from 0 to 4, with 0 being normal behavior and 4 representing severe abnormal behavior with respect to the assessment. Motor-UPDRS refers to the value of motor assessments in the UPDRS scale. Total UPDRS score refers to the sum of all the motor-UPDRS features.

Our goal is to estimate the conditional distribution - $P(par|\mathbf{x})$ where $\mathbf{x}$ represents the set of motor and non-motor assessments (i.e., features) and $par$ denotes the incidence of Parkinson's disease for the particular patient. One could apply any machine learning algorithm for learning this distribution. We focus on the gradient-boosting technique which has had success in medical applications [6]. Gradient-boosting is a gradient-ascent technique performed on the functional space. For probabilistic models, gradient-boosting represents the conditional distributions using a functional representation, for instance a *sigmoid* function. Then the gradients are calculated w.r.t to this function. For instance, one could represent $P(par|\mathbf{x}) = \frac{e^{\psi(y|\mathbf{x})}}{1+e^{\psi(y|\mathbf{x})}}$. Friedman [2] suggested that instead of obtaining the gradients w.r.t the global function $\psi$, one could obtain the gradient for each example $\langle \mathbf{x}_i, y_i \rangle$, where $y$ denotes the target i.e., presence of Parkinson's. The key idea is that these gradients are good approximations of the overall gradients because the direction of these gradients point to the same direction as that of the true gradient. This functional gradient of likelihood w.r.t. $\psi$ for each example $i$ is $\frac{\partial log(P(y_i|\mathbf{x}_i))}{\partial \psi(y_i|\mathbf{x}_i)} = I(y_i = 1) - P((y_i = 1|\mathbf{x}_i)$ where I is the indicator function (equal to 1 if $y_i = 1$ and equal to 0 if $y_i = 0$). The gradient is the difference between the true label and the current prediction and is positive for positive examples and negative for negative examples. In simpler terms, the negative examples are pushed towards 0 and the positive examples are pushed to 1 resulting in a well defined decision boundary. The $\psi$ function is represented by relational regression trees (RRT) [3] which uses the relational features as input to the trees that can be easily guided by the domain expert who could provide preference information.

## 3    Proposed Approach

Since the aim of the PPMI study is to determine the most predictive features for the disease progression, the involvement of a domain expert during the feature selection process becomes beneficial. Our specific aim is to predict the incidence of Parkinson's in a patient. Since the data is longitudinal in nature, it is important that we model time faithfully. We view time as a special type of relation and hence we create features in the predicate logic format as *feature_name (patient id, time, feature value)*. This allows the number of days since the start of the study (time) to be an argument of the predicate.

The first step of the process starts with the correlation analysis of the raw data obtained. The raw clinical data consists of 81 features. A correlation matrix is constructed with each entry representing the correlation coefficient (we use the *Pearson correlation coefficient*) between each variable and the others. The 50 features with low mutual correlation and high correlation to the class label are selected. The second step consists of the expert evaluating the obtained features giving us a further pruned set of 37 features and thus the final data to be used for the classification task. The key reason for considering a relational representation is two fold: First, relational models allow learning using the natural representation of the data and second, data is longitudinal i.e. a patient has multiple entries in the data. Propositional classifiers have limitations in learning such data (require aggregators).

The learner is provided with the training data which it uses to learn a relational regression tree. The last step is the prediction phase where the learned model can be queried to return the probability of the target being true given the evidence. Since all the evidence is observed, inference requires simply querying all the relational regression trees, summing up their regression values and returning the posterior estimates i.e. the probability that the given test example belongs to the positive class.

**Interpretability:** One key limitation of the proposed approach is the interpretability of the final model. While each boosted tree in itself is interpretable, given that they are not learned independently of each other makes the model difficult to interpret. To make the model comprehensible, we take an approximate approach that we call the Craven approach [5] which was originally developed for making neural networks interpretable. The key idea is to relabel the training data based on the boosted model that we have learned and then train an overfitted tree to this labeled data. The intuition is that this new large tree will represent the decisions made by the original set of trees due to its performance on the training data. Recall that our original training data consists of Boolean labels (Parkinson's vs negative). But the relabeled data consists of regression values that are being learned in the new tree. Hence, the resulting tree is closer to the original learned model as we show in our experiments.

**Table 1.** Results for BoostPark with and without expert advice

| Classifier | With expert | | | | Without expert | | | |
|---|---|---|---|---|---|---|---|---|
| | Accuracy | AUC-ROC | AUC-PR | F1 score | Accuracy | AUC-ROC | AUC-PR | F1 score |
| BoostPark10 | 0.889 | 0.973 | 0.937 | 0.808 | 0.854 | 0.932 | 0.9 | 0.797 |
| BoostPark20 | 0.901 | 0.977 | 0.947 | 0.851 | 0.881 | 0.94 | 0.87 | 0.832 |

## 4   Experiments

In our empirical evaluations, we aim to explicitly ask the following questions:

**Q1:** How effective is the feature selection with expert in predicting Parkinson's?
**Q2:** Given the longitudinal nature of the data, is our method more effective than using standard classifiers in this prediction task?

Table 1 shows the result of learning from 50 features obtained after correlation and 37 features after the expert advice. This helps us in answering **Q1** affirmatively. Across all scoring metrics, expertly selected features outperforms models built using the larger feature set.

We compare our method, BoostPark, to three propositional classifiers: Logistic Regression, Gradient-Boosting and Support Vector Machines. The propositional data is aggregated using three aggregator functions: min, max and mean over time. Our data consists of records for 1194 patients, with 378 positive examples (i.e. Parkinson's patients) and 816 negative examples. Regression trees are learned on the given data which form the training model. We perform 10-fold cross validation and present the results.

**Table 2.** Classifier results. Only the best classifiers among the aggregators are shown.

| Classifier | Accuracy | AUC-ROC | AUC-PR | F1 score |
|---|---|---|---|---|
| BoostPark10 (10 relational regression trees) | 0.889 | **0.973** | **0.937** | 0.808 |
| BoostPark20 (20 relational regression trees) | 0.901 | **0.977** | **0.947** | 0.851 |
| Gradient Boosting (aggregator = mean) | **0.920** | 0.914 | 0.904 | **0.885** |
| Logistic Regression (aggregator = max) | 0.903 | 0.896 | 0.884 | 0.862 |
| Support Vector Machine (kernel = linear, aggregator = max) | 0.897 | 0.892 | 0.877 | 0.855 |
| Support Vector Machine (kernel = polynomial, aggregator = max) | 0.895 | 0.883 | 0.875 | 0.848 |
| Support Vector Machine (kernel = RBF, aggregator = mean) | 0.757 | 0.806 | 0.784 | 0.734 |

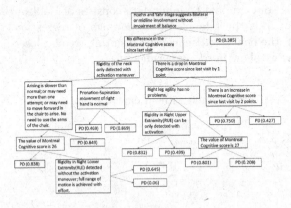

**Fig. 1.** Combined tree learnt with BoostPark10.

Since we aggregate the propositional data using 3 aggregators, the best performing aggregator for all the propositional classifiers is selected and compared to our methods BoostPark10 and BoostPark20 as shown in Table 2. Our methods perform considerably better than the propositional counterparts in terms of AUC-ROC and AUC-PR and performs equally well in terms of accuracy. This helps answer **Q2** positively. Our method is more effective than the standard classifiers in this prediction task (Fig. 1).

## 5    Conclusion

Identifying important features responsible for the progression in Parkinson's disease in a patient remains a compelling challenge. We use a human domain expert to guide our method with identifying a relatively large set of influencing risk factors. We then present a learning method that can consider this large set of weak influences in learning a probabilistic model. We evaluated our results on the PPMI data and demonstrated that our learning approach outperforms standard machine learning classifiers. Since Parkinson's is a progressive disease, developing robust temporal models for this task remains an interesting challenge. Extending our learning algorithm to handle hybrid temporal models will allow for modeling the continuous data more faithfully. Finally, scaling up the learning algorithm to learn from a broader set of population data rather than be limited by a single study remains an interesting open problem.

## References

1. Marek, K., et al.: The parkinson progression marker initiative (PPMI). Prog. Neurobiol. **95**(4), 629–635 (2011)

2. Friedman, J.H.: Greedy function approximation: a gradient boosting machine. Ann. Stat. **29**, 1189–1232 (2001)
3. Blockeel, H., De Raedt, L.: Top-down induction of first-order logical decision trees. Artif. Intell. **101**(1), 285–297 (1998)
4. Goetz, C.G., et al.: Movement Disorder Society sponsored revision of the Unified Parkinson's Disease Rating Scale (MDSUPDRS): process, format, and clinimetric testing plan. Mov. Disord. **22**(1), 41–47 (2007)
5. Craven, M.W., Jude, W.S.: Extracting tree-structured representations of trained networks. In: Advances in Neural Information Processing Systems (1996)
6. Natarajan, S., et al.: Early prediction of coronary artery calcification levels using machine learning. In: IAAI (2013)

# Echo State Networks as Novel Approach for Low-Cost Myoelectric Control

Cosima Prahm[1(✉)], Alexander Schulz[2], Benjamin Paaßen[2], Oskar Aszmann[1], Barbara Hammer[2], and Georg Dorffner[1]

[1] Medical University of Vienna, Vienna, Austria
cosima.prahm@meduniwien.ac.at
[2] Bielefeld University, Bielefeld, Germany
aschulz@techfak.uni-bielefeld.de

**Abstract.** Myoelectric signals (EMG) provide an intuitive and rapid interface for controlling technical devices, in particular bionic arm prostheses. However, inferring the intended movement from a surface EMG recording is a non-trivial pattern recognition task, especially if the data stems from low-cost sensors. At the same time, overly complex models are prohibited by strict speed, data parsimony and robustness requirements. As a compromise between high accuracy and strict requirements we propose to apply Echo State Networks (ESNs), which extend standard linear regression with (1) a memory and (2) nonlinearity. Results show that both features, memory and nonlinearity, independently as well as in conjunction, improve the prediction accuracy on simultaneous movements in two degrees of freedom (hand opening/closing and pronation/supination) recorded from four able-bodied participants using a low-cost 8-electrode-array. However, it was also shown that the model is still not sufficiently resistant to external disturbances such as electrode shift.

## 1 Introduction

Robotic arm prostheses support upper limb amputees in everyday life tasks [3]. Such prostheses are controlled through myoelectric signals derived from the patient's muscles in their residual limb. However, commercially established products are restricted to activating only a single degree of freedom (DoF) at a time and require tiresome mode-switching to execute movements in multiple DoFs [3]. More advanced prosthetic control systems acquire the user's muscle signals via an array of surface electrodes, infer the intended movement as well as the intended intensity via machine learning models and translate it to more natural prosthetic movements [7,9,16]. However, the requirements imposed on such a machine learning model are high: It should be complex enough to provide accurate prediction, but also require few patient training data, few electrodes, be

---

Funding by the DFG under grant number HA 2719/6-2, the CITEC center of excellence (EXC 277), and the Christian Doppler Research Foundation of the Austrian Federal Ministry of Science, Research and Economy is gratefully acknowledged.

A. ten Teije et al. (Eds.): AIME 2017, LNAI 10259, pp. 338–342, 2017.
DOI: 10.1007/978-3-319-59758-4_40

robust to outside disturbances and provide real-time predictions with low delay [3]. Especially electrode shift has been found to be a limiting factor in reliable prosthetic control, as it occurs every time the device is re-attached. Therefore, even though nonlinear regression techniques promise better recognition results, the strict requirements prohibit overly complex models [4]. So far, no machine learning model has yielded reliable enough predictions to be realized within a commercially available prosthesis.

In this contribution we analyzed whether extending a simple linear regression model [4] with two key features, namely memory (in time) and nonlinearity, offers significant improvement with respect to recognition accuracy without sacrificing too much in terms of speed and robustness. We applied Extreme Learning Machines (ELMs) [5] and Echo State Networks (ESNs) [6] to assess the benefit of nonlinearity and memory respectively. We evaluated the test recognition accuracy on undisturbed data as well as data which was disturbed by electrode shift. Our data set consisted of simultaneous movements in two DoFs from four able-bodied participants, recorded using the Thalmic Myo (Thalmic Labs, Canada) armband, a commercially availabe, low-cost 8-electrode array, which has already shown promising applications in the area of prosthetic training and control [8,13].

## 2   Methods

### 2.1   From Linear Regression to Echo State Networks

We introduce Echo State Networks (ESNs) as a systematic extension of linear regression (LR) by nonlinearity and memory. Let $x_t \in \mathbb{R}^K$ denote the input at time step $t$ (i.e. the $K$ features computed on from the myoelectric signal at time step $t$) and let $y_t \in \mathbb{R}^n$ be the desired output at time $t$ (i.e. the desired movement in each of the $n$ degrees of freedom). We can write these as matrices of the form $X = (x_1, \ldots, x_T) \in \mathbb{R}^{K \times T}$ and $Y = (y_1, \ldots, y_T) \in \mathbb{R}^{n \times T}$. Then, linear regression assumes a direct, linear relationship between input- and output of the form $Y = W \cdot X$ for some matrix $W \in \mathbb{R}^{n \times K}$ which is set to $W = Y \cdot X^T \cdot (X \cdot X^T)^{-1}$ in order to minimize the squared error [2].

We can equivalently describe linear regression as a feedforward neural network with a single hidden layer. Let $h_t \in \mathbb{R}^m$ denote the value of the $m$ hidden layer neurons at time $t$. This value is computed as $h_t = W^{\text{in}} \cdot x_t$ for some input weight matrix $W^{\text{in}} \in \mathbb{R}^{m \times K}$. Similarly, the output of the network is computed as $W^{\text{out}} \cdot h_t$ for some output weight matrix $W^{\text{out}} \in \mathbb{R}^{n \times m}$. Due to linearity, the network output is fully described by the product matrix $W = W^{\text{out}} \cdot W^{\text{in}}$ which we can optimize as before.

This analytical optimal solution is no longer possible if we introduce a nonlinearity in the hidden layer; that is, we compute the value of the hidden neurons as $h_t = \sigma\left(W^{\text{in}} \cdot x_t\right)$ where $\sigma()$ denotes the component-wise application of some nonlinear function $\sigma$, e.g. the tanh function. Here, one relies on gradient-based optimization methods such as backpropagation, which are prone to local optima. An alternative strategy is to not train the input weights $W^{\text{in}}$ at all, but to keep them fixed at some random value. Then, one can preprocess the input data via $H = \sigma(W^{\text{in}} \cdot X)$ and obtain an optimal solution for the output weights as in linear regression by setting $W^{\text{out}} = Y \cdot H^T \cdot (H \cdot H^T)^{-1}$. This is the *Extreme Learning Machine (ELM)* model, which has been shown to be a universal approximator with good generalization properties [5] and to be competitive with nonlinear classification models on EMG data [1].

Additionally, one can introduce a memory to the model, such that the value of the hidden layer neurons is computed dependent on the previous value of the hidden layer neurons giving rise to the equations

$$h_t = \sigma\left(W^{\text{in}} \cdot x_t + W^{\text{hid}} \cdot h_{t-1}\right) \tag{1}$$

$$y_t = W^{\text{out}} \cdot h_t \tag{2}$$

for some recurrent weight matrix $W^{\text{hid}} \in \mathbb{R}^{m \times m}$. Approximate analytical solutions for the weight matrices are provided by [12]. Alternatively, one can again set the recurrent weights $W^{\text{hid}}$ independently of the input (e.g. randomly) and train only the output weights via linear regression. Note that one also has to ensure that the recurrent weights $W^{\text{hid}}$ are set such that initial conditions vanish over time (*echo state property* [17]). Such models are called *Echo State Networks (ESNs)* [6]. If $\sigma$ is the identity, we call the model a *linear Echo State Network (lESN)*. Here, we apply *cycle reservoirs with jumps (CRJs)* as introduced by [15]. In CRJs, input and reservoir weights are set deterministically to a single fixed value with varying signs according to an aperiodic, deterministic sequence. Hidden neurons are connected sparsely in a large cycle and additional *jump* connections between distant neurons within the cycle.

Given the strict runtime constraints of our application domain, fast models are desirable. Fortunately, ESNs have little complexity overhead compared to linear regression (provided that the number of hidden neurons $m$ is sufficiently small). For all models, a matrix inversion is required for training which has cubic complexity in the matrix dimension, i.e. $\mathcal{O}(K^3)$ for linear regression and $\mathcal{O}(m^3)$ for all other models. Prediction requires one matrix multiplication in $\mathcal{O}(K \cdot n)$ for linear regression, two matrix multiplications in $\mathcal{O}(K \cdot m + m \cdot n)$ for ELMs and three matrix multiplications in $\mathcal{O}(K \cdot m + m^2 + m \cdot n)$ for lESNs and ESNs, which reduces to $\mathcal{O}(K \cdot m + m \cdot n)$ in our case due to sparsity in $W^{\text{hid}}$.

## 2.2 Experimental Protocol

Four able-bodied participants executed a sequence of eight movements (hand open, hand close, pronation, supination, hand open + pronation, hand open + supination, hand close + pronation, and hand close + supination) ten times using their non-dominant hand. Each movement was executed between 3–5 s. Afterwards, the electrode array was shifted for one electrode in medial direction around the forearm, and the movement sequence was recorded four times.

We recorded the myoelectric data with the Thalmic Myo armband using all 8 channels. For each channel, we computed the log-variance and the squared log-variance on windows of 120 ms with 40 ms overlap, resulting in $K = 16$

**Table 1.** The average classification error (between 0 and 1) across all movements for each degree of freedom (listed as rows) and each model (listed as columns). The standard deviation is provided in brackets. The top two rows show the results on the original data, the bottom two rows on the shifted data.

| condition | DoF \model | LR | ELM | lESN | ESN |
|---|---|---|---|---|---|
| unshifted | hand close/open | 0.158 | 0.072 | 0.105 | 0.042 |
| | | (0.171) | (0.102) | (0.203) | (0.112) |
| | pronation/supination | 0.186 | 0.069 | 0.116 | 0.050 |
| | | (0.176) | (0.097) | (0.183) | (0.126) |
| shifted | hand close/open | 0.446 | 0.359 | 0.437 | 0.394 |
| | | (0.226) | (0.168) | (0.231) | (0.248) |
| | pronation/supination | 0.448 | 0.336 | 0.430 | 0.383 |
| | | (0.214) | (0.129) | (0.225) | (0.245) |

features. We generated a separate output signal for both DoFs ($n = 2$) with three possible values each ($-1, 0$, and $1$, i.e.: the output vector $(1, 1)$ codes hand open + supination, the output vector $(1, 0)$ codes hand open). We obtained a classification output from the regression output using simple thresholding $((-\infty, -0.5] \mapsto -1, (-0.5, 0.5) \mapsto 0,$ and $[0.5, \infty) \mapsto 1)$.

We evaluated the models in a leave-one-out crossvalidation over the 80 movements, optimizing hyperparameters via 1000 trials of random-search on the respective training data. The resulting model was then applied to the shifted data.

## 3   Results

The average classification error for the unshifted data is shown in the top two rows of Table 1. Using a Bonferrroni corrected Wilcoxon signed-rank test with $\alpha = 0.001$ over all movements of all subjects ($N = 320$) we obtained the following significant differences: ELM is superior to linear regression (LR), lESN is superior to LR, ESN is superior to all three (LR, ELM and lESN).

After the electrode shift, a significant degradation of classification accuracy can be observed for all models, as given in the bottom two rows of Table 1.

## 4   Discussion and Conclusion

We have demonstrated that extending a linear regression model with nonlinear features and memory in the form of an Echo State Network (ESN) leads to significant improvements in terms of recognition accuracy on sensor data obtained from a low-cost electrode array for simultaneous movements in multiple degrees of freedom. Further, we have demonstrated that combining both features is superior to each single feature. As such, ESNs present a promising method for future research as well as clinical applications. However, we have also shown that ESNs are not sufficiently robust regarding external disturbances, namely electrode shift. Further work would be required to address this problem, e.g. using resistant features [11] or transfer learning [10, 14].

# References

1. Anam, K., Al-Jumaily, A.: Evaluation of extreme learning machine for classification of individual and combined finger movements using electromyography on amputees and non-amputees. Neural Netw. **85**, 51–68 (2017)
2. Bishop, C.M.: Pattern Recognition and Machine Learning. Springer-Verlag New York Inc., Secaucus (2006)
3. Farina, D., Jiang, N., Rehbaum, H., Holobar, A., Graimann, B., Dietl, H., Aszmann, O.C.: The extraction of neural information from the surface EMG for the control of upper-limb prostheses: emerging avenues and challenges. IEEE Trans. Neural Syst. Rehabil. Eng. **22**(4), 797–809 (2014)
4. Hahne, J.M., Biebmann, F., Jiang, N., Rehbaum, H., Farina, D., Meinecke, F.C., Müller, K.R., Parra, L.C.: Linear and nonlinear regression techniques for simultaneous and proportional myoelectric control. IEEE Trans. Neural Syst. Rehabil. Eng. **22**(2), 269–279 (2014)
5. Huang, G.B., Zhu, Q.Y., Siew, C.K.: Extreme learning machine: theory and applications. Neurocomputing **70**(1–3), 489–501 (2006)
6. Jaeger, H., Haas, H.: Harnessing nonlinearity: predicting chaotic systems and saving energy in wireless communication. Science **304**(5667), 78–80 (2004)
7. Jiang, N., Englehart, K.B., Parker, P.A.: Extracting simultaneous and proportional neural control information for multiple-DOF prostheses from the surface electromyographic signal. IEEE Trans. Biomed. Eng. **56**(4), 1070–1080 (2009)
8. Masson, S., Fortuna, F., Moura, F., Soriano, D., do ABC, S.B.d.C.: Integrating Myo Armband for the control of myoelectric upper limb prosthesis. In: Proceedings of the XXV Congresso Brasileiro de Engenharia Biomédica (2016)
9. Ortiz-Catalan, M., Brånemark, R., Håkansson, B.: BioPatRec: a modular research platform for the control of artificial limbs based on pattern recognition algorithms. Sour. Code Biol. Med. **8**(1), 1–18 (2013)
10. Paaßen, B., Schulz, A., Hahne, J.M., Hammer, B.: An EM transfer learning algorithm with applications in bionic hand prostheses. In: Verleysen, M. (ed.) Proceedings of the 25th European Symposium on Artificial Neural Networks, Computational Intelligence and Machine Learning (ESANN 2017), Bruges, pp. 129–134. i6doc.com (2017). ISBN: 978-2-87587-038-4
11. Pan, L., Zhang, D., Jiang, N., Sheng, X., Zhu, X.: Improving robustness against electrode shift of high density EMG for myoelectric control through common spatial patterns. J. NeuroEng. Rehabil. **12**(1), 1–16 (2015)
12. Pasa, L., Sperduti, A.: Pre-training of recurrent neural networks via linear autoencoders. In: NIPS, pp. 3572–3580 (2014)
13. Phelan, I., Arden, M., Garcia, C., Roast, C.: Exploring virtual reality and prosthetic training. In: 2015 IEEE of the Virtual Reality (VR), pp. 353–354. IEEE (2015)
14. Prahm, C., Paassen, B., Schulz, A., Hammer, B., Aszmann, O.: Transfer learning for rapid re-calibration of a myoelectric prosthesis after electrode shift. In: Ibáñez, J., González-Vargas, J., Azorín, J., Akay, M., Pons, J. (eds.) Converging Clinical and Engineering Research on Neurorehabilitation II. Biosystems & Biorobotics, vol. 15, pp. 153–157. Springer, Cham (2017). doi:10.1007/978-3-319-46669-9_28
15. Rodan, A., Tiňo, P.: Simple deterministically constructed cycle reservoirs with regular jumps. Neural Comput. **24**(7), 1822–1852 (2012)
16. Vujaklija, I., Farina, D., Aszmann, O.: New developments in prosthetic arm systems. Orthop. Res. Rev. **8**, 31–39 (2016)
17. Yildiz, I.B., Jaeger, H., Kiebel, S.J.: Re-visiting the echo state property. Neural Netw. **35**, 1–9 (2012)

# Demo's

# Semi-automated Ontology Development and Management System Applied to Medically Unexplained Syndromes in the U.S. Veterans Population

Stéphane M. Meystre[1(✉)] and Kristina Doing-Harris[2]

[1] Medical University of South Carolina, Charleston, SC, USA
meystre@musc.edu
[2] Westminster College, Salt Lake City, UT, USA

**Abstract.** Terminologies or ontologies to describe patient-reported information are lacking. The development and maintenance of ontologies is usually a manual, lengthy, and resource-intensive process. To support the development of medical specialty-specific ontologies, we created a semi-automated ontology development and management system (SEAM). SEAM supports ontology development by automatically extracting terms, concepts, and relations from narrative text, and then offering a streamlined graphical user interface to edit and create content in the ontology and finally export it in OWL format. The graphical user interface implements card sorting for synonym grouping and concept laddering for hierarchy construction. We used SEAM to create ontologies to support medically unexplained syndromes detection and management among veterans in the U.S.

**Keywords:** Ontology · Terminology · Natural language processing

## 1 Introduction and Background

Automated processing, capture, extraction, comparison, and analysis of clinical information require a structured conceptualization, which includes a collection of ideas with their definitions and inter-relationships. In information sciences, an explicit specification of a domain conceptualization in a machine-readable format is called an ontology [1]. Ontologies typically include concepts, their properties, and the relationships between these concepts. Several ontologies for clinical information already exist such as SNOMED-CT, the Foundational Model of Anatomy, the Human Disease ontology, and the Symptom ontology. These ontologies do not include lay language used in patient-reported symptoms. Patients suffering from medically unexplained syndromes (MUS; e.g., fibromyalgia, irritable bowel syndrome, chronic fatigue) often report a large variety of symptoms [2]. These symptoms are central to the MUS diagnosis process. Therefore, processing clinical documents related to MUS will require either expanding the existing ontologies or creating new ones.

Developing and maintaining ontologies can be a manual, lengthy, and resource-intensive process. Several tools have been developed to ease this process by

© Springer International Publishing AG 2017
A. ten Teije et al. (Eds.): AIME 2017, LNAI 10259, pp. 345–350, 2017.
DOI: 10.1007/978-3-319-59758-4_41

providing graphical interfaces and functionalities for ontology alignment and comparison, consistency testing, and collaborative development. A good example is Protégé [3]. The manual ontology creation process requires that the user generate all of the terms and relationships to be used in the ontology. Therefore, semi-automated methods to find terms in corpora and integrate them into ontologies have also been developed. These methods typically employ an iterative process of automated potential new element extraction from text. Extracted terms are manually integrated into the knowledgebase. Examples of such systems include Text2Onto [4] or ASIUM [5]. The main advantage of these semi-automated methods is the significant reduction in human effort, allowing for faster and less resource-intensive ontology development and maintenance. A limitation shared by most systems implementing these semi-automated methods is the absence or very limited graphical user interface making interactions with users difficult. Some require technical experts rather than domain experts to developing ontologies. Protégé includes many plug-ins for ontology maintenance, but not many that help the process of building the initial set of concepts for the ontology. In 2006, Wang et al. built a Protégé plug-in that implemented card-sorting and laddering for ontology creation [6]. Card sorting [7] elicits ideas from domain experts about which concepts to include in the ontology. Concept names and sometimes definitions are written on cards that can be selected and grouped according to the expert's domain understanding. Laddering [8] is used to compare ideas and build concept hierarchies. KASO was another Protégé plug-in built in 2011, adding a semi-automated approach within Protégé [9]. This combination of semi-automated methods with a graphical user interface to control card sorting and laddering methods was a significant achievement and well-liked by users. Unfortunately, the Protégé user-interface can be complex and require significant training before users can efficiently use it. Neither plug-in is currently available on the Protégé plug-in wiki page [3]. We adopted a similar approach to address these limitations and provide Veterans Health Administration (VHA) clinical research with methods and instruments for improving detection and mitigation of health problems in deployed veterans, and more specifically detection of MUS, a challenging family of syndromes because of their complex and sometimes ill-defined and overlapping diagnostic criteria. This paper focuses on the novel graphical user interface developed for this purpose.

## 2 Methods

Development of the semi-automated ontology development and management (SEAM) application was guided by an extensive review of existing methods and tools for terms, concepts, and relation discovery in unstructured text. SEAM is built as a Java application with modular processing steps and separate processing streams for terms and relationships. The *backend* component uses natural language processing (NLP) to automatically extract prominent terms, create synonymous groups, which represent concepts, hierarchical ('is_a'), and mereological ('part_of') relations between these concepts. More details are available in other publications [10, 11]. The extracted concepts and relations are stored in a triple store repository (Eclipse RDF4 J [12], formerly known as Sesame). The *frontend graphical user interface* is loosely coupled with the backend component, enabling control of the latter and implementing ontology development and management functionalities. It accesses the triple store repository and displays the automatically extracted terms and relations for manual editing and validation. We implemented two methods with demonstrated success: card sorting and laddering. As mentioned above, the former is a method to elicit groups of synonymous terms to create concepts. Terms are rendered as 'cards,' which can be sorted into concepts and given names. The latter is a method to organize concepts into hierarchies. Concepts and terms are presented to be organized to reflect the relationships between them. The *frontend graphical user interface* was built with JavaFX [13] graphics and media packages, allowing for customizable appearance defined in separate Cascading Style Sheets, advanced graphics, compatibility with all major desktop platforms (Windows, MacOS, Linux), and possible deployment as a self-contained application package.

**Medically Unexplained Syndromes Use Case.** We used SEAM to find new terms and relationships for an existing partial ontology of medically unexplained syndromes (MUS). The existing ontology was created using part of the body systems ontology in ICD-11 [14], consultation with domain experts on inclusion and exclusion criteria for the three most common MUS (irritable bowel syndrome, chronic fatigue syndrome and fibromyalgia), literature and chart reviews.

## 3   Results

**Graphical User Interface.** The semi-automated ontology development and management (SEAM) application supports ontology development and management by automatically extracting terms, concepts (i.e., groups of synonymous terms), and relations from narrative text. Users can then edit this automatically extracted information and export semi-automatically built ontologies.

For the frontend graphical user interface, we implemented the card sorting and laddering approaches as a novel user-friendly streamlined interface where concepts and relations are created by simply dragging and dropping extracted terms or concepts on top of other terms or concepts. The user interface is composed of three tabs, for control of the application, concept creation (Fig. 1), and relationship creation. The initial application control lets the user configure the connection to the local database (triple store), define the location of the input collection of narrative text and its type (simple text or PDF), set up parameters for the backend component (e.g., cosine similarity factor), starting the backend component processing, and exporting the resulting ontology as an OWL file. As shown in Fig. 1, the concept creation process starts with a list of all terms automatically extracted from the processed collection of narrative text. Terms can be ordered alphabetically, by frequency, and by score used during the term extraction phase (i.e., C-score and termhood [10]). All passages from the text corpus containing an extracted term are also displayed to give SEAM users information about the local context of these terms. This functionality was developed to give more detailed understanding of the extracted term and to allow detecting and dealing with homographs and ambiguous abbreviations. During the concept creation phase, combining

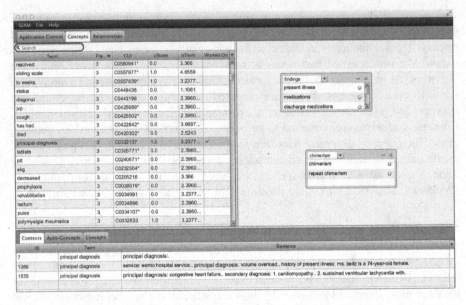

**Fig. 1.**  SEAM  user  interface  concepts  building  screenshot

extracted terms indicates that they are synonyms. The system creates a new concept (group), which the user can then name. Concepts can also be combined to create a new higher-level concept. Dragging and dropping terms and concepts on top of each other combines them. This process results in the creation of concept windows indicating all synonymous terms for the concept and the concept's preferred name. The group can then be manually edited further (e.g., changing the preferred name, editing synonyms). During the relation creation phase, combining extracted concepts creates a relation between them. The default relation type is hierarchical (i.e., is_a relation), but can also be mereological (i.e., part_of relation). Concepts or groups of concepts (i.e., portions of a taxonomy) can be placed where desired in the resulting taxonomies. Both concepts and relations can be further edited.

Concepts created are listed with their preferred name. They are placed and moved in hierarchies by dragging and dropping them where desired.

**Ontology Development Use Case.** The partial MUS ontology contained 236 concepts and 413 hierarchical relations. To help enrich it, we used a collection of 696 clinical documents from the 2009 i2b2 Medications Challenge [15] and also added 47 manually-selected journal articles focused on MUS. These documents contained 86,931 potential terms, and 4,962 full matches with UMLS Metathesaurus concepts. After processing by the backend component, 271 recommended terms were extracted, 62 of which matched concepts in the partial MUS ontology [10]. Of the remaining 209 recommended terms, without UMLS Metathesaurus matches, 67 were approved by two or more of the domain experts. 26 were approved by at least one expert. The backend component also extracted 75 synonymy relations and 14 hierarchical relations. Only 19 of the former, but 9 of the latter were approved.

## 4  Discussion

We developed and assessed SEAM as a contribution to efforts to provide Veterans Health Administration (VHA) clinical research with methods and instruments for improving detection and mitigation of health problems such as MUS in deployed veterans. SEAM was used to develop ontologies for MUS and offered significant help for this complex task. The frontend graphical user interface built on top of a backend component based on NLP allowed for user-friendly dragging and dropping of automatically extracted or manually created terms, concepts, and relations between concepts, to help build and enrich ontologies. Overall, when applied to the MUS use case, between 25% and 57% of the terms or relations automatically extracted by SEAM were approved by domain experts and added to the MUS ontology. SEAM was also successfully tested with two different use cases, to detect mentions of acute changes in mental status and enhance the pick-list of findings used to summarize echocardiogram reports [10].

**Acknowledgments.** We thank Yarden Livnat and Kristin Potter for their work with the frontend graphical user interface. Project funded under VA HSR&D contract 11RT0150. SEAM is available at http://kdh-nlp.org/Seam-project/seam-home.html.

## References

1. Gruber T.: A translation approach to portable ontology specifications. Knowledge Acquisition Stanford, CA, Technical report KSL, vol. 5(2), pp. 199–220 (1993)
2. Doing-Harris, K., Meystre, S.M., Samore, M., Ceusters, W.: Applying ontological realism to medically unexplained syndromes. Stud. Health Technol. Inform. **192**, 97–101 (2013)
3. Protégé. http://protege.stanford.edu
4. Cimiano, P., Völker, J.: Text2Onto. In: Montoyo, A., Muñoz, R., Métais, E. (eds.) NLDB 2005. LNCS, vol. 3513, pp. 227–238. Springer, Heidelberg (2005). doi:10.1007/11428817_21
5. Faure, D., Nédellec, C.: A corpus-based conceptual clustering method for verb frames and ontology acquisition. In: LREC workshop on adapting lexical and corpus resources to sublanguages and applications, pp. 707–728 (1998)
6. Wang, Y., Sure, Y., Stevens, R., Rector, A.: Knowledge elicitation plug-in for protege: card sorting and laddering. In: Mizoguchi, R., Shi, Z., Giunchiglia, F. (eds.) ASWC 2006. LNCS, vol. 4185, pp. 552–565. Springer, Heidelberg (2006). doi:10.1007/11836025_53
7. Upchurch, L., Rugg, G., Kitchenham, B.: Using card sorts to elicit web page quality attributes. IEEE Softw. **18**(4), 84–89 (2001)
8. Shadbolt, N., O'hara, K., Crow, L.: The experimental evaluation of knowledge acquisition techniques and methods: history, problems and new directions. Int. J. Hum.-Comput. Stud. **51**(4), 729–755 (1999)
9. Wang, Y., Völker, J., Haase, P.: Towards semi-automatic ontology building supported by large-scale knowledge acquisition. In: AAAI Fall Symposium on Semantic Web for Collaborative Knowledge Acquisition, vol. 6, p. 06 (2006)
10. Doing-Harris, K., Livnat, Y., Meystre, S.: Automated concept and relationship extraction for the semi-automated ontology management (SEAM) system. J. Biomed. Semant. **6**(1), 15 (2015)

11. Doing-Harris, K., Boonsirisumpun, N., Potter, K., Livnat, Y., Meystre, S.M.: Automated concept and relationship extraction for ontology development. In: AMIA, p. 344 (2013)
12. Eclipse RDF4 J. http://rdf4j.org
13. Oracle Corp. JavaFX. http://docs.oracle.com/javase/8/javase-clienttechnologies.htm
14. World Health Organization. ICD-11 Revision. http://www.who.int/classifications/icd/revision/en/
15. Uzuner, O., Solti, I., Xia, F., Cadag, E.: Community annotation experiment for ground truth generation for the i2b2 medication challenge. J. Am. Med. Inf. Assoc. 17(5), 519–523 (2010)

# pMineR: An Innovative R Library for Performing Process Mining in Medicine

Roberto Gatta[1]([envelope]), Jacopo Lenkowicz[1], Mauro Vallati[2], Eric Rojas[3],
Andrea Damiani[1], Lucia Sacchi[4][ORCID], Berardino De Bari[5], Arianna Dagliati[6],
Carlos Fernandez-Llatas[7], Matteo Montesi[8], Antonio Marchetti[8],
Maurizio Castellano[9], and Vincenzo Valentini[1]

[1] Universitá Cattólica del Sacro Cuore, Roma, Italy
roberto.gatta.bs@gmail.com
[2] University of Huddersfield, Huddersfield, UK
[3] Universidad Pontificia Católica de Chile, Santiago, Chile
[4] Universitá degli Studi di Pavia, Pavia, Italy
[5] Centre Hospitalier Régional Universitaire de Besançon, Besançon, France
[6] IRCCS ICS Maugeri Pavia, Pavia, Italy
[7] ITACA-Universitat Politecnica de Valencia, Valencia, Spain
[8] Data Warehouse, Policlinico Universitario A. Gemelli, Roma, Italy
[9] Universitá degli Studi di Brescia, Brescia, Italy

**Abstract.** Process Mining is an emerging discipline investigating tasks related with the automated identification of process models, given real-world data (Process Discovery). The analysis of such models can provide useful insights to domain experts. In addition, models of processes can be used to test if a given process complies (Conformance Checking) with specifications. For these capabilities, Process Mining is gaining importance and attention in healthcare.

In this paper we introduce pMineR, an R library specifically designed for performing Process Mining in the medical domain, and supporting human experts by presenting processes in a human-readable way.

**Keywords:** Process mining · R · Decision support system

## 1 Introduction

Process Mining [1] is an emerging discipline aiming at providing methods to automatically identify, starting from real-world data or Event Logs, accurate models of processes. This is usually defined as Process Discovery.

Among the other domains, medicine is one of the most complex from the Process Mining perspective. This is due, for instance, to the high knowledge needed to deal with corresponding data, the need to involve experts –that usually are not strongly in favor of automatic techniques–, and the lack of formal methodologies. On this matter, the interested reader is referred to a recent review about the role of Process Mining in healthcare [5].

© Springer International Publishing AG 2017
A. ten Teije et al. (Eds.): AIME 2017, LNAI 10259, pp. 351–355, 2017.
DOI: 10.1007/978-3-319-59758-4_42

It should be noted that, at the state of the art, there are few tools that can perform Process Mining. Well-known examples include DISCO [2], PROM [6] and PALIA [2]. However, there is a lack of tools focusing on Process Mining in Medicine. Moreover, we observed that a common expectation of users from the medical domain, is to be able to work with one of the most diffused statistical software (like MATLAB, R, or SPSS), to exploit the previous knowledge of a well-known environment. R [4], in particular, is one of the most common software environment for data analysis: nevertheless, with the exception of edeaR [3], which unfortunately does not support Process Discovery or Conformance Checking, there is no availability of libraries to cope with Process Mining.

In order to fill the aforementioned gap, in this work we propose pMineR,[1] an R library focused on dealing with Process Discovery and Conformance Checking in the medical domain. In such domain, it is of pivotal importance that physicians clearly understand every aspect of the elements in the generated models. In the light of that, pMineR exploits Markov Models, which are usually well-known and easy to understand for medical users.

While coping with Process Discovery, pMineR exploits some aspects taken from the Computer Interpretable Clinical-Guidelines field, in particular in terms of human-readability and representation of clinical guidelines. About this point big efforts have been made in proposing a new language, less powerful but easier to be handled from physicians to the extant ones.

## 2    The pMineR Library

The pMineR structure has been specifically designed for supporting: (i) the evolution and extensibility of the system, and (ii) to privilege improvements oriented to cope with real-world issues in medicine (it has to have a real-needs-driven improvement). The former element is reflected in a very modular architecture, and in the adoption of a strategy to govern architectural improvements by a set of development guidelines. The latter has been implemented by emphasizing a goal-driven development, by proposing scenarios of application, and by structuring the documentation in form of Case Reports.

The modular architecture exploits *collections*, specialized sets of classes built using common methods and exchanging data structures. This is done in order to maximize the re-usability of the code and the extensibility of pMineR.

The main *collections* of pMineR are described in the following.

- **Data Loader:** this *collection* includes classes handling the loading of event logs, and data pre-processing. It provides tools to translate and/or group event log terms by a given dictionary, to load arrays of *dataLoader* objects (used for storing structured information extracted from raw data) and manipulate them for creating different *views*.
- **LOG Inspection:** is a set of classes aiming to provide some descriptive statistics on event log data, such as events and processes, useful for a preliminary exploration of the data.

---

[1] https://cran.r-project.org/web/packages/pMineR/index.html.

- **Process Discovery:** classes in this *collection* implement one or more Process Discovery algorithms, given a *Data Loader* object. In the current version of pMineR there are two classes implementing, respectively, first and second order Markov Models-based algorithms. Classes in this package interact by a set of standard methods such as *load, train, play, replay,* in order to increase the possibility of interaction among objects of different classes.
- **Conformance Checking:** is a set of classes specialized in Conformance Checking. Even if also the Process Discovery classes have methods to check how a set of given processes can flow through the models, normally, the formalism for representing clinical guidelines have no algorithms to automatically generate an interpretable guideline starting from real world data: pMineR implements a class able to work with an internal formalism for representing WorkFlow-like diagrams: such formalism is called Pseudo-WorkFlow (PWF) and was designed with the contribution of our physicians in order to be human readable and to be able to represent a set of needed guidelines.
- **General Purposes:** this *collection* include classes addressing a wide range of common issues that are faced during programming, such as exception handling, and strings manipulation.

## 2.1 Process Discovery

The current version of pMineR implements two algorithms referring to the first (FOMM) and second (SOMM) order Markov Models. Both of them provide the minimum set of common methods and some different model-specific methods.

Some of the methods provided by classes of the Process Mining package are the common expected methods of PM algorithms, such as: *load, train, play,* and *replay,* which role is well described in the Manifesto [1]. Some more advanced methods include *compare,* which can be used to compare two models given an embedded default metric or a user-defined one (the simplest default metric for FOMM model is the normalized sum of absolute delta between two FOMMs). pMineR can compute the differences between generated models, and output them under the form of diagrams, that can then be analyzed by human experts.

Figure 1 (colored) shows the difference among two FOMM models, built using the *compare* method.

## 2.2 Conformance Checking

These classes are specifically designed to support Conformance Checking, by proposing schema and diagrams close to the language adopted by physician. At the moment, pMineR implements an engine which is able to parse guidelines written in the previously introduced PWF language. PWF is based on three main constructs: events, statuses, and triggers.

Given an event log, the engine reads the list of events and, for each event, it tests if one or more triggers can be fired. A trigger is an item composed by two main sections: condition and effects. The condition part can check elements of the just read event log ($ev.NOW$) or other statuses of the patient (e.g.,

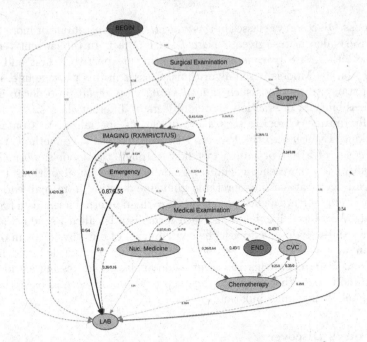

**Fig. 1.** The graph shows a FOMM related to a set of patient treated for lung cancer, compared with a set of patient treated for breast cancer: in such graph the arcs in black are the arch with a similar probability (according to a threshold). Red (green) arcs indicate that the transition probability for the lung model is lower (higher) than the one for breast. (Color figure online)

currently active statuses $st.ACTIVES$). If the condition applies, the effects listed in the subsequent section are executed. In the following example, a trigger for representing the end of a treatment is specified, using the PWF language. If the current status of the treatment is *in progress*, and a *dismission Report* event is read, the status of the patient has to be updated, according to the list of set and unset items. Using this approach, statuses are automatically updated while events are processed, sequentially, from the first to the last.

Figure 2 provides an example of the computation of a PWF for a dummy set of event log (on the left) and details about a specific patient (on the right). On the left the workflow is graphed starting from the given XML used for defining triggers (squared boxes) and statuses (rounds). On the top right an original event log, which is an input of the computation. On the bottom right, the result of the computation for the same event log, plotted under the form of the "activation time" of the different statuses.

In addition, PWF also provides a set of items for representing time and duration. It can also support different kind of *actions* on status, and allows to specify a header, on the guidelines, to indicate references, authors, validity, and information about the expected use.

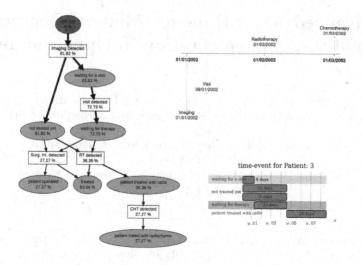

**Fig. 2.** Example output provided by pMineR after the computation of a PWF.

# 3 Conclusion

While several tools have been developed for dealing with the general Process Mining problem, only a few are able to deal with some aspects of its application to the medicine domain. Moreover, there is a lack of libraries for Process Mining in R, one of the most exploited software environment for data analysis. For filling these gaps, in this paper we presented pMineR, an innovative R library addressing all the tasks of Process Mining, with a specific focus on the medical domain. It should be noted that, while developing pMineR, we took into account insights provided by the state of the art of Computer Interpretable clinical Guideline (CIG), in order to provide easy to understand and interpret output. The system is already freely available, and can be exploited for efficiently and effectively performing Process Mining on sets of heterogeneous real world data.

# References

1. Aalst, W., et al.: Process mining manifesto. In: Daniel, F., Barkaoui, K., Dustdar, S. (eds.) BPM 2011. LNBIP, vol. 99, pp. 169–194. Springer, Heidelberg (2012). doi:10. 1007/978-3-642-28108-2_19
2. Günther, C.W., Rozinat, A.: Disco: discover your processes. BPM (Demos) **940**, 40–44 (2012)
3. Janssenswillen, G.: edeaR. cran.r-project.org/web/packages/edeaR
4. R Core Team: R: A Language and Environment for Statistical Computing. R Foundation for Statistical Computing (2016). R-project.org
5. Rojas, E., Munoz-Gama, J., Sepúlveda, M., Capurro, D.: Process mining in healthcare: a literature review. J. Biomed. Inform. **61**, 224–236 (2016)
6. Dongen, B.F., Medeiros, A.K.A., Verbeek, H.M.W., Weijters, A.J.M.M., Aalst, W.M.P.: The ProM framework: a new era in process mining tool support. In: Ciardo, G., Darondeau, P. (eds.) ICATPN 2005. LNCS, vol. 3536, pp. 444–454. Springer, Heidelberg (2005). doi:10.1007/11494744_25

# Advanced Algorithms for Medical Decision Analysis. Implementation in OpenMarkov

Manuel Arias[1], Miguel Ángel Artaso[1], Iñigo Bermejo[2],
Francisco Javier Díez[1(✉)], Manuel Luque[1], and Jorge Pérez-Martín[1]

[1] Department of Artificial Intelligence, UNED, Madrid, Spain
fjdiez@dia.uned.es
[2] School of Health and Related Research, University of Sheffield,
Sheffield, UK

**Abstract.** In spite the important advantages of influence diagrams over decision trees, including the possibility of solving much more complex problems, the medical literature still contains around 10 decision trees for each influence diagram. In this paper we analyse the reasons for the low acceptance of influence diagrams in health decision analysis, in contrast with its success in artificial intelligence. One of the reasons is the difficulty of representing asymmetric problems. Another one was the lack of algorithms for explaining the reasoning and performing cost-effectiveness analysis, as well as the scarcity of user-friendly software tools for sensitivity analysis. In this paper we review the research conducted by our group in the last 25 years, crystallised in the open-source software tool OpenMarkov, explaining how it has tried to address those challenges.

**Keywords:** Probabilistic graphical models · Bayesian networks · Influence diagrams · Markov models · Cost-effectiveness analysis · Sensitivity analysis · OpenMarkov

## 1 Introduction

Probabilistic graphical models, such as Bayesian networks (BNs) [14] and influence diagrams (IDs) [6], have been extensively used in artificial intelligence. Many of the applications have addressed medical problems, but they represent a small fraction of the immense amount of models built for medical decision analysis [13]. A quick search in PubMed,[1] the main database of medical literature, conducted in January 2017 returned 14,212 references about decision trees (DTs), 2,522 about BNs and 1,261 about IDs; i.e., there are around 10 DTs for each ID. Furthermore, in a survey of books about medical decision analysis, health economics and evidence based medicine published after 1994, we found 26 texts that explain DTs but only 3 of them describe IDs, very briefly. This is even more surprising if we consider the great advantages of IDs over DTs, which allow them to solve much more complex problems [1].

---

[1] www.pubmed.com.

© Springer International Publishing AG 2017
A. ten Teije et al. (Eds.): AIME 2017, LNAI 10259, pp. 356–360, 2017.
DOI: 10.1007/978-3-319-59758-4_43

What are then the reasons for the low acceptance of IDs in health decision analysis? In this paper we discuss some of them and review the research conducted by our research group in the last quarter of century, explaining how it has tried to overcome the limitations of IDs. Our contributions have been implemented in OpenMarkov, an open-source tool for probabilistic graphical models, which is available at www.openmarkov.org.

## 2   OpenMarkov

OpenMarkov is an open-source tool for building and evaluating several types of probabilistic graphical models, including BNs and IDs. It is implemented in Java, so it can run on any operating system. It has a powerful graphical user interface, but it can also be used as a library. The site www.openmarkov.org gives access to the source code, a compiled .jar version, a tutorial, some technical documents, lists for users and developers, etc. To our knowledge, OpenMarkov has been used at universities, research institutes, and companies in at least 26 countries. Its native format for encoding models, ProbModelXML, has also been implemented in other software tools, such as pgmpy, MADP and PILGRIM.[2]

The web page www.probmodelxml.org/networks contains several examples of probabilistic graphical models (BNs, IDs, etc.) in ProbModelXML format. Some of them were built to solve real medical problems, such as lung cancer staging, total knee arthroplasty, HPV vaccination, HIV prophylaxis, cochlear implantation, etc. They can be opened and evaluated with OpenMarkov.

## 3   Problems and Solutions

In this section we discuss the main difficulties faced when trying to solve medical decision problems with IDs, and the solutions implemented in OpenMarkov.

### 3.1   Explanation of Reasoning

Human experts are reluctant to accept the advice of a computer if they cannot understand the inference that led to the conclusions. The arc reversal algorithm [12,15], which recursively reduces the ID by eliminating its nodes one by one, does not provide an intuitive understanding of the inference. In the case of algorithms that do not operate on the graph, such as variable elimination [7], it is impossible to follow the steps performed.

As a solution, we proposed several explanation methods for BNs and IDs, such as the possibility of showing several bars at each node representing either its posterior probability (for chance nodes) or its expected value (for utility nodes), and the possibility of imposing policies for what-if reasoning [9]. These facilities have been very useful when collaborating with medical doctors in the construction of several BNs and IDs [8].

---

[2] See www.probmodelxml.org, https://github.com/pgmpy, www.fransoliehoek.net/index.php?fuseaction=software.madp and http://pilgrim.univ-nantes.fr.

## 3.2    Modelling and Solving Asymmetric Problems

In medicine virtually all problems are asymmetric. There is *structural asymmetry* when the value taken on by a variable restricts the domain of other variables; for example, the result of a test is only available when the doctor decides to order it. *Order asymmetry* occurs when the decisions can be made in different orders. IDs can only represent symmetric problems. Some asymmetries can be eliminated using different modelling tricks, but in other cases this is not possible [2].

For this reason we have proposed decision analysis networks (DANs), a new type of probabilistic graphical models that—instead of having information links, as in IDs—represent the flow of information by means of revelation links and always-observed variables [2]. This way DANs do not require a total ordering of the decisions. They can also represent restrictions among variables. OpenMarkov offers a graphical user interface for editing DANs and efficient algorithms for evaluating them.

## 3.3    Strategy Trees

Given a decision problem, the strategy consists of a rule for each decision in each context. In the case of an ID a policy for a decision consists of a decision rule for each configuration of the variables known when making it. If all those variables are discrete, the policy can be represented by a table. The strategy is implicit in the set of policies. This entails two problems: that the size of the policy for a decision grows exponentially with the number of variables known when making it, and that most columns in the last policy tables contain correspond to impossible or suboptimal scenarios.

An alternative consists in representing the strategy as a tree, which is much more compact. For example, in Mediastinet, an ID for lung cancer, the policy table for the last decision contains over 15,552 columns, but the strategy tree contains only 5 nodes [11]. We have extended the variable-elimination algorithm so that it can build the strategy tree at the same time as it evaluates the model (paper under evaluation). This method also works for DANs and has been implemented in OpenMarkov.

## 3.4    Cost-Effectiveness Analysis

The traditional approach in medical decision making was to select the most beneficial intervention for the patient. Nowadays the need to explicitly take into account the cost of interventions has made cost-effectiveness analyses more and more common. Unfortunately, standard algorithms can only evaluate unicriterion IDs.

We have developed an algorithm that can perform CEA with IDs [1] and implemented it in OpenMarkov; in the near future we will extend it to DANs. We have also implemented several types of sensitivity analysis, both deterministic (for example, tornado diagrams) and probabilistic (mainly scatter plots and acceptability curves), which are a requisite in CEAs [4].

## 3.5    Temporal Reasoning Was Not Possible with IDs

Finally, in many cases it is necessary to model the evolution of patients over time. There are dynamic IDs [5,10], but they are not used in practice because their policies grow exponentially with the time horizon.

For this reason we have proposed Markov IDs [3], which may contain both temporal and atemporal variables; in particular, in the models we have built for several medical problems[3] all the decisions are atemporal, because otherwise those models could only be evaluated for extremely short horizons.

# 4    Conclusion

We have discussed five reasons that, in our opinion, have impeded a wide acceptance of IDs in the fields of medical decision analysis and health economics. Some of these reasons are inherent to IDs, such the impossibility of representing asymmetric problems—as mentioned above, in medicine virtually all problems are asymmetric—and the difficulty of explaining its reasoning. Other reasons are related with the lack of algorithms for solving the kind of problems in which health economists are currently more interested, such as cost-effectiveness analysis. Finally, an obstacle for the use of IDs in health studies is that most software tools for probabilistic graphical models lack some facilities that medical decision analysts require; for example, probabilistic sensitivity analysis.

We have also reviewed in this paper the solutions proposed by our research group. They include new algorithms for cost-effectiveness analysis with IDs and with Markov IDs, as well as some explanation methods for BNs and IDs. We have overcome the limited expressive power of IDs by proposing decision analysis networks (DANs), a new type of probabilistic graphic especially designed for asymmetric problems. All these contributions would be of little use to medical doctors and health economists if they were only ideas proposed in journals papers and conference proceedings. We have devoted a great effort to creating a software tool with an advanced graphical user interface that allows people with no programming skills build and evaluate models. At this moment, OpenMarkov is not only a free alternative to commercial products for medical decision analysis, but a powerful software package offering new types of models and novel algorithms that are not available in any other tool.

**Acknowledgements.** This work has been supported by the Spanish Government under grants TIN2009-09158, PI13/02446, and TIN2016-77206-R, and co-financed by the European Regional Development Fund (ERDF). It has also received support from projects 262266 and 324401 (FP7-PEOPLE-2012-IAPP) of the European Union.

---

[3] See www.probmodel.xml/networks.

# References

1. Arias, M., Díez, F.J.: Cost-effectiveness analysis with influence diagrams. Methods Inf. Med. **54**, 353–358 (2015)
2. Díez, F.J., Luque, M., König, C., Bermejo, I.: Decision analysis networks. Technical report CISIAD-14-01, UNED, Madrid, Spain (2014)
3. Díez, F.J., Yebra, M., Bermejo, I., Palacios-Alonso, M.A., Arias, M., Luque, M., Pérez-Martín, J.: Markov influence diagrams: a graphical tool for cost-effectiveness analysis. Med. Decis. Mak. **37**, 183–195 (2017)
4. Drummond, M.F., Sculpher, M.J., Torrance, G.W., O'Brien, B.J., Stoddart, G.L.: Methods for the Economic Evaluation of Health Care Programmes, 3rd edn. Oxford University Press, Oxford (2005)
5. Hazen, G.B.: Dynamic influence diagrams: applications to medical decision modeling. In: Brandeau, M.L., Sainfort, F., Pierskalla, W.P. (eds.) Operations Research and Health Care, pp. 613–638. Springer, Heidelberg (2004)
6. Howard, R.A., Matheson, J.E.: Influence diagrams. In: Howard, R.A., Matheson, J.E. (eds.) Readings on the Principles and Applications of Decision Analysis, pp. 719–762. Strategic Decisions Group, Menlo Park (1984)
7. Jensen, F.V., Nielsen, T.D.: Bayesian Networks and Decision Graphs, 2nd edn. Springer, New York (2007)
8. Lacave, C., Oniśko, A., Díez, F.J.: Use of Elvira's explanation facilities for debugging probabilistic expert systems. Knowl.-Based Syst. **19**, 730–738 (2006)
9. Lacave, C., Luque, M., Díez, F.J.: Explanation of Bayesian networks and influence diagrams in Elvira. IEEE Trans. Syst. Man Cybern.-Part B: Cybern. **37**, 952–965 (2007)
10. Leong, T.Y.: Multiple perspective dynamic decision making. Artif. Intell. **105**, 209–261 (1998)
11. Luque, M., Díez, F.J., Disdier, C.: Optimal sequence of tests for the mediastinal staging of non-small cell lung cancer. BMC Med. Inform. Decis. Mak. **16**, 1–14 (2016)
12. Olmsted, S.M.: On representing and solving decision problems. Ph.D. thesis, Department Engineering-Economic Systems, Stanford University, CA (1983)
13. Pauker, S., Wong, J.: The influence of influence diagrams in medicine. Decis. Anal. **2**, 238–244 (2005)
14. Pearl, J.: Probabilistic Reasoning in Intelligent Systems: Networks of Plausible Inference. Morgan Kaufmann, San Mateo (1988)
15. Shachter, R.D.: Evaluating influence diagrams. Oper. Res. **34**, 871–882 (1986)

# A Platform for Targeting Cost-Utility Analyses to Specific Populations

Elisa Salvi$^{(\boxtimes)}$, Enea Parimbelli, Gladys Emalieu, Silvana Quaglini, and Lucia Sacchi$^{(\boxtimes)}$ (iD)

Department of Electrical, Computer and Biomedical Engineering, University of Pavia, Pavia, Italy
lucia.sacchi@unipv.it

**Abstract.** Quality-adjusted life years (QALYs) are a popular measure employed in cost-utility analysis (CUA) for informing decisions about competing healthcare programs applicable to a target population.

CUA is often performed using decision trees (DTs), i.e. probabilistic models that allow calculating the outcome related to different decision options (e.g., two different therapeutic strategies) considering all their expected effects. DTs may in fact include a measure of the quality of life, namely a utility coefficient (UC), for every health state patients might experience as a result of the healthcare interventions. Eliciting reliable UCs from patients poses several challenges, and it is not a common procedure in clinical practice.

We recently developed UceWeb, a tool that supports users in that elicitation process. In this paper we describe the public repository where UceWeb collects the elicited UCs, and how this repository can be exploited by researchers interested in performing DT-based CUAs on a specific population. To this aim, we also describe the UceWeb integration with a commercial software for DTs management, which allows to automatically run the models quantified with the mean value of the target population UCs.

**Keywords:** Patient preferences · Utility coefficients · Cost-utility analysis · Decision trees

## 1 Introduction

Quality-adjusted life years (QALYs) are a popular measure for valuing the outcome of healthcare interventions in terms of resulting patients' expected life and perceived quality of life (QOL) [1]. To compute QALYs for a specific patient, his/her expected life is split in time intervals, each one ($t_i$) presumably spent experiencing a specific health condition. For each condition, a utility coefficient (UC) $u_i$ can be elicited to measure the QOL perceived by the patient in relation to such condition. QALYs are then computed according to the following formula: $\sum_i ti * ui$.

Obtaining reliable UCs is fundamental, since the elicited values may be used in clinical practice for shared decision making (SDM) procedures guiding the personalization of care for a specific patient. Besides the use for the individual, UCs can be exploited at a population level, for cost-utility analysis (CUA). In the health economics

© Springer International Publishing AG 2017
A. ten Teije et al. (Eds.): AIME 2017, LNAI 10259, pp. 361–365, 2017.
DOI: 10.1007/978-3-319-59758-4_44

context, CUAs are evaluations that compare alternative healthcare programs in terms of "cost per gained QALY" in the target population [2]. Eliciting UCs from a patient is not straightforward, and it is not a common procedure in clinical practice. Usually, UC values for specific diseases are collected during clinical studies, using either paper-based elicitation instruments or questionnaires designed for the specific circumstances [3]. Elicitation procedures usually require the presence of trained interviewers, to properly complete the process and obtain reliable UC values from the examined patient. CUA is often performed using decision trees (DTs), probabilistic models that allow calculating the outcome related to multiple decision options (e.g., two alternative therapeutic strategies, or whether or not to apply a preventive healthcare measure to a specific population) considering all their expected effects [4]. DTs may include among their parameters one UC for every health state patients might experience as a result of the compared healthcare interventions. Since collecting UCs is challenging and time-consuming, CUAs usually rely on UCs that are already available in the literature. Unfortunately, it is not straightforward to re-use such values, since they might have been elicited from a population that consistently differs from the one defined for the CUA. Using a sub-optimal set of UCs in CUA would introduce a bias effect in the analysis, and may consequently lead to sub-optimal decisions.

Few tools are available for eliciting UCs, and the majority of them has a limited range of application. For example, some of the tools described in the literature are embedded in disease-specific instruments designed to support targeted SDM processes [5], and are not exploitable for different decision analysis problems. On the other hand, general purpose elicitation tools are usually not well integrated with functionalities that allow the actual exploitation of the obtained values in either SDM processes or CUAs [6–10].

To support elicitation processes, we had already developed UceWeb [11], a web-application to elicit patient-specific UCs through three state of the art methods (time trade-off, standard gamble, and rating scale), and to collect them in a public repository for further use. In this paper we present a new functionality of UceWeb that has been recently implemented for supporting users in a more comprehensive decision analysis workflow. The new functionality allows the re-use of the UC values collected through our tool for targeting DT-based CUAs to a population of interest.

## 2 Methods

Researchers who perform DT-based CUAs on specific populations have two needs. First, to compute QALYs, they need to elicit a population-specific UC, and its confidence interval, for every health condition included in the model. Second, they need to effectively run the DT model quantified with such UCs.

For the first need, they could exploit the UceWeb public database where UC values are collected. In this repository, each UC is stored along with additional information on the related elicitation procedure, including the elicitation date, the elicitation method, the SNOMED [12] code identifying the health condition, and an anonymous identifier for the patient who has elicited the coefficient. For each patient, UceWeb collects a profile gathering relevant anonymous information such as age, sex, ethnicity, education, marital status, parental status, and occupation. Since patient profiles may be updated over time, the repository keeps track of any occurred modification. Thanks to the described framework, it is possible to query the UCs repository for retrieving the coefficients elicited from patients who fit a user-defined population (e.g. caucasian female aged from 35 to 55, employed in executive professions and having children). For targeting a DT-based CUA to such population, it is possible to use the mean value of the retrieved UCs as a population-specific parameter, which is in turn used to quantify the DT model.

To support users in employing the obtained population-specific UCs in DTs, we integrated UceWeb with TreeAge Pro [13], a widely used software for the formalization and analysis of DTs. The integration with the TreeAge Pro Suite was possible thanks to the TreeAge Pro Object Interface, which enables to open, update, and analyze DTs using programming languages (in our case, Java) and, consequently, to embed these functionalities in custom applications. To the best of our knowledge, this is the first attempt to integrate TreeAge Pro into a web-based application. To achieve this goal, we had to explore the Object Interface resources to assess which TreeAge Pro objects (e.g. variables and properties of the DT model) could be successfully managed via programming language. Thanks to the obtained integration, users are now able to run DTs formalized with TreeAge Pro and collected in a dedicated repository, directly from the UceWeb interface. The DTs repository currently contains two models, specifically formalized to address two decision problems: the optimization of the antiarrhythmic therapy for patients affected by atrial fibrillation (AF)("AAT model"), and the optimization of the anticoagulant treatment to prevent thromboembolic complications in the same patients ("OAT model") [14].

# 3   Results

This section describes the UceWeb workflow allowing users to perform DT-based population-specific CUAs.

First, the user can select the model from the ones available in the DTs repository. Then, thanks to the described integration with TreeAge Pro, UceWeb provides the user with the list of the health states considered in the selected model, whose UCs are required to quantify and run the DT. As an example, Fig. 1 illustrates the response returned by UceWeb when the user selects the "OAT" model. For each health condition, the user can decide to use the default UC values set in the model by the DT author ("Use default" buttons in Fig. 1). As an alternative, the user can decide to exploit the UceWeb UCs repository for "eliciting" population-specific UCs ("Elicit" buttons in Fig. 1). In this case, he/she is asked to fill in a dedicated form for defining the characteristics of the population to consider (e.g. age range, sex, type of occupation, ethnicity). On the basis of those parameters, UceWeb retrieves from the repository all the UCs elicited from patients who fit the specified population. It then computes the mean value, and presents it as the population-specific UC assessment required for the DT quantification. In the form, the user can also define a time frame of interest for the desired QOL assessment, limiting the retrieval to UCs elicited during a selected time period (e.g. consider only UCs elicited from January 2010).

Decision tree : Mobiguide_models/OAT.trex

**Please elicit utility coefficients for the following health states:**

| Health state fully specified name | SNOMED ID | Elicitation method | Utility coeff. value |
|---|---|---|---|
| 25133001 Completed stroke (disorder), 255604002 Mild (qualifier value) | 25133001 | ... | Elicit / Use default |
| 25133001 Completed stroke (disorder), 371924009 Moderate to severe (qualifier value) | 25133001 | ... | Elicit / Use default |
| 49436004 Atrial fibrillation (disorder) | 49436004 | ... | Elicit / Use default |

Use the elicited values in the decision tree and view results    Submit

**Fig. 1.** List of the UCs values required for quantifying the "OAT" model.

When all the coefficients required for the selected model have been collected, UceWeb exploits the TreeAge Pro Object Interface for setting the obtained values into the selected DT, and to run the model. In this way, the analysis will be targeted according to the population assessed through our tool. Finally, for each decision option the results of the analysis in terms of QALYs and costs are presented to the user through the UceWeb interface.

## 4 Discussion and Conclusion

The aim of the described work is to provide an innovative tool for empowering researchers engaged in DT-based healthcare decision making processes. The proposed framework for bridging the gap between UCs elicitation and UCs exploitation in DT analyses has multiple advantages. First, it promotes the sharing and re-use of available UCs assessments, facilitating CUAs targeted to specific populations. Second, thanks to the ability to run any DT model formalized with TreeAge Pro, the UceWeb tool is domain-independent and thus exploitable in a wide range of applications. On the other hand, it allows users to perform disease-specific analyses once a specific model is selected from the DT repository.

Further steps are necessary to fully exploit UceWeb potentiality. First, we must promote the use of UceWeb in order to enrich its UC repository. In addition, we aim to expand the DT repository with models that are already described in the literature. Future work will also be focused on implementing a functionality to allow users to add their own models to the DTs repository. Since in healthcare decision making TreeAge Pro is widely used for DTs formalization, we believe that a significant number of researchers interested in CUAs have already used this software to formalize their own models. With a minor modification (i.e. the insertion of structured comments into the formalized DT), these models could be easily added into our DTs repository. This possibility should encourage researchers to use UceWeb as a collaborative platform for collecting DT models, increasing the value of the efforts dedicated to modeling disease-specific decision problems. Finally, we will define a validation procedure for assessing the usefulness of the described framework in practice.

# References

1. Kind, P., Lafata, J.E., Matuszewski, K., Raisch, D.: The use of QALYs in clinical and patient decision-making: issues and prospects. Value Health. **12**(Suppl. 1), S27–S30 (2009)
2. Gold, M.R.: Cost-Effectiveness in Health and Medicine. Oxford University Press, Oxford (1996)
3. Dijkers, M.P.J.M.: Quality of life of individuals with spinal cord injury: a review of conceptualization, measurement, and research findings. J. Rehabil. Res. Dev. **42**, 87 (2004)
4. Weinstein, M.C., Fineberg, H.V.: Clinical Decision Analysis. Saundersm, Philadelphia (1980)
5. Thomson, R., Robinson, A., Greenaway, J., Lowe, P.: Development and description of a decision analysis based decision support tool for stroke prevention in atrial fibrillation. Qual. Saf. Health Care **11**, 25–31 (2002)
6. Bansod, A., Skoczen, S., Lenert, L.A: IMPACT4: a framework for rapid, modular construction of web-based patient decision support systems and preference measurement tools. In: AMIA Annual Symposium Proceedings, p. 782. American Medical Informatics Association (2003)
7. Sims, T.L., Garber, A.M., Miller, D.E., Mahlow, P.T., Bravata, D.M., Goldstein, M.K.: Multimedia quality of life assessment: advances with FLAIR. In: AMIA Annual Symposium Proceedings, pp. 694–698 (2005)
8. PROSPEQT: A New Program for Computer-Assisted Utility Elicitations. https://smdm. confex.com/smdm/2004ga/techprogram/P1523.HTM
9. Goldstein, M.K., Miller, D.E., Davies, S., Garber, A.M.: Quality of life assessment software for computer-inexperienced older adults: multimedia utility elicitation for activities of daily living. In: Proceedings of the AMIA Symposium, p. 295. American Medical Informatics Association (2002)
10. Sumner, W., Nease, R., Littenberg, B.: U-titer: a utility assessment tool. In: Proceedings Annual Symposium Computer Application in Medical Care, pp. 701–705 (1991)
11. Parimbelli, E., Sacchi, L., Rubrichi, S., Mazzanti, A., Quaglini, S.: UceWeb: a web-based collaborative tool for collecting and sharing quality of life data. Methods Inf. Med. **53**, 156–163 (2014)
12. SNOMED CT. https://www.nlm.nih.gov/healthit/snomedct/index.html
13. TreeAge Pro. https://www.treeage.com/
14. Rognoni, C., Marchetti, M., Quaglini, S., Liberato, N.L.: Apixaban, dabigatran, and rivaroxaban versus warfarin for stroke prevention in non-valvular atrial fibrillation: a cost-effectiveness analysis. Clin. Drug Investig. **34**, 9–17 (2014)

# Author Index

Printed in the United States
By Bookmasters